Methods in Molecular Biology

Series Editor
John M. Walker
School of Life and Medical Sciences
University of Hertfordshire
Hatfield, Hertfordshire, UK

For further volumes:
http://www.springer.com/series/7651

For over 35 years, biological scientists have come to rely on the research protocols and methodologies in the critically acclaimed *Methods in Molecular Biology* series. The series was the first to introduce the step-by-step protocols approach that has become the standard in all biomedical protocol publishing. Each protocol is provided in readily-reproducible step-by-step fashion, opening with an introductory overview, a list of the materials and reagents needed to complete the experiment, and followed by a detailed procedure that is supported with a helpful notes section offering tips and tricks of the trade as well as troubleshooting advice. These hallmark features were introduced by series editor Dr. John Walker and constitute the key ingredient in each and every volume of the *Methods in Molecular Biology* series. Tested and trusted, comprehensive and reliable, all protocols from the series are indexed in PubMed.

Cytoskeleton Dynamics

Methods and Protocols

Edited by

Helder Maiato

i3S University of Porto, Porto, Portugal

Humana Press

Editor
Helder Maiato
i3S University of Porto
Porto, Portugal

ISSN 1064-3745 ISSN 1940-6029 (electronic)
Methods in Molecular Biology
ISBN 978-1-0716-0218-8 ISBN 978-1-0716-0219-5 (eBook)
https://doi.org/10.1007/978-1-0716-0219-5

Cover Caption: Microtubule-labeled U2OS cell imaged using CH-STED microscopy. Raw images were projected into a single plane using the 'Temporal-color code' with the 'Magenta-Hot' look-up table.

Image Credits: António Pereira, Ana Luísa Ferreira and Helder Maiato

This Humana imprint is published by the registered company Springer Science+Business Media, LLC part of Springer Nature.
The registered company address is: 233 Spring Street, New York, NY 10013, U.S.A.

Preface

The cytoskeleton accounts for critical cellular processes from cell division and motility to axonal transport and vesicle trafficking. They all rely on the individual or combined dynamic properties of the two main cytoskeletal components: actin microfilaments and microtubules. Given their essential role, the molecular mechanisms behind their structure, dynamic properties, and functions have been extensively scrutinized over several decades, but new knowledge is constantly emerging from more recent studies. This book provides a comprehensive overview of the state-of-the-art methodology to investigate how actin (and actin-like) microfilaments and microtubules work to accomplish different cellular functions in different cellular contexts from bacteria to humans. The book covers extensive and exciting methodologies for protein purification and in vitro reconstitution of several cytoskeleton properties, as well as the functional dissection of microtubule- and actin-based structures, such as centrioles, mitotic spindles, axons, and cytokinetic contractile rings. It also provides a methodology for the study and manipulation of microtubules and actin, including the study of their nucleation, ultrastructural and/or dynamic properties (e.g., using optogenetics), and the functional dissection of post-translational modifications. There is a focus on specific biological processes, inevitably on mitosis, which lies very close to my heart, but also on bacterial division, meiosis, cytokinesis, axonal transport, and nuclear migration. For all these great contributions, I am extremely indebted to all colleagues that accepted to write a chapter (or more!) to this book. These chapters provide step-by-step protocols and rationale that can be followed not only by experts, but especially by training students and postdocs that only recently joined the field. A special mention to Prof. John Walker for his guidance during the preparation and revision of all the chapters. I am confident that the content of this book provides a breadth of experimental approaches that can be made standard in any laboratory and will represent an invaluable resource for present and future generations of cytoskeleton researchers.

Porto, Portugal *Helder Maiato*

Contents

Preface . . . . *v*
Contributors . . . . *ix*

1 Purification of Mammalian Tubulins and Tubulin-Associated Proteins Using a P2A-Based Expression System 1
Nuo Yu and Niels Galjart

2 Expression and Purification of Microtubule-Associated Proteins from HEK293T Cells for In Vitro Reconstitution 19
Shasha Hua and Kai Jiang

3 In Vitro Reconstitution Assays of Microtubule Amplification and Lattice Repair by the Microtubule-Severing Enzymes Katanin and Spastin 27
Annapurna Vemu, Ewa Szczesna, and Antonina Roll-Mecak

4 In Vitro Microtubule Dynamics Assays Using Dark-Field Microscopy 39
Jeffrey O. Spector, Annapurna Vemu, and Antonina Roll-Mecak

5 In Vitro Reconstitution of Dynamic Co-organization of Microtubules and Actin Filaments in Emulsion Droplets 53
Kim J. A. Vendel, Celine Alkemade, Nemo Andrea, Gijsje H. Koenderink, and Marileen Dogterom

6 Studying Tau-Microtubule Interaction Using Single-Molecule TIRF Microscopy 77
Virginie Stoppin-Mellet, Nassiba Bagdadi, Yasmina Saoudi, and Isabelle Arnal

7 In Vitro FRET- and Fluorescence-Based Assays to Study Protein Conformation and Protein-Protein Interactions in Mitosis 93
Stephanie C. Ems-McClung and Claire E. Walczak

8 Methods for Studying Membrane-Associated Bacterial Cytoskeleton Proteins In Vivo by TIRF Microscopy 123
Charlène Cornilleau, Arnaud Chastanet, Cyrille Billaudeau, and Rut Carballido-López

9 Processing TIRF Microscopy Images to Characterize the Dynamics and Morphology of Bacterial Actin-Like Assemblies 135
Cyrille Billaudeau, Arnaud Chastanet, and Rut Carballido-López

10 Studying Centriole Duplication and Elongation in Human Cells 147
Catarina Peneda, Carla A. M. Lopes, and Mónica Bettencourt-Dias

11 Assaying Microtubule Nucleation 163
Artur Ezquerra, Ricardo Viais, and Jens Lüders

12 Detection of Microtubule Nucleation Hotspots at the Golgi 179
Roslin J. Thoppil, Anna A. W. M. Sanders, and Irina Kaverina

13 Ultrastructural Analysis of Microtubule Ends 191
J. Richard McIntosh, Eileen O'Toole, Cynthia Page, and Garry Morgan

14 Optogenetic Control of Microtubule Dynamics 211
Jeffrey van Haren, Lauren S. Adachi, and Torsten Wittmann

15 Measurement of Microtubule Half-Life and Poleward Flux in the Mitotic Spindle by Photoactivation of Fluorescent Tubulin 235
Hugo Girão and Helder Maiato

16 Functional Dissection of Mitosis Using Immortalized Fibroblasts from the Indian Muntjac, a Placental Mammal with Only Three Chromosomes 247
Ana C. Almeida, Danica Drpic, Naoyuki Okada, Joana Bravo, Marta Madureira, and Helder Maiato

17 Visualization and Functional Analysis of Spindle Actin and Chromosome Segregation in Mammalian Oocytes 267
Binyam Mogessie

18 Using the Four-Cell *C. elegans* Embryo to Study Contractile Ring Dynamics During Cytokinesis 297
Fung Yi Chan, Ana Marta Silva, and Ana Xavier Carvalho

19 Knocking Out Multiple Genes in Cultured Primary Neurons to Study Tubulin Posttranslational Modifications.............................. 327
Satish Bodakuntla, Carsten Janke, and Maria M. Magiera

20 Measuring the Impact of Tubulin Posttranslational Modifications on Axonal Transport 353
Satish Bodakuntla, Maria M. Magiera, and Carsten Janke

21 Methods to Measure Perinuclear Actin Dynamics During Nuclear Movement in Migrating Cells.................................... 371
Cátia S. Janota, Francisco J. Calero-Cuenca, and Edgar R. Gomes

Correction to: Visualization and Functional Analysis of Spindle Actin and Chromosome Segregation in Mammalian Oocytes........................... C1

Index *387*

Contributors

LAUREN S. ADACHI • *Department of Cell and Tissue Biology, University of California San Francisco, San Francisco, CA, USA*

CELINE ALKEMADE • *Department of Bionanoscience, Kavli Institute of Nanoscience, Delft University of Technology, Delft, The Netherlands; Department of Living Matter, AMOLF, Amsterdam, The Netherlands*

ANA C. ALMEIDA • *Chromosome Instability & Dynamics Laboratory, IBMC—Instituto de Biologia Molecular e Celular, Universidade do Porto, Porto, Portugal; i3S—Instituto de Investigação e Inovação em Saúde, Universidade do Porto, Porto, Portugal; Graduate Program in Biomedicine, Faculdade de Medicina, Universidade do Porto, Porto, Portugal*

NEMO ANDREA • *Department of Bionanoscience, Kavli Institute of Nanoscience, Delft University of Technology, Delft, The Netherlands*

ISABELLE ARNAL • *Univ. Grenoble Alpes, Inserm, U1216, CNRS, Grenoble Institut Neurosciences, Grenoble, France*

NASSIBA BAGDADI • *Univ. Grenoble Alpes, Inserm, U1216, CNRS, Grenoble Institut Neurosciences, Grenoble, France*

MÓNICA BETTENCOURT-DIAS • *Instituto Gulbenkian de Ciência, Oeiras, Portugal*

CYRILLE BILLAUDEAU • *Micalis Institute, INRA, AgroParisTech, Université Paris-Saclay, Jouy-en-Josas, France*

SATISH BODAKUNTLA • *Institut Curie, PSL Research University, CNRS UMR3348 Universitaire, Orsay, France; Université Paris Sud, Université Paris-Saclay, CNRS UMR3348, Orsay, France*

JOANA BRAVO • *Graduate Program in Areas of Basic and Applied Biology (GABBA), Instituto de Ciências Biomédicas Abel Salazar, Universidade do Porto, Porto, Portugal*

FRANCISCO J. CALERO-CUENCA • *Faculdade de Medicina, Instituto de Medicina Molecular, Universidade de Lisboa, Lisbon, Portugal*

RUT CARBALLIDO-LÓPEZ • *Micalis Institute, INRA, AgroParisTech, Université Paris-Saclay, Jouy-en-Josas, France*

ANA XAVIER CARVALHO • *Instituto de Investigação e Inovação em Saúde (i3S), Universidade do Porto, Porto, Portugal; Instituto de Biologia Molecular e Celular (IBMC), Universidade do Porto, Porto, Portugal*

FUNG YI CHAN • *Instituto de Investigação e Inovação em Saúde (i3S), Universidade do Porto, Porto, Portugal; Instituto de Biologia Molecular e Celular (IBMC), Universidade do Porto, Porto, Portugal*

ARNAUD CHASTANET • *MICALIS, INRA, AgroParisTech, Université Paris-Saclay, Jouy-en-Josas, France*

CHARLÈNE CORNILLEAU • *Micalis Institute, INRA, AgroParisTech, Université Paris-Saclay, Jouy-en-Josas, France*

MARILEEN DOGTEROM • *Department of Bionanoscience, Kavli Institute of Nanoscience, Delft University of Technology, Delft, The Netherlands*

DANICA DRPIC • *Chromosome Instability & Dynamics Laboratory, IBMC—Instituto de Biologia Molecular e Celular, Universidade do Porto, Porto, Portugal; i3S—Instituto de Investigação e Inovação em Saúde, Universidade do Porto, Porto, Portugal*

STEPHANIE C. EMS-MCCLUNG • *Indiana University School of Medicine—Bloomington, Medical Sciences, Bloomington, IN, USA*
ARTUR EZQUERRA • *Institute for Research in Biomedicine (IRB Barcelona), The Barcelona Institute of Science and Technology (BIST), Barcelona, Spain*
NIELS GALJART • *Department of Cell Biology, Erasmus University Medical Center, Rotterdam, The Netherlands*
HUGO GIRÃO • *Chromosome Instability & Dynamics Laboratory, IBMC—Instituto de Biologia Molecular e Celular, Universidade do Porto, Porto, Portugal; i3S—Instituto de Investigação e Inovação em Saúde, Universidade do Porto, Porto, Portugal*
EDGAR R. GOMES • *Faculdade de Medicina, Instituto de Medicina Molecular, Universidade de Lisboa, Lisbon, Portugal*
SHASHA HUA • *The State Key Laboratory Breeding Base of Basic Science of Stomatology (Hubei-MOST) and Key Laboratory of Oral Biomedicine (Ministry of Education), School and Hospital of Stomatology, Wuhan University, Wuhan, China; Medical Research Institute, Wuhan University, Wuhan, China*
CARSTEN JANKE • *Institut Curie, PSL Research University, CNRS UMR3348, Centre Universitaire, Orsay, France; Université Paris Sud, Université Paris-Saclay, CNRS UMR3348, Orsay, France*
CÁTIA S. JANOTA • *Faculdade de Medicina, Instituto de Medicina Molecular, Universidade de Lisboa, Lisbon, Portugal*
KAI JIANG • *The State Key Laboratory Breeding Base of Basic Science of Stomatology (Hubei-MOST) and Key Laboratory of Oral Biomedicine (Ministry of Education), School and Hospital of Stomatology, Wuhan University, Wuhan, China; Medical Research Institute, Wuhan University, Wuhan, China*
IRINA KAVERINA • *Department of Cell and Developmental Biology, Vanderbilt University, Nashville, TN, USA*
GIJSJE H. KOENDERINK • *Department of Living Matter, AMOLF, Amsterdam, The Netherlands*
CARLA A. M. LOPES • *Instituto Gulbenkian de Ciência, Oeiras, Portugal*
JENS LÜDERS • *Institute for Research in Biomedicine (IRB Barcelona), The Barcelona Institute of Science and Technology (BIST), Barcelona, Spain*
MARTA MADUREIRA • *Graduate Program in Areas of Basic and Applied Biology (GABBA), Instituto de Ciências Biomédicas Abel Salazar, Universidade do Porto, Porto, Portugal*
MARIA M. MAGIERA • *Institut Curie, PSL Research University, CNRS UMR3348, Centre Universitaire, Orsay, France; Université Paris Sud, Université Paris-Saclay, CNRS UMR3348, Orsay, France*
HELDER MAIATO • *Chromosome Instability & Dynamics Laboratory, IBMC—Instituto de Biologia Molecular e Celular, Universidade do Porto, Porto, Portugal; i3S—Instituto de Investigação e Inovação em Saúde, Universidade do Porto, Porto, Portugal; Experimental Biology Unit, Faculdade de Medicina, Cell Division Group, Department of Biomedicine, Universidade do Porto, Porto, Portugal*
BINYAM MOGESSIE • *School of Biochemistry, University of Bristol, Bristol, UK*
GARRY MORGAN • *Department of Molecular, Cellular, and Developmental Biology, University of Colorado, Boulder, CO, USA*
EILEEN O'TOOLE • *Department of Molecular, Cellular, and Developmental Biology, University of Colorado, Boulder, CO, USA*

Naoyuki Okada • *Chromosome Instability and Dynamics Laboratory, IBMC—Instituto de Biologia Molecular e Celular, Universidade do Porto, Porto, Portugal; i3S—Instituto de Investigação e Inovação em Saúde, Universidade do Porto, Porto, Portugal*

Cynthia Page • *Department of Molecular, Cellular, and Developmental Biology, University of Colorado, Boulder, CO, USA*

Catarina Peneda • *Instituto Gulbenkian de Ciência, Oeiras, Portugal; Instituto de Ciências Biomédicas Abel Salazar, Universidade do Porto, Porto, Portugal*

J. Richard McIntosh • *Department of Molecular, Cellular, and Developmental Biology, University of Colorado, Boulder, CO, USA*

Antonina Roll-Mecak • *Cell Biology and Biophysics Unit, Porter Neuroscience Research Center, National Institute of Neurological Disorders and Stroke, Bethesda, MD, USA; Biochemistry and Biophysics Center, National Heart, Lung and Blood Institute, Bethesda, MD, USA*

Anna A. W. M. Sanders • *Department of Cell and Developmental Biology, Vanderbilt University, Nashville, TN, USA*

Yasmina Saoudi • *Univ. Grenoble Alpes, Inserm, U1216, CNRS, Grenoble Institut Neurosciences, Grenoble, France*

Ana Marta Silva • *Instituto de Investigação e Inovação em Saúde (i3S), Universidade do Porto, Porto, Portugal; Instituto de Biologia Molecular e Celular (IBMC), Universidade do Porto, Porto, Portugal*

Jeffrey O. Spector • *Cell Biology and Biophysics Unit, Porter Neuroscience Research Center, National Institute of Neurological Disorders and Stroke, Bethesda, MD, USA*

Virginie Stoppin-Mellet • *Univ. Grenoble Alpes, Inserm, U1216, CNRS, Grenoble Institut Neurosciences, Grenoble, France*

Ewa Szczesna • *Cell Biology and Biophysics Unit, Porter Neuroscience Research Center, National Institute of Neurological Disorders and Stroke, Bethesda, MD, USA*

Roslin J. Thoppil • *Department of Cell and Developmental Biology, Vanderbilt University, Nashville, TN, USA*

Jeffrey van Haren • *Department of Cell and Tissue Biology, University of California San Francisco, San Francisco, CA, USA; Department of Cell Biology, Erasmus MC, Rotterdam, The Netherlands*

Annapurna Vemu • *Cell Biology and Biophysics Unit, Porter Neuroscience Research Center, National Institute of Neurological Disorders and Stroke, Bethesda, MD, USA*

Kim J. A. Vendel • *Department of Bionanoscience, Kavli Institute of Nanoscience, Delft University of Technology, Delft, The Netherlands*

Ricardo Viais • *Institute for Research in Biomedicine (IRB Barcelona), The Barcelona Institute of Science and Technology (BIST), Barcelona, Spain*

Claire E. Walczak • *Indiana University School of Medicine—Bloomington, Medical Sciences, Bloomington, IN, USA*

Torsten Wittmann • *Department of Cell and Tissue Biology, University of California San Francisco, San Francisco, CA, USA*

Nuo Yu • *Department of Cell Biology, Erasmus University Medical Center, Rotterdam, The Netherlands*

Chapter 1

Purification of Mammalian Tubulins and Tubulin-Associated Proteins Using a P2A-Based Expression System

Nuo Yu and Niels Galjart

Abstract

The microtubule cytoskeleton plays a crucial role in a myriad of cellular events, including mitosis, cell differentiation, migration, and the maintenance of cell shape. Microtubules are assembled from α- and β-tubulin heterodimers, whose biosynthesis is a complex process requiring the balanced production of α- and β-tubulin subunits. This chapter focuses on a method for the combined expression of tagged α- and β-tubulin dimers, their purification, and the isolation of co-purifying tubulin-associated proteins (TAPs) in mammalian cells. This approach is currently used in our laboratory to study tubulin function and to identify and characterize TAPs.

Key words Microtubule, Tubulin, Transient transfection, Affinity-based purification, Tubulin-associated proteins

1 Introduction

Microtubules (MTs) are cylindrical filamentous structures, which are composed of 13 protofilaments of head-to-tail stacked tubulins that associate laterally to form a hollow tube with a diameter of 25 nm. MTs provide shape and strength to cells, position the nucleus and organelles, and support long-range intracellular transport. They also form the inner core of flagella and cilia and are part of the centrosome that nucleates MTs in interphase and mitosis. The building block of MTs is tubulin, a heterodimer composed of two closely related globular proteins designated α- and β-tubulin. Both proteins are about 55 kDa, and each can bind GTP. However, upon dimerization, the β-tubulin subunit binds tightly "on top" of an α-tubulin monomer, at the position of the GTP-binding pocket in α-tubulin. This causes GTP bound to α-tubulin to be "non-exchangeable," whereas GTP bound to β-tubulin is "exchangeable." MT assembly is in fact accompanied by GTP hydrolysis on β-tubulin; the energy input is stored in the MT lattice and released upon depolymerization. The stochastic switching from

Helder Maiato (ed.), *Cytoskeleton Dynamics: Methods and Protocols*, Methods in Molecular Biology, vol. 2101,
https://doi.org/10.1007/978-1-0716-0219-5_1,

polymerization to depolymerization and vice versa is termed dynamic instability [1]. In cells, it mainly occurs at MT plus ends, as minus ends are often embedded in some kind of protective structure and are therefore not dynamic.

In vertebrates, a small multigene family encoding multiple α- and β-tubulin isotypes has been identified (mice have 7 α-tubulin and 8 β-tubulin encoding genes, whereas humans have 8 α- and 9 β-tubulin encoding genes [2]). Tubulin isotypes show extensive sequence homology, although they differ somewhat at their extreme C-terminal region. They have distinct expression patterns, which corroborates the original "multi-tubulin hypothesis" which proposed that the individual tubulin genes encode functionally distinct polypeptides that confer unique properties to MTs [3].

The folding of newly made α- and β-tubulin dimers is elaborate [4], with chaperones (e.g., hsp70) first binding to nascent tubulin polypeptides, followed by cytosolic chaperonins (CCTs), which facilitate productive folding. After release of tubulin-folding intermediates into the cytoplasm, further folding and dimerization occurs with the help of tubulin-specific folding cofactors (TBCA-TBCE). More recently, a modification to the long-standing tubulin-folding model has been proposed, with a role for ADP ribosylation factor-like protein 2 (Arl2) [5]. Proper tubulin folding appears to require a balanced production of both α- and β-tubulin subunits; overexpression of one subunit is thought to "clog" the folding pathway, leading to a less efficient dimerization [6]. This explains in part why it is difficult to express, purify, and study tubulin dimers via conventional overexpression methods.

Functional tubulin can be purified from porcine or calf brain [7, 8]. However, due to unknown isotype composition and post-translational modifications (PTMs), it is difficult to dissect the effects of a specific tubulin isotype on in vitro MT dynamic instability using brain tubulin. Based on the original observation by Minoura et al. [9], it has recently been possible to produce functional tubulin isotypes with the use of the baculovirus insect cell system (e.g., [10]). This type of tubulin is very well suited for in vitro studies; however, analysis of tubulin function in mammalian cells needs other approaches. To circumvent imbalance problems upon overexpression of a single exogenous α- or β-tubulin subunit in mammalian cells, we utilized a "polycistronic" or "dual" mRNA approach, where we encoded both the α- and the β-tubulin subunits on a single mRNA and used the P2A self-cleavage peptide [11] in between the subunits to synthesize α-tubulin and β-tubulin at equimolar levels in transfected mammalian cells [12]. Since we are interested in neuronal functions of MTs, we chose to express human *TUBA1A*, a commonly expressed α-tubulin isotype, in combination with *TUBB3*, a β-tubulin isotype that is enriched in neurons. We furthermore tagged the N-terminus of α-tubulin with a small biotin recognition sequence, which is recognized and efficiently biotinylated by the bacterial BirA enzyme [13]. We then used magnetic streptavidin-coated beads to pull down recombinant biotinylated tubulin. The biotin-streptavidin interaction allows stringent purification of tubulin using a single-step approach; however, since the biotin-streptavidin interaction is very tight, we also introduced a Sumo* sequence in the tubulin construct after the biotin tag. The SUMOstar protease recognizes the Sumo* motif in a highly specific manner and cleaves at the C-terminus of this motif, releasing recombinant α-tubulin. In this chapter, we provide a detailed practical overview of the methods used in our laboratory to clone constructs and express and purify tubulin from mammalian cells (Subheading 1) and to study tubulin-associated proteins (TAPs) (Subheading 2).

2 Materials

2.1 Tubulin Expression and Purification

2.1.1 Generation of "Dual" Tubulin Expression Construct

1. PCR tube 0.2 mL.
2. Thermocycler.
3. Owl™ EasyCast™ B2 Minigel-Elektrophoresis system.
4. 1% Agarose gel: Weigh 1.5 g of agarose in 150 mL of Tris-borate-EDTA (TBE) gel electrophoresis buffer. Depending on the size of the agarose gel, one can use less (or more) than 1.5 g and 150 mL TBE. We use a 10× TBE stock to prepare TBE buffer. The 10× stock is prepared by dissolving 1090 g Tris–HCl, 556 g boric acid, and 93 g of EDTA into H_2O and bringing the volume to 10 L by addition of H_2O (*see* **Note 1**). Dissolve agarose by heating in a microwave for 5–10 min. Swirl solution regularly so as to dissolve the agarose well. After heating, the solution should be brought back to 150 mL using distilled water.
5. Phusion High-Fidelity DNA Polymerase (NEB).
6. 5× Phusion buffer (NEB; this buffer comes with the Phusion enzyme).
7. QiAquick PCR purification kit (Qiagen).
8. 6× DNA loading dye. We either use a commercial dye (NEB) or a homemade one (dissolve 0.6 g Orange G and 15 g Ficoll into sterile H_2O, and bring volume to 100 mL with sterile H_2O).
9. DNA molecular weight marker (e.g., Lambda-PstI from NEB).
10. dNTPs: Mix of dATP, dCTP, dGTP, and dTTP, each nucleotide at 10 mM concentration.
11. Nanodrop.
12. UV transilluminator.

2.1.2 Vector Preparation

1. 37 °C Water bath.
2. Owl™ EasyCast™ B2 Minigel-Elektrophoresis system.
3. XhoI.
4. Asp718.
5. Enzyme buffer.
6. QiAquick gel extraction kit (Qiagen).
7. 1% Agarose gel (prepared as described in Subheading 2.1.1).
8. Nanodrop.
9. UV transilluminator.

2.1.3 Gibson Assembly

1. Thermoshaker.
2. Centrifuge for Eppendorf tube.
3. Gibson assembly master mix (NEB).

2.1.4 Transformation and Selection

1. Thermoshaker with temperature adjustment function.
2. 37 °C Incubator.
3. Centrifuge for Eppendorf tube.
4. Bacteria culture shaker.
5. Owl™ EasyCast™ B2 Minigel-Elektrophoresis system.
6. DH5α chemical competent cells (Thermo Fisher Scientific).
7. LB medium without antibiotic: For 1 L, dissolve 10 g tryptone, 10 g NaCl, and 5 g yeast extract in 900 mL distilled H_2O. Adjust pH to 7.0 by using 1 M NaOH. Once the pH is adjusted, complete the volume with distilled H_2O. Autoclave to sterilize and store at 4 °C.
8. LB medium containing 50 μg/mL ampicillin.
9. LB agar plate containing 50 μg/mL ampicillin.
10. Glass pasteur pipettes with long tips of 120 mm.
11. QIAprep Spin Miniprep Kit (Qiagen).
12. Sma I.
13. Enzyme buffer.
14. 5× DNA loading dye.
15. Agarose.

2.2 Cell Culture

1. 15 cm Plastic dish for mammalian cell culture for transfection of HEK293T cells.
2. Complete growth medium: A mixture of Dulbecco's modified Eagle's medium (DMEM) and Ham's F-10 medium (1:1) supplemented with 10% fetal bovine serum (FBS), penicillin, and streptomycin.
3. Sterile laminar flow hood.
4. 37 °C, 5% CO_2 culture incubator.

2.3 Transient Transfection of HEK293T Cells

1. Eppendorf tube 1.5 mL.
2. OptiMEM (Thermo Fisher Scientific).
3. X-tremeGENE HP DNA transfection reagent (Roche).
4. DNA constructs: BirA, Bio-Sumo*-GFP, TUBB3-P2A-Bio-Sumo*-GFP-TUBA1A.

2.4 One-Step Purification of Tubulin

2.4.1 Magnetic Beads Blocking

1. 1.5 mL Eppendorf tubes.
2. Dynabeads™ MPC™-S Magnetic Particle Concentrator (Thermo Fisher Scientific).
3. Stuart Tube Rotator (Cole-Parmer).
4. Dynabeads™ M-280 streptavidin (Thermo Fisher Scientific).
5. Blocking buffer: 20 mM Tris–HCl pH 6.8, 150 mM KCl, 20% glycerol, 0.02% Triton X-100, 1 mM EGTA, 4 mM $MgCl_2$, 1 μg/μL chicken egg albumin. The Tris–HCl is made up from a stock of 1 M Tris–HCl pH 6.8, which is prepared as follows: dissolve 121 g Tris in H_2O, adjust to pH 6.8 with concentrated HCl (in fume hood), bring to 1 L with H_2O, and sterilize by autoclaving. Be careful to pipette the right amount of glycerol, as the solution is very viscous. To make 0.02% Triton, we use a 10% stock, which is not viscous. However, the 10% stock is made from the original Triton solution, which is very viscous.

2.4.2 Cell Harvesting

1. Suction pump.
2. Centrifuge for falcon tubes.
3. Dulbecco's phosphate-buffered saline (DPBS).

2.4.3 Cell Lysis

1. 1.5 mL Eppendorf tubes.
2. Centrifuge for Eppendorf tubes with precooling function.
3. Lysis buffer: 20 mM Tris–HCl pH 6.8, 150 mM KCl, 0.2% Triton X-100, 10% glycerol, 1 mM EGTA, 4 mM $MgCl_2$, 4% Protease Inhibitor Cocktail (Roche, 1 tablet in 2 mL sterile H_2O).

2.4.4 Protein Immunoprecipitation

1. Stuart Tube Rotator in cold room.
2. Dynabeads™ MPC™-S Magnetic Particle Concentrator (Thermo Fisher Scientific).
3. Wash buffer: 20 mM Tris–HCl pH 6.8, 150 mM KCl, 0.2% Triton X-100, 10% glycerol, 1 mM EGTA, 4 mM $MgCl_2$.

2.4.5 Protein Cleavage

1. Thermoshaker.
2. 30 kDa Cutoff Amicon Ultra-0.5 mL Centrifuge Filter.
3. Cleavage buffer: 20 mM Tris–HCl pH 6.8, 150 mM KCl, 0.2% Triton X-100, 10% glycerol, 1 mM EGTA, 4 mM $MgCl_2$, 2 mM DTT.
4. SUMOstar protease (LifeSensors).

2.5 Isolation of Tubulin-Associated Proteins

2.5.1 Protein Separation on Clean Gels

1. XCell SureLock™ Mini-Cell Electrophoresis System (Thermo Fisher Scientific).
2. 4–12% NuPAGE tris-acetate gels (Thermo Fisher Scientific).
3. Sample buffer (4×, Invitrogen): We use a commercial supplier for this buffer because these buffers need to be as clean as possible in terms of contaminating proteins.
4. NuPAGE 20× SDS MOPS buffer (Thermo Fisher Scientific).
5. MilliQ water.

2.5.2 Staining and Destaining of "Mass Spectrometry" Gels

1. 15 cm Plastic culture dish.
2. Ultraclean pipettes.
3. Suction pump.
4. Mini orbital shaker.
5. Invitrogen™ Novex™ Colloidal Blue Staining Kit (Thermo Fisher Scientific).
6. MilliQ water.

3 Methods

3.1 Tubulin Expression and Purification

3.1.1 Generation of "Dual" Tubulin Expression Construct

1. Prepare inserts (*see* Fig. 1 for graphical explanation). *TUBB3* and *TUBA1A* cDNAs can be synthesized and amplified from mRNA derived from human cell lines (or from cell lines from other species, depending on which tubulin one wants to study). Alternatively, tubulin cDNAs can be purchased from Addgene or be purchased as a gene block from a commercial vendor.

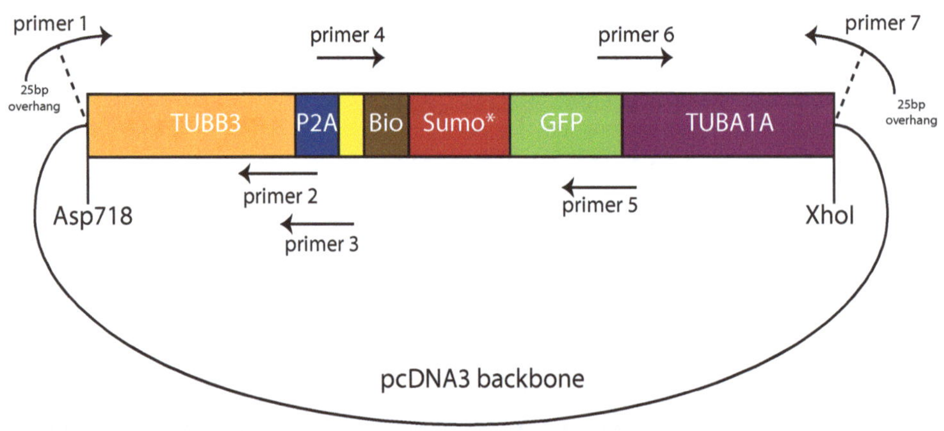

Fig. 1 Schematic representation of final "dual" plasmid construct encoding β-tubulin (TUBB3 isoform), α-tubulin (TUBA1A isoform), and the other indicated proteins. The position of the primers used for PCR and Gibson assembly is indicated

Table 1
PCR primer sequences

Primer	Sequence
Primer 1	5′-TGGCTAGCGTTTAAACTTAAGCTTGGTACCA TGAGGGAGATCGTGCACATCCAGG-3′
Primer 2	5′-GCTTCAGCAGGCTGAAGTTAGTAGCTCCGCTT CCCTTGGGCCCCTGGGCTTCCGATTCCT-3′
Primer 3	5′-TGCAGGTCCAGGGTTCTCCTCCACGTCTCCA GCCTGCTTCAGCAGGCTGAAGTTAGTAGC-3′
Primer 4	5′-CGTGGAGGAGAACCCTGGACCTGCAGGAGC TGGAGCGATGGCTTCCGGCCTGAACGACAT-3′
Primer 5	5′-AGATGCACTCACGCATAAGCTTGAGCTCGTC CATGCCGAGAGTGATCCCGGCGGCGGTCA-3′
Primer 6	5′-CTCAAGCTTATGCGTGAGTGCATCTCCATCC ACGTTGGCCAGGCTGGTGTCCAGATTGGC-3′
Primer 7	5′-GTTTAAACGGGCCCTCTAGACTCGAGTTAGTA TTCCTCTCCTTCTTCCTCACCCTCTCCT-3′

Similarly, the Bio-Sumo* sequence can be purchased as a gene block and be hooked up to GFP by PCR to generate biotin-Sumo*-GFP, or the whole Bio-Sumo*-GFP sequence can be ordered. Note that while we provide our own primer sequences in Table 1 based on our own cDNAs and protocols, sequences should be carefully checked when using other cDNAs or modifications of our protocol. Please *see* Fig. 1 for graphical explanation of the PCR procedure.

2. Introduce P2A sequence by PCR by pipetting the components listed below in the order from top to bottom (stock concentration is between brackets):
   (a) TUBB3 construct: 5 μL (10 ng/μL).
   (b) Primer 1 (Table 1): 1 μL (10 μM).
   (c) Primer 2 (Table 1): 1 μL (10 μM).
   (d) dNTP: 1 μL (10 mM).
   (e) 5× Phusion buffer: 10 μL.
   (f) Phusion polymerase: 0.5 μL.
   (g) H_2O: 31.5 μL.

3. PCR program:

98 °C	30 s
98 °C	15 s
68 °C	30 s
72 °C	30 s
Repeat last three steps 10×	
72 °C	5 min

4. Mix 5 μL of the PCR sample with 1 μL 6× DNA loading dye on a 1% agarose gel, alongside marker DNA, to confirm fragment sizes. Visualize DNA using a UV transilluminator. Use QiAquick PCR purification kit to clean the rest of the PCR product, and measure the DNA concentration.

5. Take the PCR product from above reaction after PCR cleanup, and prepare the following mix:
   (a) PCR product: 5 μL (10 ng/μL).
   (b) Primer 1 (Table 1): 1 μL (10 μM).
   (c) Primer 3 (Table 1): 1 μL (10 μM).
   (d) dNTP: 1 μL (10 mM).
   (e) 5× Phusion buffer: 10 μL.
   (f) Phusion polymerase: 0.5 μL.
   (g) H_2O: 31.5 μL.

6. PCR program:

98 °C	30 s
98 °C	15 s
64 °C	30 s
72 °C	30 s
Repeat last three steps 15×	
72 °C	5 min

7. Mix 5 μL of the PCR sample with 1 μL 6× DNA loading dye on a 1% agarose gel, alongside marker DNA, to confirm fragment sizes. Visualize DNA using a UV transilluminator. Use QiAquick PCR purification kit to clean the rest of the PCR product, and measure the DNA concentration.

8. Hook up appropriate primer sequences to the Bio-Sumo*-GFP construct by preparing the following mix:

(a) Bio-Sumo*-GFP: 2 μL (10 ng/μL).
(b) Primer 4 (Table 1): 1 μL (10 μM).
(c) Primer 5 (Table 1): 1 μL (10 μM).
(d) dNTP: 1 μL (10 mM).
(e) 5× Phusion buffer: 10 μL.
(f) Phusion polymerase: 0.5 μL.
(g) H_2O: 34.5 μL.

9. PCR program:

98 °C	30 s
98 °C	15 s
65 °C	30 s
72 °C	30 s
Repeat last three steps 25×	
72 °C	5 min

10. Mix 5 μL of the PCR sample with 1 μL 6× DNA loading dye on a 1% agarose gel, alongside marker DNA, to confirm fragment sizes. Visualize DNA using a UV transilluminator. Use QiAquick PCR purification kit to clean the rest of the PCR product, and measure the DNA concentration.
11. Prepare the following mix to amplify TUBA1A sequence:
    (a) TUBA1A construct: 2 μL (10 ng/μL).
    (b) Primer 6 (Table 1): 1 μL (10 μM).
    (c) Primer 7 (Table 1): 1 μL (10 μM).
    (d) dNTP: 1 μL (10 mM).
    (e) 5× Phusion buffer: 10 μL.
    (f) Phusion polymerase: 0.5 μL.
    (g) H_2O: 34.5 μL.
12. PCR program:

98 °C	30 s
98 °C	15 s
64 °C	30 s
72 °C	30 s
Repeat last three steps 25×	
72 °C	5 min

13. Mix 5 μL of the PCR sample with 1 μL 6× DNA loading dye on a 1% agarose gel, alongside marker DNA, to confirm fragment sizes. Visualize DNA using a UV transilluminator. Use QiAquick PCR purification kit to clean the rest of the PCR product, and measure the DNA concentration.

3.1.2 Vector Preparation

1. Prepare the following mix:
   (a) pcDNA3.1: 3 μL (500 ng/μL).
   (b) Asp718: 1 μL.
   (c) XhoI: 1 μL.
   (d) Enzyme buffer: 2.5 μL.
   (e) H_2O: 17.5 μL.
2. Mix components by ticking against the tube with your finger; do a quick spin in a microfuge to bring sample down to the bottom of the tube and digest at 37 °C for 2 h.
3. Load digested samples (mixed with 5 μL 6× DNA loading dye) alongside marker DNA on a 1% agarose gel. Visualize DNA using a UV transilluminator (wear protective glasses to protect eyes from UV light). Cut out the proper-sized band from the gel using a clean scalpel, purify the linearized vector using QiAquick gel extraction kit, and measure the DNA concentration on a Nanodrop.

3.1.3 Gibson Assembly

1. Prepare the following mix:
   (a) pcDNA3.1 (Asp718 and XhoI): 100 ng.
   (b) TUBB3-P2A PCR product: 26 ng.
   (c) P2A-Bio-Sumo*-GFP PCR product: 23 ng.
   (d) TUBA1A PCR product : 26 ng.
   (e) 2× Gibson assembly master mix: 10 μL.
   (f) H_2O: Fill up to 20 μL final volume.
2. Incubate at 50 °C for 1 h.
3. Gibson assembly mixture is collected by centrifugation at 300 × *g* for 15 s at room temperature.

3.1.4 Transformation and Selection

1. Transform 5 μL of the Gibson assembly mixture into DH5α chemical competent cells, mix gently, and incubate on ice for 30 min.
2. Heat-shock the cells for 90 s at 42 °C followed by 2-min incubation on ice. Add 1 mL of pre-warmed LB medium without antibiotic, and shake at 37 °C for 30 min at 200 rpm in a shaking incubator.

3. Collect competent cells by centrifugation at 14,000 × *g* for 30 s, resuspend bacteria with 100 μL LB, and spread bacteria out on a pre-warmed LB agar plate using a bent glass pasteur pipette. Sterilize pipette by passing through a flame, and then flame the tip of the pipette to melt it and close it. Turn pipette constantly in your hand to spread the heat equally. Then bend the pipette by putting it into the flame at ~5 cm distance from the tip. The moment the pipette starts to bend, take it out of the flame and cool it in the air for ~30 s. It can then be used to spread bacteria. Incubate plate overnight at 37 °C.
4. Pick colonies into 5 mL LB medium supplemented with ampicillin, and grow at 37 °C overnight in a shaking incubator with 200 rpm.
5. Perform DNA purification using QIAprep Spin Miniprep Kit.
6. Digest with SmaI at 37 °C in water bath for 30 min, according to the following reaction mix:
   (a) Purified DNA: 5 μL (100 ng/μL).
   (b) SmaI: 0.5 μL.
   (c) Enzyme buffer: 1.0 μL.
   (d) H_2O: 3.5 μL.
7. Mix digestion product with 2.5 μL 6× DNA loading dye, and run the digestion reaction on 1% agarose gel (alongside marker DNA) to confirm the separation of 1884, 2259, and 5099 bp fragments.
8. Digest plasmid to test for known restriction enzyme patterns (*see* Fig. 2 for an example of such a digestion with SmaI).
9. If digestion pattern is correct, send plasmid for sequencing.

3.2 Cell Culture

HEK293T cells are cultured in complete growth medium at 37 °C and 5% CO_2. Cell maintenance, transfection, and harvesting are performed in a sterile laminar flow hood.

3.3 Transient Transfection of HEK293T Cells

In order to biotinylate recombinant tubulins in the HEK293T cells, a plasmid encoding BirA needs to be co-transfected with the tubulin constructs. In addition, a negative control experiment is always performed; we use a Bio-Sumo*-GFP-encoding plasmid for this purpose. These plasmids are available from our laboratory upon request.

1. Seed 6 × 10^6 HEK293T cells in 20 mL complete growth medium on a 15 cm dish for each transfection.
2. Next day, prepare the following conditions in 1.5 mL Eppendorf tubes (*see* **Note 2**):
   (a) 1 mL opti-MEM + 10 μg Bio-Sumo*-GFP + 10 μg BirA +20 μL X-tremGENE HP DNA Transfection reagent.

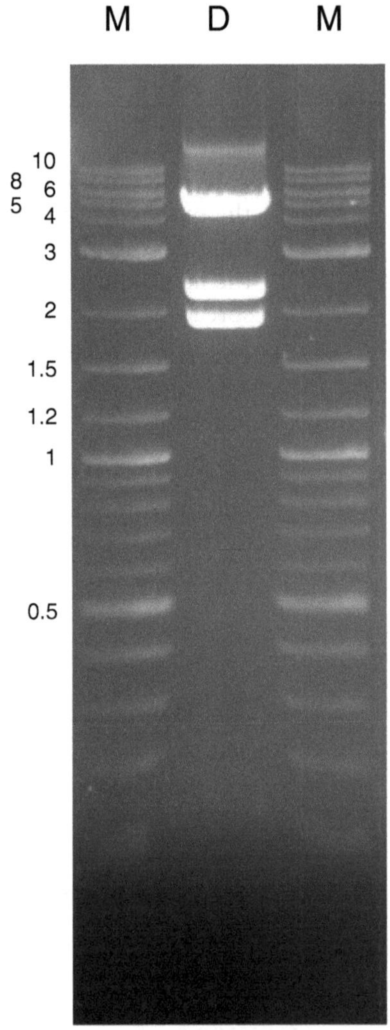

Fig. 2 Digestion of "dual" plasmid DNA (P) with SmaI. To the left and right are marker lanes (M), with sizes of important marker fragments indicated to the left (in kb)

(b) 1 mL opti-MEM + 10 μg TUBB3-P2A-Bio-Sumo*-GFP-TUBA1A + 10 μg BirA +20 μL X-tremGENE HP DNA Transfection reagent.

3. Mix transfection reagent well with DNA and incubate for 20 min at room temperature.
4. Add the transfection mixture to the dish dropwise, and shake the plate gently to distribute transfection mixture (*see* **Note 3**).
5. As protein purification is performed after 72 h, no medium change is required.

3.4 One-Step Purification of Tubulin

3.4.1 Magnetic Bead Blocking

1. Mix streptavidin magnetic beads by shaking the bottle (*see* **Note 4**).
2. Take 80 μL magnetic beads for each 15 cm dish and transfer to 1.5 mL Eppendorf tube.
3. Leave the Eppendorf tube in the magnetic particle concentrator for 1 min.
4. Remove liquid using a 200 μL pipette tip (*see* **Note 5**).
5. Add 400 μL of blocking buffer per tube (i.e., for each 15 cm dish) to the magnetic beads, and gently resuspend beads with 200 μL pipette tips (*see* **Note 4**).
6. Incubate beads at room temperature for 1 h on the rotator.

3.4.2 Cell Harvesting

1. Remove complete growth medium and wash cells once gently with 20 mL DPBS.
2. Add 20 mL DPBS to each 15 cm dish and collect cells by pipetting up and down (*see* **Note 6**).
3. Cells are collected by centrifugation at 200 × *g* for 10 min at room temperature.
4. Use suction pump and attached pipette to remove DPBS entirely.

3.4.3 Cell Lysis

1. Resuspend cell pellet with 400 μL ice-cold lysis buffer per 15 cm dish, and transfer to 1.5 mL Eppendorf tube.
2. Leave Eppendorf tube on a rotator in the cold room for 30 min (4 °C). This helps cells to lyse more efficiently.
3. Lysates are clarified by centrifugation at 14,000 × *g* for 10 min at 4 °C.
4. 10% of the supernatant can be saved as cell lysate control for Western blots and Coomassie staining analysis.
5. Wash magnetic beads twice with 800 μL lysis buffer to remove excess chicken egg albumin.
6. Transfer the lysate to the washed magnetic beads and mix gently (avoid introducing bubbles).

3.4.4 Protein Immunoprecipitation

1. Incubate magnetic beads with lysate on a rotator for 2 h in the cold room (4 °C).
2. Leave Eppendorf tube in the magnetic particle concentrator for 1 min to allow magnetic beads to sink to the bottom of the tubes, and collect 10% supernatant as unbound fraction (rest can be discarded).
3. Add 800 μL ice-cold wash buffer to Eppendorf tube, and resuspend the beads gently with 200 μL pipette tips.

4. Leave Eppendorf tubes on the rotator for 5 min in the cold room.
5. Repeat the last two steps seven times to wash the beads sufficiently, and after the last wash save 10% bead solution as bound fraction.

3.4.5 Protein Cleavage

1. Remove the wash buffer entirely, and resuspend with 40 μL cleavage buffer per 15 cm dish (*see* **Note 7**).
2. Add 2 units of SUMOstar protease in cleavage buffer, and leave Eppendorf tube in a thermoshaker at 30 °C and 1200 rpm for 1 h.
3. Put Eppendorf tube in the magnetic particle concentrator for 1 min, and take supernatant (purified tubulin) with 200 μL pipette tips.
4. Save 10% supernatant as your purified tubulin for Western blot and Coomassie staining analysis.
5. The rest of the supernatant can be concentrated further for in vitro assays by using 30 kDa cutoff Amicon Ultra-0.5 mL Centrifuge Filter.
6. Resuspend beads with 40 μL cleavage buffer per 15 cm dish, and save 10% as bead sticky fraction for Western blot and Coomassie staining analysis.

3.5 Isolation of Tubulin-Associated Proteins

Co-expression of tagged tubulins in mammalian cells with the P2A system yields more dimeric recombinant tubulin than when α- or β-tubulin are expressed as single constructs [12]. Thus, the "dual" biotin-streptavidin purification method provided us with the opportunity to identify tubulin-associated proteins (TAPs) binding to dimeric recombinant tubulins. In this case, milder washing conditions are used to retain TAPs as much as possible. In order to improve the confidence in the HEK293T cell "tubulome," we examine two types of purified material: proteins that are present on the streptavidin-coated beads before SUMOstar protease cleavage and proteins that are found in the supernatant after SUMOstar protease cleavage.

3.5.1 Protein Separation on Clean Gels

1. To identify TAPs, we execute all steps described above, but we only perform 3–5 washes (*see* **Note 8**). Both the proteins present on the streptavidin-coated beads before the SUMOstar protease cleavage are kept, and, of course, the proteins are released after the SUMOstar protease cleavage.
2. Add sample buffer to samples (if it is a 2× sample buffer solution, add 1 part sample buffer and 1 part sample), and boil at 95 °C for 5 min.

3. 4–12% NuPAGE tris-acetate gels (Invitrogen) are used to separate proteins, as these gels are very clean. For mass spectrometry, we spare 2–3 empty lanes in between protein samples loaded onto the gels, to prevent spillover and subsequent cross-contamination. For investigation of the experiment prior to mass spectrometry, we do not use spare lanes. For the mass spectrometry experiment, we only load samples containing streptavidin-coated beads before SUMOstar protease cleavage and after SUMOstar protease cleavage and use Bio--Sumo*-GFP as negative control for our experiments. Running buffer is made from 20× SDS MOPS buffer diluted into 1× SDS MOPS buffer by sterile MilliQ water.
4. Run 4–12% NuPAGE tris-acetate gels with 170 mV for 1.5 h.

3.5.2 Staining and Destaining of "Mass Spectrometry" Gels

1. Transfer each gel to a sterile 15 cm culture dish, and shake in fixing solution for 10 min at room temperature (*see* **Note 9**).
2. Remove fixing solution using ultraclean pipettes and a suction device, and then incubate (i.e., shake) the gel in staining solution without Stainer B for 10 min at room temperature (*see* **Note 10**).
3. Add 2.5 mL Stainer B to the 15 cm dish and shake the gel for 3 h.
4. Remove staining solution as described above, and add 50 mL sterile milliQ water to the 15 cm dish. Leave overnight for destaining.
5. Next day, refresh MilliQ water and send gels for mass spectrometry. Figure 3 shows an example of a purification from transiently transfected HEK293T cells.

4 Notes

1. We use deionized H_2O for experiments that do not require MilliQ H_2O. The deionized H_2O is autoclaved depending on the application, e.g., the 10× TBE has regular deionized H_2O, whereas for specific molecular biology experiments (e.g., PCR), we use autoclaved deionized H_2O.
2. Even though our "dual" tubulin expression approach yields nice quantities of dimeric tubulin, one should not forget that we are collecting these tubulins from transiently transfected HEK293T cells. We imagine that in this system, individual cells are starting to express recombinant tubulins at different time points after transfection and will express tubulins at different levels. We harvest cells after 72 h, and we collect all cells, including those that are on their way to express and fold tubulins as well as cells that have just started to express them

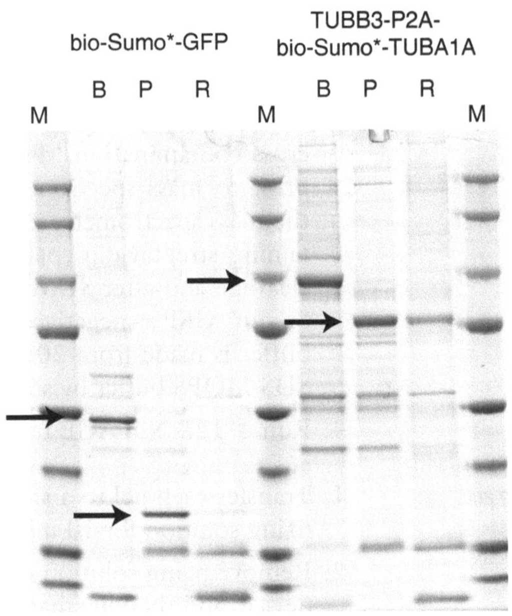

Fig. 3 Coomassie-stained gel showing protein samples after purification on streptavidin-coated magnetic beads and cleavage from the beads. We transfected Bio-Sumo*-GFP (negative control (left samples)) and "dual" tubulin (i.e., the construct 5′-TUBB3-P2A-Bio-Sumo*-GFP-TUBA1A-3′, right samples) in HEK293T cells and harvested lysates after 72 h of transfection. B, beads, after extensive washing; P, soluble proteins released from beads after Sumo* protease treatment; R, retained material on beads; M, marker lane (molecular weights from top to bottom: 250, 150, 100, 75, 50, 37, 25 kDa). The arrows indicate biotin-tagged fusion proteins prior to cleavage (in B lanes) and released principal proteins after cleavage (in P lanes)

and cells that are overexpressing them in very high quantities. It is therefore logical that we not only isolate fully folded recombinant dimeric tubulins (and their associated network of TAPs) but also tubulins that are at some step in the folding pathway, as well as tubulins that are on their way to be degraded, and perhaps even some aggregated tubulins. We therefore advice to experiment with different time points of harvesting cells to obtain the optimal point where most of the recombinant tubulin is dimeric and functional. In our hands, optimal experiments in HEK293T cells yield approximately 70% functional dimeric tubulin. From one transfected 15 cm dish of HEK293T cells, we obtain about 50 μg of purified functional tubulin. To obtain 1 mg of purified functional protein, about 30 dishes need to be transfected. Note that for Hela cell transfection, one should add 40 μL X-tremeGENE HP DNA Transfection reagent in each condition.

3. Do not pipette transfection mixture up and down after incubation.
4. Avoid introducing bubbles.
5. Avoid touching the beads.
6. HEK293T cells do not attach very well to a culture dish and can be resuspended without the use of trypsin. If you want to isolate tubulin from other cells, e.g., Hela cells, these are often much more tightly attached to the plate, and you will need trypsin-EDTA to harvest such cells.
7. You can concentrate your protein by using less cleavage buffer.
8. It is wise to experiment with washing buffers and conditions for purifying tubulins and TAPs. The most optimal pH for isolating tubulin is ~6.8; however, components (and perhaps even pH) of the buffer could be varied, certainly if one is after TAPs rather than tubulins.
9. At all times, wear gloves and avoid touching the gel.
10. Staining solution listed below does contain Stainer B.

References

1. Mitchison T, Kirschner M (1984) Dynamic instability of microtubule growth. Nature 312 (5991):237
2. Breuss M, Keays DA (2014) Microtubules and neurodevelopmental disease: the movers and the makers. Adv Exp Med Biol 800:75–96. https://doi.org/10.1007/978-94-007-7687-6_5
3. Fulton C, Simpson PA (1976) Selective synthesis and utilization of flagellar tubulin. The multi-tubulin hypothesis. Cell Motil 3:987–1005
4. Lewis SA, Tian G, Cowan NJ (1997) The α-and β-tubulin folding pathways. Trends Cell Biol 7(12):479–484
5. Nithianantham S, Le S, Seto E, Jia W, Leary J, Corbett KD, Moore JK, Al-Bassam J (2015) Tubulin cofactors and Arl2 are cage-like chaperones that regulate the soluble αβ-tubulin pool for microtubule dynamics. Elife 4:e08811
6. Gasic I, Mitchison TJ (2019) Autoregulation and repair in microtubule homeostasis. Curr Opin Cell Biol 56:80–87. https://doi.org/10.1016/j.ceb.2018.10.003
7. Hinman ND, Morgan JL, Seeds NW, Cann JR (1973) Isolation of brain tubulin by affinity chromatography. Biochem Biophys Res Commun 52(3):752–758
8. Williams RC Jr, Lee JC (1982) Preparation of tubulin from brain. Methods Enzymol 85:376–385
9. Minoura I, Hachikubo Y, Yamakita Y, Takazaki H, Ayukawa R, Uchimura S, Muto E (2013) Overexpression, purification, and functional analysis of recombinant human tubulin dimer. FEBS Lett 587(21):3450–3455
10. Ti S-C, Alushin GM, Kapoor TM (2018) Human β-tubulin isotypes can regulate microtubule protofilament number and stability. Dev Cell 47(2):175–190. e175
11. Yan F, Doronina VA, Sharma P, Brown JD (2010) Orchestrating ribosomal activity from inside: effects of the nascent chain on the peptidyltransferase centre. Biochem Soc Trans 38 (6):1576–1580. https://doi.org/10.1042/BST0381576
12. Yu N, Signorile L, Basu S, Ottema S, Lebbink JH, Leslie K, Smal I, Dekkers D, Demmers J, Galjart N (2016) Isolation of functional tubulin dimers and of tubulin-associated proteins from mammalian cells. Curr Biol 26 (13):1728–1736
13. de Boer E, Rodriguez P, Bonte E, Krijgsveld J, Katsantoni E, Heck A, Grosveld F, Strouboulis J (2003) Efficient biotinylation and single-step purification of tagged transcription factors in mammalian cells and transgenic mice. Proc Natl Acad Sci U S A 100(13):7480–7485. https://doi.org/10.1073/pnas.1332608100

Chapter 2

Expression and Purification of Microtubule-Associated Proteins from HEK293T Cells for In Vitro Reconstitution

Shasha Hua and Kai Jiang

Abstract

In vitro reconstitution has been an invaluable tool to elucidate the roles and mechanisms of microtubule-associated proteins (MAPs). Like all biochemical assays, the quality of the proteins is vital for success. In the microtubule field, proteins produced in bacteria and insect cells have been widely used for in vitro reconstitution. Recently, we applied the mammalian HEK293T cell expression system to our research on several MAPs. We find that such system is especially suitable for quick functional studies and can produce active proteins that sometimes are difficult for either bacteria or insect cell expression systems. Here, we provide a detailed protocol to express and purify microtubule-associated proteins from HEK293T cells using a Strep-tag strategy. The method described here can be adopted for preparation of other proteins and protein complexes for reconstitution studies.

Key words Microtubule, Microtubule-associated protein, In vitro reconstitution, Mammalian cell expression system, Strep-tag, StrepTactin

1 Introduction

In cells, microtubule dynamics are regulated by a variety of microtubule-associated proteins (MAPs). Understanding the behavior of MAPs on microtubules and the underlying mechanisms requires careful functional analyses both in vivo and in vitro, which often are complementary to each other. With the discovery of tubulin and subsequent development of the method of microtubule assembly using purified tubulin [1–3], in vitro reconstitution has played a pivotal role in uncovering the key proteins in the regulation of microtubule polymerization [4, 5], depolymerization [6], nucleation [7], severing [8], etc. In the early days, MAPs were identified and isolated, either through their ability to co-purify with tubulin or by displaying specific activities toward microtubules. Thus, the amount of proteins used for further characterization was limited. This has been improved dramatically with the availability of recombinant proteins in sufficient quantity and validated

Helder Maiato (ed.), *Cytoskeleton Dynamics: Methods and Protocols*, Methods in Molecular Biology, vol. 2101,
https://doi.org/10.1007/978-1-0716-0219-5_2,

quality. In the past two decades, benefitting from the advances in labelling proteins with fluorescent tags and imaging them under total internal reflection fluorescence (TIRF) microscopy, a vast knowledge of how different MAPs act on microtubules has been gained [9–11]. Moreover, the success is not limited to MAPs with a single polypeptide chain. Huge multi-subunit MAP complexes, such as the yeast Dam1 complex [12, 13] and the human dynein complex [14], have also been studied using recombinant proteins.

In the microtubule field, MAPs currently used for in vitro reconstitution assays are generally purified from bacteria or insect cells. Relatively small proteins and protein fragments can be easily expressed in *E. coli*, while for large protein expression, insect cells give in most cases better results than *E. coli*. Recently, instead of these systems, we applied the mammalian HEK293T cell expression system to our research on several MAPs [15–17]. Mammalian cell expression systems have been widely used for structural studies [18] and protein production by biotech companies [19]. It has several advantages over *E. coli* or insect cells, such as proper protein folding, posttranslational modifications, and flexible exogenous gene delivery methods. It's noteworthy that we found that in few cases, only proteins purified from HEK293T, but not insect cells, show microtubule binding activity in vitro (our unpublished observations). Because of its increasing popularity, several cloning strategies have also been developed to express multi-subunit protein complexes in mammalian cells [20, 21]. Foreign genes can be introduced to HEK293T cells through transient transfection, stable integration into the genome, and virus infection. For in vitro work, we found that transient transfection is more convenient and less time consuming than the other two, especially when it comes to test a panel of deletion or point mutation mutants.

To drive the overexpression of the protein of interest in HEK293T cells, a strong promoter is often chosen. CAG [22], CMV, and its optimized version CMV5 [23, 24] are commonly used promoters. While CMV promoter has been widely used for expressing exogenous genes in human cell lines, CAG or CMV5 promoter is often the better choice when yield is the critical factor, as they always give higher yields compared to CMV promoter [24, 25]. However, as in the case of all expression systems, the more does not always mean the better. It's advisable to use CMV promoter when the target protein has a tendency to aggregate, as higher expression often has adverse effects. In order to purify the overexpressed protein, the protein of interest should be conjugated with an affinity tag. We prefer Strep-tag to His-tag and Flag-tag as it offers a good compromise between yield, purity, and cost. With one-step purification, the proteins can achieve up to 95% purity and be eluted at mild conditions to be used directly for in vitro reconstitution experiments [26].

In this chapter, we provide a detailed procedure to express and purify MAPs from HEK293T cells with the CMV5 promoter-based pTT5 expression vector, including guidance for troubleshooting.

2 Materials

1. Expression vector: We prefer vector with CMV5 promoter to CAG one because of the smaller size. The original vector was obtained from Addgene (#55188) and subsequently modified to express Strep-tag fusion proteins with GFP at N- or C-terminus of the proteins (*see* Fig. 1 for vector maps and multiple cloning sites (MCS)). These vectors are available upon request (*see* **Note 1**).
2. 1 mg/mL Polyethylenimine: Dissolve 200 mg polyethylenimine (PEI) "MAX" in 200 mL MilliQ-grade water, and adjust pH to 7.0. Sterilize the solution by passing it through 0.22 μm membrane filter. Prepare 2 mL aliquots and store at −20 °C. The working stock is kept at 4 °C.
3. StrepTactin Sepharose (#28-9355-99, GE Healthcare).

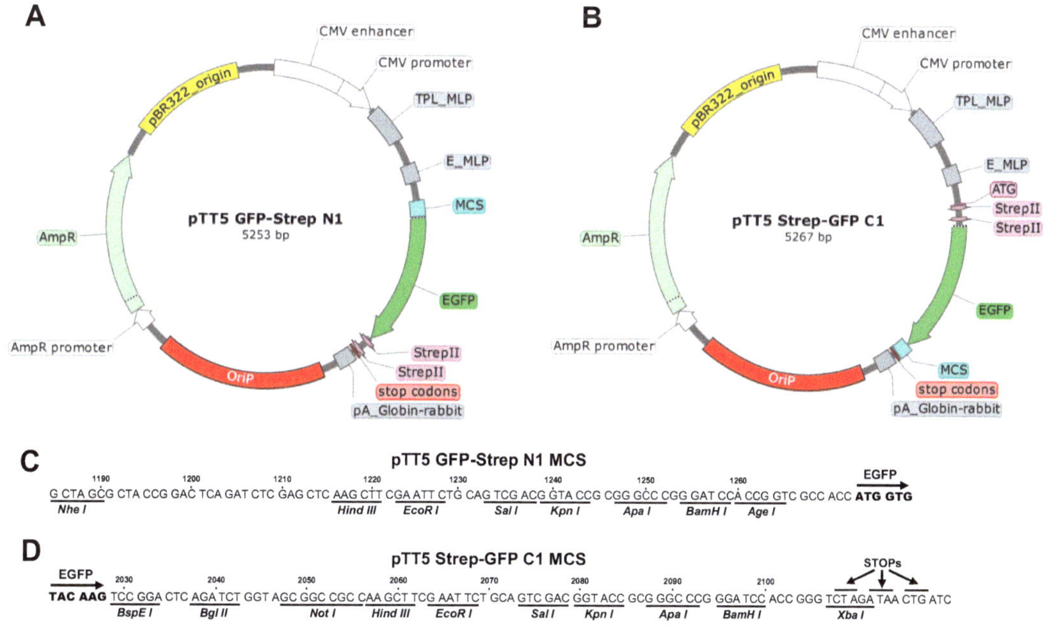

Fig. 1 Map of pTT5 GFP-Strep N1 (**A, C**) and Strep-GFP C1 (**B, D**) vector. For expression in mammalian cells, cDNAs encoding the target proteins are cloned into the multiple cloning sites. The tripartite leader (TPL_MLP) and enhancer (E_MLP) of adenovirus major late promoter that are inserted downstream of CMV promoter can significantly increase the expression level of target proteins. The OriP element allows the construct undergoing episomal replication in Epstein-Barr Nuclear Antigen 1 (EBNA1) containing cell line, such as HEK293E, which can further enhance the protein expression. However, the proteins purified from HEK293E cells didn't work as well as the ones from HEK293T cells in our hands

4. Lysis buffer: 50 mM HEPES, 300 mM NaCl, 0.5% Triton X-100, protease inhibitors, pH 7.4.
5. Wash buffer A: 50 mM HEPES, 300 mM NaCl, 0.5% Triton X-100, protease inhibitors, pH 7.4.
6. Wash buffer B: 50 mM HEPES, 1 M NaCl, 0.5% Triton X-100, pH 7.4.
7. Wash buffer C: 50 mM HEPES, 150 mM NaCl, 0.01% Triton X-100, pH 7.4.
8. Elution buffer: 2.5 mM Desthiobiotin, 50 mM HEPES, 150 mM NaCl, 0.01% Triton X-100, pH 7.4.
9. Costar® Spin-X® Centrifuge Tube Filters (Corning).
10. Eppendorf Centrifuge 5810R with Rotor S-4-104.
11. Eppendorf Centrifuge 5424R with Rotor FA-45-24-11.

3 Methods

3.1 Generation of the Strep-Tag Fusion Protein Expression Construct

For in vitro reconstitution, in most cases, we would like to image the protein of interest under the microscope. Thus, the target protein should be fused with a fluorescent tag like GFP for imaging, alongside with a Strep-tag for purification. Care should be taken when determining which terminus (N or C) of the target protein to insert the Strep-GFP tag. The tag should not impair the localization and function of the target protein. We always test the expression constructs for their cellular localization before starting the purification. The cDNA encoding the target protein can be cloned in our modified pTT5 vectors by traditional restriction enzyme/ligase strategy or Gibson assembly. Once the correct Strep-GFP fusion construct has been obtained, use commercial kits to prepare plasmid DNA with high quality (*see* **Note 2**).

3.2 Transfection of HEK293T Cells with PEI

As described below, we provide the detailed procedure to transfect one 15-cm dish HEK293T cells with PEI. In most cases, the yield is about 10–100 μg of protein for one 15-cm dish. If the expression level of the target protein is very low, use more dishes. Our expression and purification of Strep-GFP-CAMSAP1mini [16] is shown as an example in Fig. 2.

3.2.1 Day 1

Seed the cells on one 15-cm dish at about 40% confluency.

3.2.2 Day 2

1. Transfect the cells with DNA/PEI complex. The ratio of DNA (μg) to PEI(μL) is 1:4. Dilute 15–20 μg DNA and 60–80 μL PEI in 1.5 mL serum-free medium separately (*see* **Note 3**). Add diluted PEI dropwise to diluted DNA with shaking to mix well, and incubate the mixture for 20 min at room temperature.
2. During the incubation, refresh the medium.

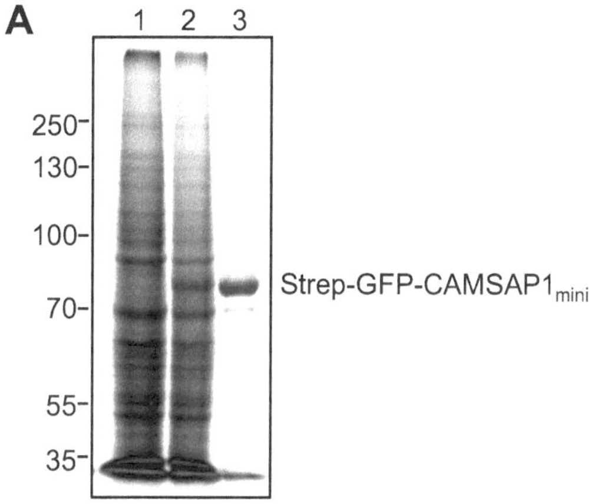

Fig. 2 Expression and purification of Strep-GFP-CAMSAP1$_{mini}$ in HEK293T cells. Coomassie blue-stained protein gel loaded with lysates from control cells (lane 1), lysates from cells transfected with Strep-GFP-CAMSAP1$_{mini}$ (lane 2), and eluted proteins (lane 3). The position of Strep-GFP-CAMSAP1$_{mini}$ is indicated on the right. Molecular weight markers are shown on the left

3. Add DNA/PEI mixture dropwise to the cells. Gently rotate the dish to mix well.

3.2.3 Day 3 (Optional)

For MAPs that localize to the nucleus during interphase, it's essential to release the nuclear pool of proteins into the cytoplasm to increase the yields by synchronizing the cells in mitosis. Twenty-four hours after transfection, treat the cells with 200 ng/mL nocodazole for 16–20 h (*see* **Note 4**).

3.3 StrepTactin Affinity Purification

1. 36–48 hours post-transfection, wash the cells once with ice-cold PBS (*see* **Note 5**).
2. Add 10 mL ice-cold PBS to the dish and scrape the cells into PBS using a cell scraper. Transfer the cells into one 15 mL tube.
3. Centrifuge the cells at 1000 rpm (94 × *g*, Rotor S-4-104) for 5 min at 4 °C.
4. Remove the supernatant and resuspend the cell pellets with 0.9 mL ice-cold lysis buffer. Transfer the lysate to a 1.5 mL Eppendorf tube. Let the tube on ice for 10 min.
5. Clear the debris by centrifuging the lysate at 14,000 rpm (18,407 × *g*, Rotor FA-45-24-11) for 20 min at 4 °C.
6. Wash 100 μL StrepTactin beads twice with lysis buffer to remove ethanol in the storage buffer.
7. Transfer the supernatant to the tube with StrepTactin beads, and incubate on rotator for 45 min at 4 °C.
8. Wash the beads three times with 1 mL wash buffer A.

9. Wash the beads three times with 1 mL wash buffer B (*see* **Note 6**).
10. Wash the beads twice with 1 mL wash buffer C. In the final wash step, remove the buffer as much as possible.
11. Add 80 μL elution buffer to elute the proteins. Gently tap the tube and keep it on ice for 2 min. Centrifuge at 3000 rpm (845 × *g*, Rotor FA-45-24-11) for 30 s at 4 °C. Transfer the supernatant to a Spin-X centrifuge tube to remove the beads that are accidently aspirated out (*see* **Note 7**).
12. Repeat **step 11**.
13. Normally, the amounts of proteins eluted in **step 11** (E1) are a bit higher than in **step 12** (E2). If lower concentration is not an issue for the subsequent assays, combine E1 with E2 to get higher yields; otherwise, only keep E1.
14. Prepare 2 μL aliquots of the protein, snap freeze with liquid N_2, and store at −80 °C.
15. Run SDS-PAGE, stain with Coomassie blue solution, and compare the intensity of the target protein with bovine serum albumin (BSA) standard. The concentration of target protein can also be measured with Nanodrop if the purified protein is very good and has no major degradation bands.
16. Determine the purity of proteins by mass spectrometry (*see* **Note 8**).

3.4 Conclusion

Here, we describe the protocol to purify proteins from HEK293T cells for in vitro reconstitution assays. Since the final concentration of proteins needed for an in vitro assay is relatively low, the amount of proteins purified from one 15-cm dish is enough for more than fifty in vitro reactions. We have successfully applied this method to purify several MAPs for in vitro work, such as the minus-end binding proteins, ASPM and CAMSAPs, and microtubule severing protein katanin. As the Strep-GFP fusion construct for purification can also be used for live-cell imaging experiment, we find it a very useful tool to compare the in vivo and in vitro results. We hope that the method described here will be broadly applied in future studies.

4 Notes

1. ATUM (formerly DNA2.0) offers a collection of mammalian expression vectors (https://www.atum.bio/catalog/expression-vectors/mammalian). Some of them are claimed to be better than CMV5 and CAG promoter-based vectors.
2. The quality of the DNA prepared with miniprep kits is usually good enough. However, DNA prepared with midiprep or maxiprep kits gives higher transfection efficiency.

3. For proteins with low expression levels, the amount of DNA can be increased up to 60 μg per 15-cm dish. Excessive amounts of DNA/PEI complexes will decrease the viability of the cells. For one 15-cm dish transfection, we normally mix the diluted DNA and PEI in 35-mm dish.
4. We find that nocodazole treatment sometimes can increase the yield and solubility of difficult MAPs.
5. Skip this step if too many cells are floating after nocodazole treatment.
6. It is essential to include this extra-high-salt wash step to get rid of strong binding partners of the target protein [27].
7. In the case of poor elution efficiency, increasing the ion strength and the concentration of Triton X-100 and adding 10% glycerol might help.
8. It is always advisable to verify the purity of proteins by mass spectrometry analysis. Common contaminations are several endogenously biotinylated proteins and heat-shock proteins.

Acknowledgment

This work is supported by 1000 Youth Talents Program to K.J. and an NSFC grant (31871356) to K.J.

References

1. Borisy GG, Taylor EW (1967) The mechanism of action of colchicine: binding of colchincine-3H to cellular protein. J Cell Biol 34 (2):525–533. https://doi.org/10.1083/jcb.34.2.525
2. Weisenberg RC, Borisy GG, Taylor EW (1968) The colchicine-binding protein of mammalian brain and its relation to microtubules. Biochemistry 7(12):4466–4479
3. Weisenberg RC (1972) Microtubule formation in vitro in solutions containing low calcium concentrations. Science 177(54):1104–1105
4. Cleveland DW, Hwo SY, Kirschner MW (1977) Purification of tau, a microtubule-associated protein that induces assembly of microtubules from purified tubulin. J Mol Biol 116(2):207–225
5. Gard DL, Kirschner MW (1987) A microtubule-associated protein from Xenopus eggs that specifically promotes assembly at the plus-end. J Cell Biol 105(5):2203–2215
6. Desai A, Verma S, Mitchison TJ, Walczak CE (1999) Kin I kinesins are microtubule-destabilizing enzymes. Cell 96(1):69–78. pii: S0092-8674(00)80960-5
7. Zheng Y, Wong ML, Alberts B, Mitchison T (1995) Nucleation of microtubule assembly by a gamma-tubulin-containing ring complex. Nature 378(6557):578–583. https://doi.org/10.1038/378578a0
8. McNally FJ, Vale RD (1993) Identification of katanin, an ATPase that severs and disassembles stable microtubules. Cell 75(3):419–429. https://doi.org/10.1016/0092-8674(93)90377-3
9. Helenius J, Brouhard G, Kalaidzidis Y, Diez S, Howard J (2006) The depolymerizing kinesin MCAK uses lattice diffusion to rapidly target microtubule ends. Nature 441 (7089):115–119. https://doi.org/10.1038/nature04736
10. Bieling P, Laan L, Schek H, Munteanu EL, Sandblad L, Dogterom M, Brunner D, Surrey T (2007) Reconstitution of a microtubule plus-end tracking system in vitro. Nature 450 (7172):1100–1105. https://doi.org/10.1038/nature06386

11. Brouhard GJ (2008) XMAP215 is a processive microtubule polymerase. Cell 132:79–88
12. Westermann S, Wang HW, Avila-Sakar A, Drubin DG, Nogales E, Barnes G (2006) The Dam1 kinetochore ring complex moves processively on depolymerizing microtubule ends. Nature 440(7083):565–569. https://doi.org/10.1038/nature04409
13. Westermann S, Avila-Sakar A, Wang HW, Niederstrasser H, Wong J, Drubin DG, Nogales E, Barnes G (2005) Formation of a dynamic kinetochore-microtubule interface through assembly of the Dam1 ring complex. Mol Cell 17(2):277–290. https://doi.org/10.1016/j.molcel.2004.12.019
14. Schlager MA, Hoang HT, Urnavicius L, Bullock SL, Carter AP (2014) In vitro reconstitution of a highly processive recombinant human dynein complex. EMBO J 33(17):1855–1868. https://doi.org/10.15252/embj.201488792
15. Jiang K, Hua S, Mohan R, Grigoriev I, Yau KW, Liu Q, Katrukha EA, Altelaar AF, Heck AJ, Hoogenraad CC, Akhmanova A (2014) Microtubule minus-end stabilization by polymerization-driven CAMSAP deposition. Dev Cell 28(3):295–309. https://doi.org/10.1016/j.devcel.2014.01.001
16. Atherton J, Jiang K, Stangier MM, Luo Y, Hua S, Houben K, van Hooff JJE, Joseph AP, Scarabelli G, Grant BJ, Roberts AJ, Topf M, Steinmetz MO, Baldus M, Moores CA, Akhmanova A (2017) A structural model for microtubule minus-end recognition and protection by CAMSAP proteins. Nat Struct Mol Biol 24(11):931–943. https://doi.org/10.1038/nsmb.3483
17. Jiang K, Rezabkova L, Hua S, Liu Q, Capitani G, Altelaar AFM, Heck AJR, Kammerer RA, Steinmetz MO, Akhmanova A (2017) Microtubule minus-end regulation at spindle poles by an ASPM-katanin complex. Nat Cell Biol 19(5):480–492. https://doi.org/10.1038/ncb3511
18. Goehring A, Lee CH, Wang KH, Michel JC, Claxton DP, Baconguis I, Althoff T, Fischer S, Garcia KC, Gouaux E (2014) Screening and large-scale expression of membrane proteins in mammalian cells for structural studies. Nat Protoc 9(11):2574–2585. https://doi.org/10.1038/nprot.2014.173
19. Geisse S, Fux C (2009) Recombinant protein production by transient gene transfer into Mammalian cells. Methods Enzymol 463:223–238. https://doi.org/10.1016/S0076-6879(09)63015-9
20. Kriz A, Schmid K, Baumgartner N, Ziegler U, Berger I, Ballmer-Hofer K, Berger P (2010) A plasmid-based multigene expression system for mammalian cells. Nat Commun 1:120. https://doi.org/10.1038/ncomms1120
21. Lu P, Bai XC, Ma D, Xie T, Yan C, Sun L, Yang G, Zhao Y, Zhou R, Scheres SHW, Shi Y (2014) Three-dimensional structure of human gamma-secretase. Nature 512(7513):166–170. https://doi.org/10.1038/nature13567
22. Niwa H, Yamamura K, Miyazaki J (1991) Efficient selection for high-expression transfectants with a novel eukaryotic vector. Gene 108(2):193–199
23. Massie B, Dionne J, Lamarche N, Fleurent J, Langelier Y (1995) Improved adenovirus vector provides herpes simplex virus ribonucleotide reductase R1 and R2 subunits very efficiently. Biotechnology (NY) 13(6):602–608
24. Massie B, Mosser DD, Koutroumanis M, Vitte-Mony I, Lamoureux L, Couture F, Paquet L, Guilbault C, Dionne J, Chahla D, Jolicoeur P, Langelier Y (1998) New adenovirus vectors for protein production and gene transfer. Cytotechnology 28(1–3):53–64. https://doi.org/10.1023/A:1008013211222
25. Qin JY, Zhang L, Clift KL, Hulur I, Xiang AP, Ren BZ, Lahn BT (2010) Systematic comparison of constitutive promoters and the doxycycline-inducible promoter. PLoS One 5(5):e10611. https://doi.org/10.1371/journal.pone.0010611
26. Schmidt T, Skerra A (2015) The Strep-tag system for one-step affinity purification of proteins from mammalian cell culture. Methods Mol Biol 1286:83–95. https://doi.org/10.1007/978-1-4939-2447-9_8
27. Hooikaas PJ, Martin M, Muhlethaler T, Kuijntjes GJ, Peeters CAE, Katrukha EA, Ferrari L, Stucchi R, Verhagen DGF, van Riel WE, Grigoriev I, Altelaar AFM, Hoogenraad CC, Rudiger SGD, Steinmetz MO, Kapitein LC, Akhmanova A (2019) MAP7 family proteins regulate kinesin-1 recruitment and activation. J Cell Biol 218(4):1298–1318. https://doi.org/10.1083/jcb.201808065

Chapter 3

In Vitro Reconstitution Assays of Microtubule Amplification and Lattice Repair by the Microtubule-Severing Enzymes Katanin and Spastin

Annapurna Vemu, Ewa Szczesna, and Antonina Roll-Mecak

Abstract

Microtubules are non-covalent dynamic polymers essential for the life of all eukaryotic cells. Their dynamic behavior is regulated by a large array of cellular effectors. In vitro microtubule assays have been instrumental in dissecting the mechanism of microtubule-associated proteins. In this chapter, we focus on microtubule-severing enzymes katanin and spastin. They are AAA ATPases that generate internal breaks in microtubules by extracting tubulin dimers out of the microtubule lattice. We present protocols for TIRF microscopy-based assays that were instrumental in proving that these enzymes not only sever microtubules but also remodel the microtubule lattice by promoting the exchange of lattice GDP-tubulin with GTP-tubulin from the soluble pool. This activity can modulate microtubule dynamics and support microtubule-dependent microtubule amplification in the absence of a nucleating factor.

Key words Cytoskeleton, Microtubule, Microtubule severing, Microtubule repair, Spastin, Katanin, Microtubule dynamics, GTP-tubulin, Rescue, AAA ATPase

1 Introduction

Microtubules are dynamic cytoskeletal polymers that are essential for basic cellular functions such as cell division, intracellular transport, and differentiation. Microtubules stochastically grow and depolymerize from their ends, a process known as dynamic instability [1]. Microtubules grow through the addition of GTP-tubulin at their ends. Lattice incorporation triggers the hydrolysis of GTP to GDP. The GDP-tubulin lattice is labile and is protected from depolymerization by the GTP-tubulin cap which results from the lag between the rate of GTP-tubulin incorporation and GTP hydrolysis [1–3]. Microtubule end dynamics are regulated in cells by a vast array of microtubule-associated proteins [4, 5].

Microtubule-severing enzymes are ATPases associated with various cellular activities (AAA ATPases) that use the energy of ATP hydrolysis to break microtubules in the middle [6]. Microtubule

Helder Maiato (ed.), *Cytoskeleton Dynamics: Methods and Protocols*, Methods in Molecular Biology, vol. 2101, https://doi.org/10.1007/978-1-0716-0219-5_3, © Springer Science+Business Media, LLC, part of Springer Nature 2020

severing activity was initially observed in metaphase *Xenopus laevis* extracts [7]. The protein responsible for this activity was later isolated and identified through biochemical purification and named katanin after the Japanese word for sword "katana" [8]. Since this original discovery, several other microtubule-severing enzymes have been identified: spastin [9, 10], fidgetin [11], and katanin-like proteins [12, 13]. Microtubule-severing enzymes are essential for cilia biogenesis, cell division, and morphogenesis, and their mutation is associated with debilitating neurodevelopmental and neurodegenerative diseases [6].

Light microscopy-based in vitro assays were instrumental in establishing the activity of microtubule-severing enzymes and proving that they were breaking microtubules along their lengths and not depolymerizing them from their ends [7, 9, 10]. However, these assays used stabilized microtubules (Taxol or GMPCPP stabilized) immobilized to glass using antibodies, Neutravidin-biotin linkages, or a rigor kinesin and thus missed two factors that characterize in vivo microtubules: (1) their dynamic behavior and (2) the presence of μM concentrations of soluble GTP-tubulin. Using a combination of negative-stain electron microscopy (EM) and total internal reflection fluorescence microscopy (TIRFM), we recently showed that microtubule-severing enzymes, katanin and spastin, sever microtubules by removing tubulin dimers from the microtubule lattice and that these nanodamage sites are repaired by the spontaneous incorporation of GTP-tubulin from the soluble tubulin pool [14], thus creating GTP islands along the microtubule. These transient GTP-tubulin islands can protect the microtubule from depolymerization and promote microtubule rescue [14]. When a microtubule is severed, the severed plus end emerges with a higher density of GTP-tubulin that can stabilize it against instantaneous depolymerization. The synergy between the increased rescue frequency and increased stability of the newly severed plus ends results in an amplification of microtubule number and mass [14]. Severing-mediated microtubule amplification explains the paradoxical in vivo findings that loss of severing enzymes leads to a reduction in microtubule mass in *C. elegans* meiotic spindles, plant microtubule cortical arrays, and *Drosophila* neurons [15–19].

In this chapter, we describe protocols to observe severing enzyme-mediated repair of stabilized and non-stabilized microtubules and severing enzyme-mediated microtubule amplification. These assays can be used to decipher the biophysical mechanism of microtubule severing enzymes and their effects on microtubule dynamics as well as for the identification and characterization of cellular factors that regulate their activity.

2 Materials

2.1 Equipment

1. Multicolor TIRF motorized inverted microscope (e.g., Nikon Ti-E 2000 with perfect focus).
2. A high NA TIRF objective (e.g., 1.49 NA Nikon CFI Apo TIRF 100×).
3. 488 and 640 nm solid-state lasers (Sapphire, Coherent).
4. Appropriate filters and dichroic mirrors.
5. Cooled electron-multiplying charge-coupled device (EMCCD) camera (e.g., iXON 897, Andor).
6. Controller software (e.g., MicroManager [20]).
7. Objective heater (Bioptechs).

2.2 Materials and Solutions

1. Plasma-cleaned/piranha-cleaned and silanized slides and coverslips (*see* **Note 1**).
2. Neutravidin (Thermo Fisher Cat# 31000).
3. BRB80: 80 mM PIPES, 1 mM $MgCl_2$, and 1 mM EGTA, pH 6.8 (adjusted with KOH).
4. Severing enzyme buffer: 20 mM HEPES, 300 mM KCl, 10 mM $MgCl_2$, 15% glycerol, 5 mM DTT or 1 mM TCEP, pH 7.0; pass through a 0.1 μm filter and keep on ice before use; use to dilute severing enzyme.
5. BB: BRB80 with 0.1% 2-mercaptoethanol.
6. BBC: BB with 2 mg/mL casein.
7. 1% Methylcellulose 4000 cP.
8. Oxygen scavengers (catalase, glucose oxidase, and 2 M glucose) prepared as described previously [21] (*see* **Note 2**).
9. Purified spastin or katanin (*see* Subheading 2.3).
10. Porcine brain tubulin (*see* Subheading 2.4).
11. GMPCPP-stabilized microtubules (*see* Subheading 2.5) (*see* **Note 3**).
12. GMPCPP-stabilized unmodified microtubule seeds (*see* Subheading 2.5).
13. Pre-perfusion mixture: BBC supplemented with 50 mM KCl, 1 mM ATP, 1% Pluronic F127, 20 mM glucose, glucose oxidase, and catalase (*see* **Note 4**).
14. Tubulin solution: BRB80, 0.5 mM GTP, 1% Pluronic F127, 2.5 mg/mL casein, 1 μM HiLyte 488-labeled tubulin.
15. Wash solution: BRB80 with 1.5 mg/mL casein, 10 mM 2-mercaptoethanol, 1% Pluronic F127, 20 mM glucose, glucose oxidase, and catalase.
16. Enzyme solution: Severing enzyme at desired concentration in pre-perfusion mixture.

17. Severing assay buffer: 50 mM KCl, 0.5% Pluronic F127, 0.2 mg/mL casein, 1.5% glycerol, 0.1% methylcellulose 4000 cP, 1 mM GTP, 1 mM ATP, 20 mM glucose, glucose oxidase, and catalase in BB.

2.3 Purified Severing Enzymes

1. Purify spastin or katanin according to [21, 22], and store in small aliquots at −80 °C.
2. Thaw an aliquot of enzyme, and spin for 10 min at 4 °C and 279,000 × *g* to remove any aggregates (*see* **Note 5**).
3. Remove the supernatant, and measure the concentration using absorbance at 280 nm or Bradford assay.
4. Enzymes are stable on ice for several hours at 20 μM.

2.4 Tubulin (See Note 6)

1. Porcine brain unlabeled tubulin, HiLyte 488 tubulin, HiLyte 647 tubulin, and biotin tubulin (Cytoskeleton, Inc.). Tubulin labeled with other fluorophores can also be used.
2. Quickly thaw an aliquot of unlabeled porcine brain tubulin and place on ice.
3. Resuspend 20 μg of lyophilized, labeled tubulin in ice-cold BRB80.
4. Spin the unlabeled and labeled tubulin in separate tubes for 10 min at 4 °C and 279,000 × *g* to remove any aggregates (*see* **Note 5**).
5. Carefully remove the top portion of the supernatant without touching the bottom of the tube that may contain tubulin aggregates. Measure the concentration by Bradford assay.
6. Keep the tubulin on ice. Use as quickly as possible as tubulin aggregates over time.

2.5 Double-Cycled, GMPCPP-Stabilized Microtubules and Unmodified Microtubule Seeds (Adapted from [23])

For dynamic and GMPCPP-capped GDP microtubules, we used GMPCPP microtubule seeds polymerized from unmodified tubulin purified from tsa201 cells as described previously [24]. Unmodified microtubules are severed at a lower rate compared to brain microtubules [25]. Therefore, by using unmodified microtubule seeds, severing enzymes preferentially sever the dynamic or GDP-brain microtubule extensions and not the unmodified GMPCPP microtubule seeds at the spastin or katanin concentrations used in these assays. Alternatively, sea urchin axonemes can also be used [26] to grow dynamic brain microtubule extensions.

1. For fluorescently labeled, GMPCPP-stabilized brain microtubules, mix 1 μL of 1 mg/mL biotin tubulin, 10 μL of 2 mg/mL HiLyte 647 tubulin, and 39.5 μL of 2 mg/mL unlabeled brain tubulin. For unmodified microtubule seeds, mix 2 μL of 2 mg/mL biotin tubulin and 98 μL of 2 mg/mL unmodified tubulin. Incubate on ice for 5 min.

2. Ultracentrifuge the tubulin mixture to remove any tubulin aggregates at 279,000 × *g* for 10 min at 4 °C (*see* **Note 5**).
3. Add GMPCPP to a final concentration of 0.5 mM, and incubate at 37 °C in a water bath (or heat block) for 1 h.
4. Pellet the microtubules using an ultracentrifuge at 126,000 × *g* for 15 min at 37 °C.
5. Remove the supernatant. Gently wash the pellet with warm BRB80.
6. Resuspend the microtubule pellet in cold BRB80 such that the final tubulin concentration is ~2 mg/mL.
7. Incubate on ice for 30 min, and mix by pipetting gently every 10 min.
8. Add GMPCPP to a final concentration of 0.5 mM. For GMPCPP-stabilized brain microtubules, incubate in 37 °C water bath (or heat block) between 4 and 16 h. For GMPCPP-stabilized unmodified seeds, incubate in 37 °C water bath (or heat block) for 30 min to obtain short microtubules.
9. Spin microtubules at 126,000 × *g* for 15 min at 37 °C.
10. Remove the supernatant without touching the pellet. Gently wash the pellet with warm BRB80.
11. Gently resuspend the microtubule seeds in warm BRB80 with a cut tip to prevent microtubule shearing.
12. GMPCPP-stabilized microtubules are stored at 30 °C and should be used within 2–3 days. For GMPCPP-stabilized seeds, flash-freeze in small aliquots. On the day of performing the assays, dilute the seed aliquot 1:10 in warm BB and place at 37 °C for 1 h before use and then keep at room temperature.

3 Methods

All steps are performed at room temperature unless stated otherwise.

3.1 Generation and Detection of Microtubule Nanoscale Damage on GMPCPP-Stabilized Microtubules Using Spastin or Katanin (Fig. 1)

1. Assemble chamber as described previously [21].
2. Perfuse room temperature BRB80 into the chamber.
3. Perfuse cold 0.1 mg/mL Neutravidin and incubate for 5–10 min.
4. Wash the chamber with BRB80 and BBC.
5. Dilute the microtubules with warm BB, and mix with a cut tip to avoid shearing. Perfuse the microtubules into the chamber, and incubate for 10 min (*see* **Note 7**).
6. Wash chamber three times with BBC (*see* **Note 8**).

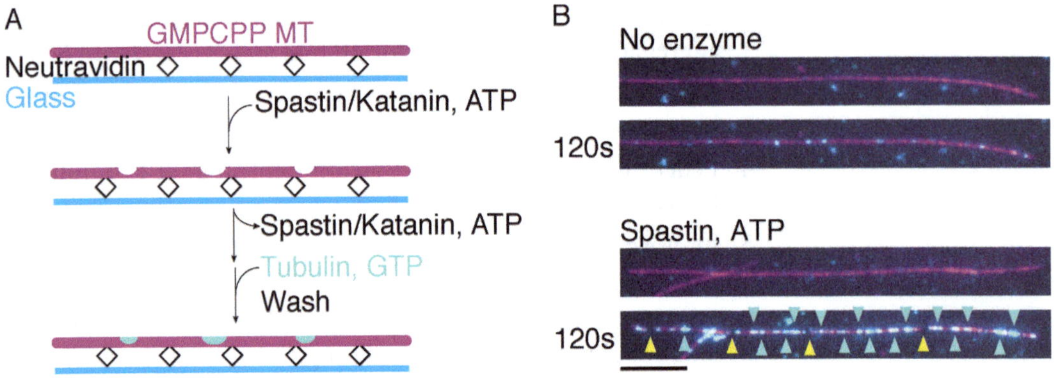

Fig. 1 (**a**) Schematic of assay for visualizing severing enzyme-mediated nanodamage and healing of GMPCPP-stabilized microtubules. (**b**) Top panels, GMPCPP-stabilized microtubule (magenta) not treated with severing enzymes shows no tubulin incorporation (cyan) along the microtubule lattice. Bottom panels, GMPCPP-stabilized microtubule incubated 120 s with 2 nM spastin and ATP shows tubulin incorporation (cyan) along the microtubule lattice. Cyan arrows indicate sites of tubulin incorporation; yellow arrows indicate severing events. Scale bar, 5 μm

7. Perfuse pre-perfusion mix (*see* **Notes 8–10**).
8. Acquire images of multiple areas and save the coordinates for these areas. To avoid photodamage, use short exposures and low laser power. The purpose of this step is to quickly visualize the microtubules before starting the assay with the severing enzymes and to count the number of severing events after introducing the enzyme.
9. Perfuse enzyme solution into the chamber and incubate for 30–120 s. The nanodamage extent varies steeply with incubation time, so the time has to be well controlled. We typically use enzyme concentrations between 2 and 20 nM.
10. Stop here if you want to obtain nanodamaged microtubules for subsequent assays (for example with motor proteins to investigate the effects of the nanodamage on motility) and do not want to visualize repair.
11. Perfuse tubulin solution and incubate for 5 min. Tubulin labeled with a different dye than the microtubules will allow visualization of the incorporation of tubulin at sites of nanodamage introduced by the enzyme in **step 9** (*see* **Note 11**).
12. Wash chamber three times with the wash solution, and visualize microtubules and incorporated tubulin.

3.2 Two-Color Severing Assays with GMPCPP-Capped GDP Microtubules to Monitor Lattice Nanodamage and Repair (Fig. 2)

The following protocol describes the generation of GMPCPP-capped microtubule extensions (adapted from [27, 28]). GMPCPP-capped microtubules are needed in order to be able to exchange the tubulin concentration in the chamber without microtubule depolymerization.

1. Perfuse BRB80 into the chamber.

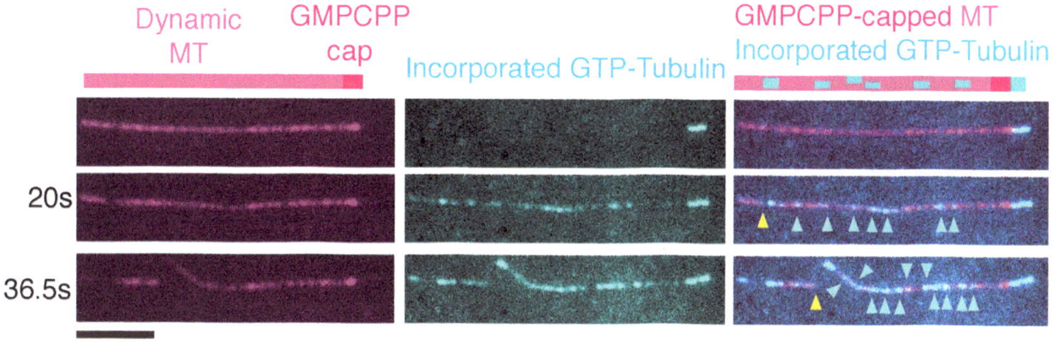

Fig. 2 Time course of GMPCPP-capped dynamic microtubule extensions (magenta) exposed to 10 nM spastin in the presence of 7 μM GTP-tubulin (cyan). Cyan arrows indicate sites of tubulin incorporation; yellow arrows indicate severing events. Scale bar, 5 μm

2. Incubate the chamber with 0.1 mg/mL Neutravidin for 5–10 min.
3. Wash the chamber with BBC.
4. Incubate the chamber with unmodified microtubule seeds or axonemes for 10 min (*see* **Notes 7** and **8**).
5. Wash the chamber with BBC.
6. Polymerize 10% or 20% HiLyte 647-labeled dynamic microtubules at 16 μM tubulin in severing assay buffer for 10–12 min on the microscope (*see* **Notes 7–12**).
7. Perfuse in 2 μM of 10% HiLyte 647-labeled tubulin in severing assay buffer supplemented with 0.5 mM GMPCPP instead of GTP (*see* **Note 13**).
8. Wash the chamber with severing assay buffer.
9. Start image acquisition as 10 nM spastin and 7 μM tubulin (or any tubulin concentration that is investigated) in severing assay buffer are perfused into the chamber. A frame rate of 2 Hz or higher is sufficient with 100 ms exposure in each channel.
10. Seal the chamber to prevent evaporation.

3.3 Severing Assays with Dynamic Microtubules (Fig. 3)

1. Perfuse BRB80 to the chamber.
2. Incubate the chamber with 0.1 mg/mL Neutravidin for 10 min.
3. Wash the chamber with BBC.
4. Incubate the chamber with seeds for 10 min (*see* **Notes 7** and **8**).
5. Wash the chamber with BBC.
6. Polymerize 10% HiLyte 647-labeled dynamic microtubules at 12 μM tubulin in severing assay buffer (*see* **Notes 7–12**).

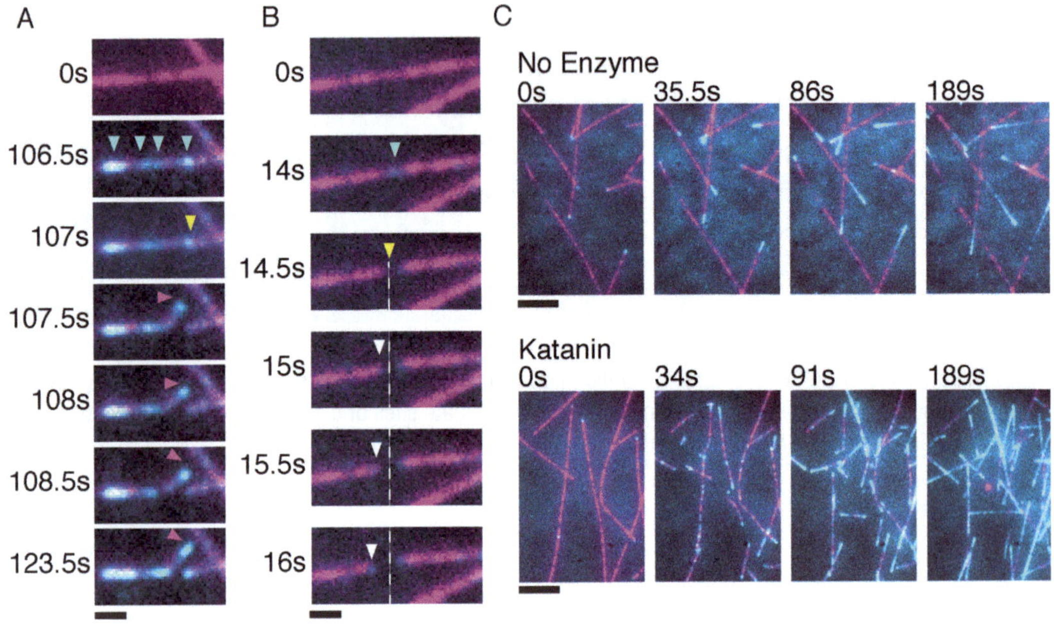

Fig. 3 Incorporation of soluble tubulin (cyan) along the lattice of dynamic microtubules (magenta) in the presence of 25 nM katanin and 12 μM GTP-tubulin. (**a**) Time course of a severing enzyme-generated plus end that is stable and grows. Cyan arrows indicate sites of tubulin incorporation, yellow arrow shows severing event, and magenta arrow follows the stable new plus end. (**b**) Time course of a severing enzyme-generated plus end that is unstable and depolymerizes. Cyan arrow indicates site of tubulin incorporation, yellow arrow indicates a severing event, and white arrow follows the depolymerizing new plus end. The dotted line marks the severing enzyme-generated new plus end. Scale bar, 1 μm. (**c**) Top panels, time lapse images showing microtubule dynamics at 12 μM tubulin in the absence of severing enzymes. The last frame is bleach corrected in the microtubule (magenta) channel. Bottom panels, time-lapse images showing microtubule number and mass amplification through katanin severing. Cyan shows newly incorporated tubulin only at the growing ends (top panel) or at the growing ends and along the microtubule shaft in the presence of enzyme and ATP (bottom panel). Scale bar, 5 μm

7. Perfuse 10% HiLyte 488-labeled tubulin at 12 μM and spastin or katanin (5–25 nM) in severing assay buffer into the chamber. Start image acquisition during perfusion. A range of tubulin and enzyme concentrations should be tested to investigate the effects of tubulin lattice incorporation.
8. Seal the chamber to prevent evaporation.

3.4 Quantification

Quantification can be performed using FIJI [29].

3.4.1 Quantification of the Intensity of Incorporated Tubulin and Microtubule Severing Rates

1. Manually select microtubules with a line selection.
2. Measure background-subtracted average intensity of the fluorescent tubulin incorporated (MatLab scripts can be used to generate background-subtracted line scans and detect the peaks on line scans, measure their full-width-at-half-maximum and intensity).

3. To measure severing rate, count the number of mesoscale severing events per microtubule length as a function of time [25].

3.4.2 Scoring the Stability of the Newly Severed Plus Ends in Dynamic Microtubule Assays

1. Determine microtubule polarity. This can be done by measuring the growth rate of the two ends or by imaging during the polymerization phase since the plus end grows faster. After severing, keep track of the polarity of the severed microtubules that are no longer connected to the coverslip through the seed.
2. Once a severing event occurs, follow the newly severed end frame by frame. If there is no depolymerization (in the 647 microtubule channel) immediately following severing, tubulin incorporation (in the 488 channel) is seen at the microtubule end, and growth persists, the newly severed end is considered stable (Fig. 3a). If the newly severed end depolymerizes, loss of fluorescent microtubule intensity (in the 647 microtubule channel) at the new plus end is observed (Fig. 3b).

4 Conclusion

In this chapter, we describe protocols to monitor severing enzyme-catalyzed tubulin exchange along the microtubule lattice as well as severing enzyme-mediated microtubule amplification. These assays lay down the foundation for understanding the effects of microtubule-severing enzymes on microtubule dynamics and organization and how their activities synergize with other microtubule-associated proteins, such as CLASP [30] or depolymerizing factors like MCAK [31] that can interact with the newly severed microtubule ends or the severing enzyme-generated GTP-tubulin islands.

5 Notes

1. Proteins adsorb nonspecifically to untreated glass. Therefore, it is essential for all glass surfaces to be clean and passivated for these assays. Otherwise, severing enzymes and tubulin will nonspecifically adsorb to the glass. This is especially important for severing enzymes as they are used at low concentrations in these assays. A detailed protocol for piranha-cleaned glass can be found in [21]. Alternatively, glass can be plasma cleaned with argon for 3 min using the Expanded Plasma Cleaner (Harrick Plasma, Ithaca, NY). Rinse the glass three times with MilliQ water, incubate for 15 min in 0.1 M KOH, and continue from the KOH step in [21].
2. If chambers have fluorescent aggregates, filter the catalase/glucose oxidase through a 0.1 μm filter.

3. For healing assays using stabilized microtubules, Taxol-stabilized microtubules can be used instead of GMPCPP-stabilized microtubules. To polymerize Taxol microtubules, *see* [21]. If Taxol microtubules are used, make sure to add Taxol to all buffers.
4. Solutions 13–17 should be prepared just before use.
5. Make sure that the ultracentrifuge tubes and Eppendorf tubes are prechilled prior to placing protein into them.
6. Severing enzyme activity is very sensitive to aggregated tubulin because the enzymes will bind to the tubulin aggregates instead of microtubules. Therefore, it is essential to remove all aggregates and work with high-quality, pure samples. To remove tubulin aggregates, ultracentrifuge tubulin at 279,000 × *g* for 10 min at 4 °C. If tubulin aggregates persist, cycle the tubulin through one cycle of polymerization and depolymerization [32, 33] or purify a new batch of tubulin.
7. The extent of nanoscale damage and/or severing depends on the density of microtubules in the chamber and how strongly they are attached to the coverslip with the Neutravidin-biotin linkage. The same applies if a rigor kinesin mutant is used to immobilize the microtubules on the glass [9, 34]. Therefore, if quantitation is attempted, it is important to use similar microtubule densities in all chambers and the same immobilization protocol.
8. Do not add any cold solutions to the chamber once microtubules are added as this will depolymerize the microtubules.
9. The perfusion volumes used in the protocol will vary depending on the width of the chambers. Typically, our chambers hold ~7 μL. We usually perfuse volumes 2–3 times the chamber volume.
10. To prevent evaporation from the edges of the chamber, leave a few μL of perfusion solution at the ends of the channel and place a humidity cover over the chamber.
11. Add tubulin after the addition of all other components.
12. From this step onwards, assays are performed on the microscope at 30 °C.
13. To avoid de novo microtubule nucleation in the chamber, add the 0.5 mM GMPCPP just before perfusing the tubulin mixture into the chamber. The incubation time with the GMPCPP tubulin will vary. Therefore, image the chamber intermittently to see if GMPCPP extensions are growing.

Acknowledgment

A.R.M. is supported by the intramural programs of the National Institute of Neurological Disorders and Stroke (NINDS) and the National, Heart, Lung, and Blood Institute (NHLBI).

References

1. Mitchison T, Kirschner M (1984) Dynamic instability of microtubule growth. Nature 312:237–242
2. Carlier M-F (1982) Guanosine-5-′-triphosphate hydrolysis and tubulin polymerization. Mol Cell Biochem 47(2):97–113
3. Carlier MF, Pantaloni D (1981) Kinetic analysis of guanosine 5′-triphosphate hydrolysis associated with tubulin polymerization. Biochemistry 20(7):1918–1924
4. Alfaro-Aco R, Petry S (2015) Building the microtubule cytoskeleton piece by piece. J Biol Chem 290(28):17154–17162. https://doi.org/10.1074/jbc.R115.638452
5. Akhmanova A, Steinmetz MO (2015) Control of microtubule organization and dynamics: two ends in the limelight. Nat Rev Mol Cell Biol 16(12):711–726. https://doi.org/10.1038/nrm4084
6. McNally FJ, Roll-Mecak A (2018) Microtubule-severing enzymes: from cellular functions to molecular mechanism. J Cell Biol 217(12):4057–4069. https://doi.org/10.1083/jcb.201612104
7. Vale RD (1991) Severing of stable microtubules by a mitotically activated protein in Xenopus egg extracts. Cell 64(4):827–839
8. McNally FJ, Vale RD (1993) Identification of katanin, an ATPase that severs and disassembles stable microtubules. Cell 75(3):419–429
9. Roll-Mecak A, Vale RD (2005) The Drosophila homologue of the hereditary spastic paraplegia protein, spastin, severs and disassembles microtubules. Curr Biol 15(7):650–655. https://doi.org/10.1016/j.cub.2005.02.029
10. Evans KJ, Gomes ER, Reisenweber SM, Gundersen GG, Lauring BP (2005) Linking axonal degeneration to microtubule remodeling by Spastin-mediated microtubule severing. J Cell Biol 168(4):599–606. https://doi.org/10.1083/jcb.200409058
11. Mukherjee S, Diaz Valencia JD, Stewman S, Metz J, Monnier S, Rath U, Asenjo AB, Charafeddine RA, Sosa HJ, Ross JL, Ma A, Sharp DJ (2012) Human Fidgetin is a microtubule severing the enzyme and minus-end depolymerase that regulates mitosis. Cell Cycle 11 (12):2359–2366. https://doi.org/10.4161/cc.20849
12. Zhang D, Rogers GC, Buster DW, Sharp DJ (2007) Three microtubule severing enzymes contribute to the "Pacman-flux" machinery that moves chromosomes. J Cell Biol 177 (2):231–242. https://doi.org/10.1083/jcb.200612011
13. Sonbuchner TM, Rath U, Sharp DJ (2010) KL1 is a novel microtubule severing enzyme that regulates mitotic spindle architecture. Cell Cycle 9(12):2403–2411. https://doi.org/10.4161/cc.9.12.11916
14. Vemu A, Szczesna E, Zehr EA, Spector JO, Grigorieff N, Deaconescu AM, Roll-Mecak A (2018) Severing enzymes amplify microtubule arrays through lattice GTP-tubulin incorporation. Science 361(6404):eaau1504. https://doi.org/10.1126/science.aau1504
15. Roll-Mecak A, Vale RD (2006) Making more microtubules by severing: a common theme of noncentrosomal microtubule arrays? J Cell Biol 175(6):849–851. https://doi.org/10.1083/jcb.200611149
16. Srayko M, O'Toole ET, Hyman AA, Müller-Reichert T (2006) Katanin disrupts the microtubule lattice and increases polymer number in C. elegans meiosis. Curr Biol 16 (19):1944–1949
17. Sherwood NT, Sun Q, Xue M, Zhang B, Zinn K (2004) Drosophila spastin regulates synaptic microtubule networks and is required for normal motor function. PLoS Biol 2(12):e429. https://doi.org/10.1371/journal.pbio.0020429
18. Lindeboom JJ, Nakamura M, Hibbel A, Shundyak K, Gutierrez R, Ketelaar T, Emons AMC, Mulder BM, Kirik V, Ehrhardt DW (2013) A mechanism for reorientation of cortical microtubule arrays driven by microtubule severing. Science 342(6163):1245533
19. Burk DH, Ye ZH (2002) Alteration of oriented deposition of cellulose microfibrils by mutation of a katanin-like microtubule-severing protein. Plant Cell 14(9):2145–2160
20. Edelstein AD, Tsuchida MA, Amodaj N, Pinkard H, Vale RD, Stuurman N (2014)

Advanced methods of microscope control using muManager software. J Biol Methods 1 (2):e10. https://doi.org/10.14440/jbm.2014.36

21. Ziółkowska NE, Roll-Mecak A (2013) In vitro microtubule severing assays. Methods Mol Biol 1046:323–334. https://doi.org/10.1007/978-1-62703-538-5_19; pmid: 23868597

22. Zehr E, Szyk A, Piszczek G, Szczesna E, Zuo X, Roll-Mecak A (2017) Katanin spiral and ring structures shed light on power stroke for microtubule severing. Nat Struct Mol Biol 24(9):717–725. https://doi.org/10.1038/nsmb.3448

23. Gell C, Bormuth V, Brouhard GJ, Cohen DN, Diez S, Friel CT, Helenius J, Nitzsche B, Petzold H, Ribbe J, Schaffer E, Stear JH, Trushko A, Varga V, Widlund PO, Zanic M, Howard J (2010) Microtubule dynamics reconstituted *in vitro* and image by single-molecule fluorescence microscopy. Methods Cell Biol 95:221–245

24. Vemu A, Garnham CP, Lee D-Y, Roll-Mecak A (2014) Generation of differentially modified microtubules using in vitro enzymatic approaches. Methods Enzymol 540:149–166

25. Valenstein ML, Roll-Mecak A (2016) Graded control of microtubule severing by tubulin glutamylation. Cell 164(5):911–921

26. Gibbons IR, Fronk E (1979) A latent adenosine triphosphatase form of dynein 1 from sea urchin sperm flagella. J Biol Chem 254 (1):187–196

27. Severin FF, Sorger PK, Hyman AA (1997) Kinetochores distinguish GTP from GDP forms of the microtubule lattice. Nature 388 (6645):888–891. https://doi.org/10.1038/42270

28. Aumeier C, Schaedel L, Gaillard J, John K, Blanchoin L, Thery M (2016) Self-repair promotes microtubule rescue. Nat Cell Biol 18 (10):1054–1064. https://doi.org/10.1038/ncb3406

29. Schindelin J, Arganda-Carreras I, Frise E, Kaynig V, Longair M, Pietzsch T, Preibisch S, Rueden C, Saalfeld S, Schmid B, Tinevez JY, White DJ, Hartenstein V, Eliceiri K, Tomancak P, Cardona A (2012) Fiji: an open-source platform for biological-image analysis. Nat Methods 9(7):676–682. https://doi.org/10.1038/nmeth.2019

30. Al-Bassam J, Kim H, Brouhard G, van Oijen A, Harrison SC, Chang F (2010) CLASP promotes microtubule rescue by recruiting tubulin dimers to the microtubule. Dev Cell 19 (2):245–258

31. Walczak CE, Mitchison TJ, Desai A (1996) XKCM1: a Xenopus kinesin-related protein that regulates microtubule dynamics during mitotic spindle assembly. Cell 84(1):37–47

32. Weisenberg RC, Borisy GG, Taylor EW (1968) The colchicine-binding protein of mammalian brain and its relation to microtubules. Biochemistry 7(12):4466–4479

33. Borisy GG, Marcum JM, Olmsted JB, Murphy DB, Johnson KA (1975) Purification of tubulin and associated high molecular weight proteins from porcine brain and characterization of microtubule assembly in vitro. Ann N Y Acad Sci 253:107–132

34. Hartman JJ, Mahr J, McNally K, Okawa K, Iwamatsu A, Thomas S, Cheesman S, Heuser J, Vale RD, McNally FJ (1998) Katanin, a microtubule-severing protein, is a novel AAA ATPase that targets to the centrosome using a WD40-containing subunit. Cell 93 (2):277–287

Chapter 4

In Vitro Microtubule Dynamics Assays Using Dark-Field Microscopy

Jeffrey O. Spector, Annapurna Vemu, and Antonina Roll-Mecak

Abstract

Microtubules are dynamic non-covalent mesoscopic polymers. Their dynamic behavior is essential for cell biological processes ranging from intracellular transport to cell division and neurogenesis. Fluorescence microscopy has been the method of choice for monitoring microtubule dynamics in the last two decades. However, fluorescent microtubules are prone to photodamage that alters their dynamics, and the fluorescent label itself can affect microtubule properties. Dark-field imaging is a label-free technique that can generate high signal-to-noise, low-background images of microtubules at high acquisition rates without the photobleaching inherent to fluorescence microscopy. Here, we describe how to image in vitro microtubule dynamics using dark-field microscopy. The ability to image microtubules label-free allows the investigation of the dynamic properties of non-abundant tubulin species where fluorescent labeling is not feasible, free from the confounding effects arising from the addition of fluorescent labels.

Key words Microtubule dynamics, Dark field, Label-free, Microscopy, Tubulin, Microtubule

1 Introduction

Microtubules are non-covalent hollow cylindrical polymers with an outer diameter of 25 nm. They are built through the addition of tubulin αβ-heterodimers at their tips. The first visualization of individual microtubules was achieved with dark-field microscopy in 1975 [1]. Microtubules oscillate stochastically between growing and shrinking states. The cycle of growth, shrinkage, and the rapid transition between growth and depolymerization (catastrophe) and depolymerization and growth (rescue) is known as dynamic instability [2]. Dynamic instability was directly confirmed using differential interference contrast (DIC) or Nomarski microscopy [3] and dark-field microscopy [4, 5]. This dynamic behavior is important for basic biological processes ranging from intracellular transport to cell division and differentiation and is highly regulated in cells by a large number of cellular effectors [6].

Helder Maiato (ed.), *Cytoskeleton Dynamics: Methods and Protocols*, Methods in Molecular Biology, vol. 2101,
https://doi.org/10.1007/978-1-0716-0219-5_4,

In vitro microtubule dynamics assays have been an essential tool for understanding the mechanism of microtubule regulators. Fluorescence microscopy, in particular total internal reflection microscopy (TIRFM), has been the dominant imaging technique for microtubules in the last two decades. This dominance over other techniques is in part due to the ease of fluorescence imaging, the low signal-to-noise ratio of DIC microscopy, and the challenge in obtaining high-quality images using dark-field microscopy. Most importantly, multiplexing fluorescent microtubules with other fluorescently labeled motors and microtubule-associated proteins (MAPs) allows the examination of microtubule-effector interactions in real time. However, there are several drawbacks to fluorescence-based microtubule dynamics assays: (1) The presence of the fluorophore can perturb microtubule dynamics. (2) Fluorescent tubulin photobleaches, and it has been shown that photodamaged microtubules acquire lattice defects [7, 8] and will even break [7, 9]. (3) Fluorescence imaging relies on the emission of photons from an excited molecule. Thus, image acquisition speed is limited by the number of photons emitted and collected per frame. This can limit the accuracy of determining depolymerization rates, which are ~10-fold faster than polymerization rates. (4) Fluorescent labeling requires large tubulin concentrations, and the process is inefficient. This is not easily achieved with tubulin from less abundant sources such as tubulin isolated from various cell lines [10, 11], recombinant tubulin [12–15], or tubulin from diverse organisms such as *C. elegans* [10, 16]. As a result, most fluorescently labeled tubulin used in microtubule dynamics assays is brain tubulin because of its abundance and ease of purification [17]. However, tubulin isolated from brain tissue is a highly heterogeneous mixture of tubulin isoforms and posttranslational modifications [11, 12] and thus not well suited for understanding how various tubulin isoforms or tubulin modifications affect intrinsic polymer properties and interactions with MAPs.

Recent years have seen a resurgence in label-free imaging of microtubules [12, 16, 18–24]. Label-free techniques rely on imaging the light scattered by the imaged object and thus are very powerful for imaging microtubules because they contain enough mass in a diffraction-limited spot. One important advantage of scattering-based techniques is the lack of photobleaching. Second, because scattered light is measured, as opposed to emitted photons, scattering techniques are limited by the speed of the camera and the amount of light hitting the sample. Therefore, the development of high-intensity LED light sources and lasers has made it possible to image microtubules label-free at high frame rates [12, 18, 21, 23].

Dark-field microscopy images the sample at an oblique angle and collects only the light that is scattered by the sample. Unlike interference reflection microscopy (IRM) and interferometric scattering microscopy (iSCAT), dark-field microscopy is not strictly

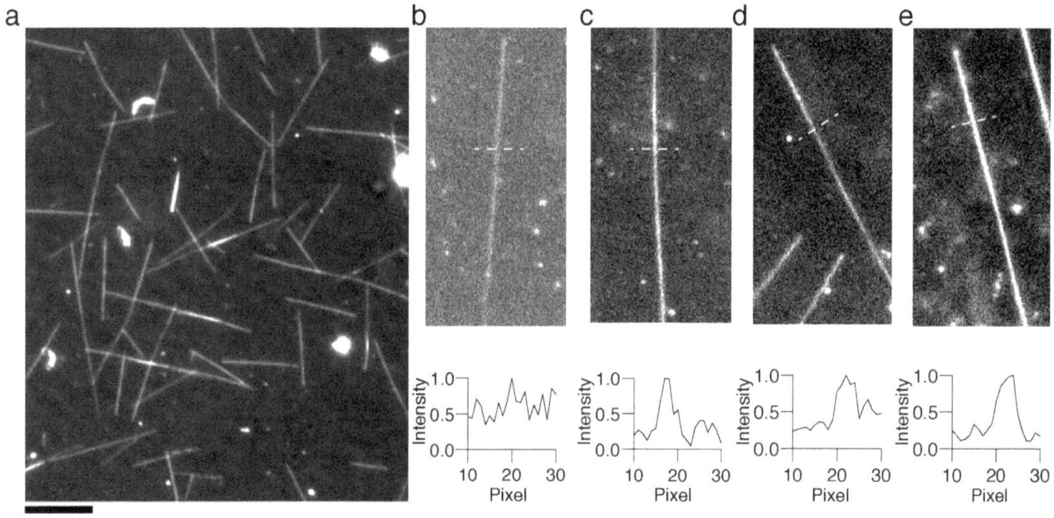

Fig. 1 Dark-field image of growing microtubules imaged at 20 Hz. Scale bar represents 10 μm. (**a**) Images of a single microtubule and normalized line scans across the dotted white line at 10, 25, 50, and 100 ms exposure times, respectively (**b–e**)

limited to near-surface imaging and the microtubule contrast is generally more uniform than in interference techniques where a small distance from the surface can lead to contrast inversion [22]. Individual microtubules can be easily visualized with exposure times of 20 ms and even as low as 1 ms. A typical dark-field image of unlabeled microtubules growing from seeds immobilized to the cover slide is shown in Fig. 1. Dark-field imaging can be performed either through the condenser or through the objective in a total internal reflection configuration [25]. We note that because dark field relies on the detection of the scattered light, it requires very clean surfaces and samples since stray light due to misalignment or scattered by even small amounts of aggregates will overwhelm the microtubule signal. The ability to image single dynamic microtubules at high frame rates and without the complication of exogenous fluorescent molecules makes dark-field microscopy a powerful technique for imaging in vitro microtubule dynamics. In addition, many commercial options are readily available and easy to integrate into existing microscopes and thus do not require custom addition of optical elements as some of the interference techniques do. The goal of this chapter is to describe how to image in vitro reconstituted microtubule dynamics using dark-field microscopy.

2 Materials

2.1 Dark-Field Microscopy

1. High-numerical-aperture (NA) (>1.4) oil immersion dark-field condenser (*see* **Note 1**).
2. Adjustable NA (1.0–1.3) 100× objective lens.

3. Low-magnification, low-NA (0.45) objective.
4. High-power LED light source.
5. sCMOS camera (for example we use a Hamamatsu Flash 4.0 v2, but there are many options available, e.g., Andor Zyla, Photometrics Prime 95B, PCO pco.edge).
6. Inverted microscope with objective heater (for example we use a Nikon Ti-E 2000 with perfect focus and an objective heater from Bioptechs).
7. Control Software (for example in our lab we use micromanager [26]).

2.2 Flow-Chamber Construction

1. Double-sided permanent tape (3 M).
2. 22 × 40 mm # 1.5 coverslip cleaned and silanized as described previously [27].
3. 3″ × 1″ × 1 mm thick microscope slide cleaned and silanized as described previously [27].

2.3 Buffers and Solutions

Prepare all buffer solutions with ultrapure water (>18 MΩ cm at 25 °C) and filter using 0.22 μm bottle-top filters.

1. BRB80: 80 mM PIPES, 1 mM $MgCl_2$, and 1 mM EGTA (pH 6.8 adjusted with KOH).
2. GMPCPP (Jena Bioscience).
3. BB buffer: BRB80 supplemented with 0.1% β-mercaptoethanol.
4. BBC buffer: BB containing a final concentration of 2 mg/mL casein (keep at room temperature).
5. BBP buffer: BB containing a final concentration of 1% Pluronic F-127 (keep at room temperature).
6. Salt buffer (SB): BB supplemented with GTP (1 mM final concentration) and KCl (100 mM final concentration).
7. Neutravidin solution: 0.1 mg/mL Neutravidin (*see* **Note 2**).
8. Oxygen scavengers (catalase, glucose oxidase, and 2 M glucose) prepared as described previously [27].
9. Tubulin: Unlabeled tubulin purified from porcine brain (Cytoskeleton, Inc.) for dynamic extensions (*see* **Note 3**) and biotin-tubulin (Cytoskeleton, Inc.) for microtubule seeds.
10. GMPCPP-stabilized biotinylated microtubule seeds (*see* Subheading 3.2).
11. Dynamics buffer: Tubulin at the desired concentration in BRB0 supplemented with 100 mM KCl, 1 mM GTP 1% Pluronic F-127, 20 mM glucose, glucose oxidase, and catalase (*see* **Note 4**).

3 Methods

3.1 Dark-Field Setup and Alignment on an Inverted Microscope

1. Before beginning dark-field alignment, first set up the microscope for Köhler illumination.
2. Once Köhler illumination is achieved, insert the dark-field condenser and switch to a low-power objective (we use a 10× 0.45 NA objective for this).
3. Swab the inside of your cheek with a pipette tip and place it on a coverslip. Press the coverslip down and seal the chamber edges. Cheek cells are barely visible in transmission but are very pronounced in dark field.
4. Place the sample on the microscope stage (coverslip down), and place a drop of immersion oil on the condenser side of the slide. Slowly bring the condenser into oil contact with the top of the slide (*see* **Note 5**).
5. Slowly focus the microscope using the focus knobs to obtain an image of the cheek cells.
6. The use of low magnification should allow you to see the dark-field stop in the bright-field image.
7. Using condenser lens centering screws, move the dark-field stop to the center of the image.
8. Once centered, slowly lower the condenser until the background is as dark as possible and the dark-field stop is a dark disk. Raise the condenser back up and the dark disk should appear bright (Fig. 2).
9. Switch to the 100× adjustable iris objective.
10. With the iris closed ~2/3 of the way, focus on the cheek sample again (Fig. 3a).
11. Now, switch to a sample of stabilized microtubule seeds (Fig. 3b) (*see* **Note 6**).
12. Adjust the light intensity, iris size, and exposure time to suit your experiment (*see* **Note 7**).

3.2 Making Double-Cycled GMPCPP-Stabilized Biotinylated Microtubule Seeds

1. Mix 5 μL of 2 mg/mL biotin-tubulin with 95 μL of 2 mg/mL unlabeled tubulin, and incubate on ice for 5 min. This results in microtubules that have ~5% of tubulin-biotin labeled.
2. Spin the tubulin mixture to remove any tubulin aggregates at 279,000 × *g* in an ultracentrifuge using a TLA-100 rotor (Beckman) for 10 min at 4 °C. Make sure to prechill your tubes.
3. Add GMPCPP to a final concentration of 0.5 mM, and incubate at 37 °C for 1 h.

   Spin the microtubules in an ultracentrifuge at 126,000 × *g* for 15 min at 37 °C.

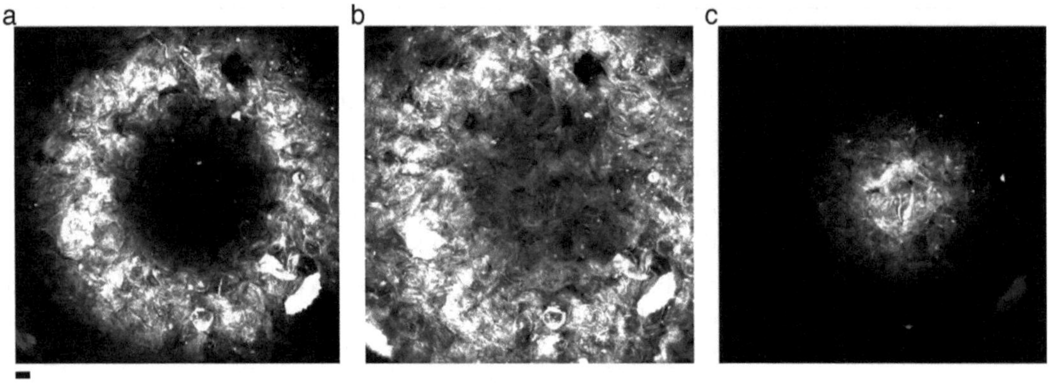

Fig. 2 Examples of low-magnification images of cheek cells. The dark-field stop will appear initially as a dark disk (**a**). As the condenser moves down, the central spot will get brighter (**b**). When the correct position is reached, the cells look bright while the background remains dark (**c**). At this point, good alignment is achieved and you can switch to higher magnification. Scale bar represents 50 μm

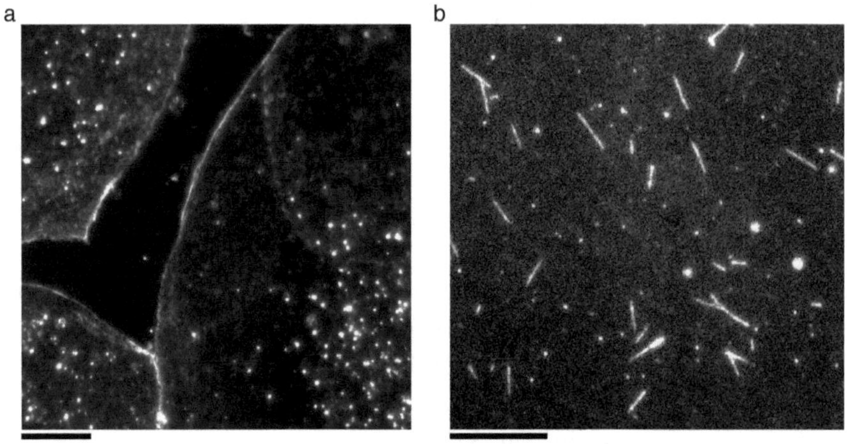

Fig. 3 Images of well-aligned cheek cells at 100× magnification. Scale bar represents 20 μm (**a**). Images of immobilized microtubule seeds. Scale bar represents 10 μm (**b**)

4. Remove the supernatant. Gently wash the pellet with warm BRB80.
5. Resuspend the microtubule pellet in cold BRB80 such that the final tubulin concentration is ~2 mg/mL (*see* **Note 8**).
6. Incubate on ice for 30 min, and mix by pipetting gently every 10 min.
7. Add GMPCPP to a final concentration of 0.5 mM, and incubate in 37 °C water bath (or heat block) for 30 min.
8. Spin microtubules at 126,000 × *g* for 15 min at 37 °C in a TLA100 ultracentrifuge rotor.
9. Remove the supernatant without touching the pellet. Gently wash the pellet with warm BRB80.

10. Gently resuspend the microtubule seeds in warm BRB80 (*see* **Note 9**).
11. Flash-freeze in small aliquots. On the day of performing the assays, dilute a seed aliquot 1:10 in warm BB and place at 37 °C for 1 h before use.

3.3 Flow-Chamber Assembly

1. Place two pieces of double-sided tape on a silanized microscope slide such that they are parallel and 3–4 mm apart.
2. Using a sharp blade, trim the excess tape off the edge of the slide.
3. Place a silanized coverslip on the tape such that its long axis is perpendicular to the long axis of the microscope slide.
4. Using gentle pressure, press down on the slide to ensure that it is sealed to the double-sided tape.

3.4 Chamber Preparation and Image Acquisition

1. On the day of performing the assay, filter all solutions through a 0.1 μm filter by spinning at 10,000 × *g* for 2 min at 4 °C (*see* **Note 10**).
2. Turn on all equipment and start heating the microscope objective (*see* **Note 11**).
3. Thaw seeds, dilute 1:10 in warm BB, and place at 37 °C for at least 1 h (*see* **Note 12**).
4. Thaw tubulin on ice, and ultracentrifuge for 10 min at 279,000 × *g* at 4 °C in a TLA100 ultracentrifuge rotor (*see* **Note 13**).
5. Perfuse the chamber with 20–40 μL of room-temperature BRB80 (*see* **Note 14**).
6. Perfuse the chamber with 20 μL of ice-cold Neutravidin solution, and wait for 5–10 min.
7. While the Neutravidin incubates, remove your tubulin from the centrifuge and check the concentration by Bradford assay.
8. Wash the chamber with 2× chamber volumes (~20 μL) of BBC and 2× chamber volume (~20 μL) of BBP (*see* **Note 15**).
9. Perfuse in 20 μL of seeds at an appropriate dilution. Wait for 5–10 min (*see* **Note 16**).
10. Check the density of seeds by imaging the chamber. If there are too many seeds, or if the chamber looks dirty, start over and prepare a new one. Save an image of the seeds in the area of interest. This can be used to demarcate where the seeds end and the dynamic microtubule extension begins. Alternatively, seeds can be prepared with fluorescent tubulin and imaged first using fluorescence microscopy.
11. Wash the chamber with 2× chamber volumes (~20 μL) of BBC and 2× chamber volume (~20 μL) of BBP.

12. Perfuse in dynamics buffer (20 μL perfusion volume) (*see* **Note 17**).
13. Seal the chamber with vacuum grease to prevent evaporation.
14. Find a suitable area and image microtubules for 30 min. We image at a frame rate of 2 Hz for measuring growth rates, catastrophe, and rescue frequencies and a rate of 20–40 Hz for depolymerization rates.

3.5 Data Analysis

Microtubule dynamic parameters are measured by tracing the microtubule tip positions over time and generating kymographs. The slope represents the growth rate. Catastrophes mark the switch from growth to depolymerization. Rescues mark the switch from depolymerization to growth (Fig. 4). To obtain reliable kymographs, movies should first be corrected for drift. This entire process can be performed in many software packages. We use ImageJ/ FIJI.

1. To correct for drift, a registration plug-in in FIJI can be used [28] (*see* **Note 18**).
2. Scroll through the drift-corrected movie to generate kymographs of each microtubule (Fig. 4). From kymographs, extract the polymerization and depolymerization rates, catastrophe frequency (number of catastrophe per time microtubule spent growing), and rescue frequency (number of rescues per time spent depolymerizing). As an alternative to analyzing kymographs manually, there are now several automated solutions available. They will either analyze the kymograph directly [29] or track the growing tips of microtubules and extract dynamic parameters form tip positions [30, 31].

4 Conclusions

This chapter describes a protocol for imaging in vitro microtubule dynamics using dark-field microscopy. Dark-field imaging is easy and cheap to integrate into an existing inverted microscope. The label-free nature of the technique allows one to image with high signal-to-background ratios, at high frame rates and without photobleaching, an unavoidable downside of fluorescence-based imaging. Dark-field microscopy and label-free imaging techniques in general allow the study of tubulin isoforms that are in low abundance and not amenable to fluorescent labeling. These techniques are making a comeback in the microtubule field, catalyzed by the recent advances in expressing and purifying single-isoform tubulin and isolating tubulin from diverse sources using affinity approaches.

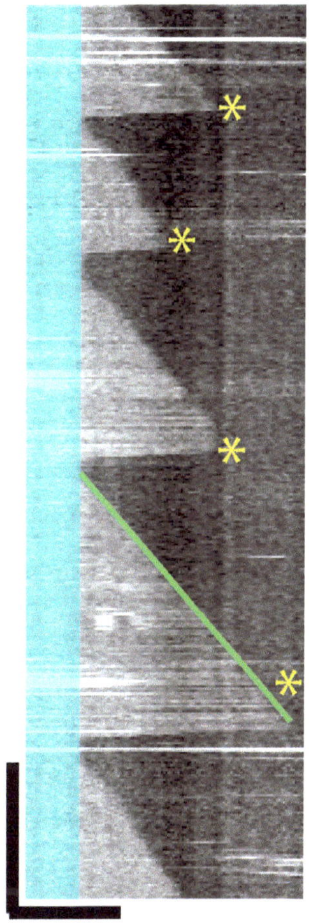

Fig. 4 A kymograph of a microtubule imaged in dark field. The bright streaks are due to scattering objects that were rapidly diffusing in solution. The stabilized GMPCPP seed has been pseudo-colored in blue. Yellow stars indicate a catastrophe event. The slope of the green line is used to calculate the growth rate. Scale bar, 5 μm horizontally and 5 min vertically

5 Notes

1. A high-NA oil immersion condenser is necessary in order to obtain high-quality images. There are many different commercially available options. Our lab uses a Nikon Dark-Field Oil Immersion NA 1.4 condenser, but several other manufacturers make dark-field condensers as well. Imaging microtubules require a well-aligned condenser and the appropriate iris size on the objective. The NA of the condenser needs to be higher than the NA of the objective to achieve dark-field imaging. We use a 1.45 NA condenser and a 1–1.3 NA adjustable 100× objective from Nikon, but there are other commercial options available (i.e., Leica, Zeiss, Olympus).

2. Prepare the Neutravidin stock solution at 1 mg/mL as *per* manufacturer's protocol. Flash-freeze and store at −80 °C in small aliquots.
3. After thawing on ice, the tubulin needs to be ultracentrifuged at 279,000 × *g* for 10 min at 4 °C to remove any tubulin aggregates. Make sure to prechill the ultracentrifuge tubes as well as the rotor. Place the tubulin in cold Eppendorf tubes, and measure tubulin concentration after the spin by Bradford or A280 absorbance.
4. Add tubulin to the perfusion mixture last. Make sure that there are no bubbles as tubulin and proteins in general denature at the air-liquid interface. Mix and use immediately.
5. Light exits above the critical angle when using a high-NA dark-field condenser; therefore, no light will exit the condenser until the slide makes contact with the oil. As the condenser approaches the sample and makes contact with the oil, a flash of light is observed. This is an indication that the condenser has made oil contact with the topside of your chamber. The condenser that we use has a working distance that is slightly larger than the thickness of the slide (~1 mm); therefore, we are imaging microtubules attached to the slide surface (in contrast to interference techniques which image microtubules that are growing from seeds immobilized to the coverslip). Because of this, the correct focal plane may be slightly higher for dark field than for TIRF imaging.
6. You will only be able to image the stabilized microtubules if the cheek cell slide was properly aligned. If you cannot see them, adjust the microscope as follows. First, check that the condenser spot is still centered on the field of view of your camera. Next, adjust all apertures to allow the maximum amount of light possible to enter the sample. Adjust the objective iris until the background is a light gray. If necessary, make small adjustments to the position of the condenser. When at the proper focal plane, the microtubules will appear bright on a dark background. It may also be necessary to adjust the size of the illumination using the field stop. Note that you might need to adjust the image histogram to obtain a good image. Software autoscaling will make it difficult to observe the microtubules if there are very bright objects near them.
7. There is a trade-off between signal-to-background, imaging speed, and size of the objective iris. We find that closing the iris ~2/3 of the way gives good signal for exposure times as low as 20 ms. If high speeds are not necessary, one can also adjust the brightness of the light source and use longer exposure times (~200 ms).

8. For brain tubulin, the typical polymerization efficiency is ~70%. Therefore, resuspend the pellet with a volume such that the concentration will be 2 mg/mL.
9. Cut the pipette tip to prevent shearing when pipetting microtubules.
10. Any scattering mass in the sample will obscure your dark-field signal. We find filtering our solutions through a 0.1 μm filter allows us to obtain high-quality dark-field images. Moreover, we would like to emphasize that the sample has to be very clean and free of aggregates. This is one of the challenges of dark-field microscopy. Unlike fluorescence microscopy where one sees only what is labeled, any scattering mass will give a signal in dark-field microscopy. It is not uncommon for a sample to look "clean" with high signal-to-background in TIRFM but have extremely high background in dark field because of scattering from a very small amount of aggregates. Thus, high sample quality is paramount.
11. We use a Bioptechs objective heater, but there are many commercial objective heaters available.
12. After incubating the seeds at 37 °C, store them at room temperature. This incubation time gives an average seed length of 1–3 μm. If longer seeds are desired, incubate at 37 °C for a longer time.
13. Tubulin can be stored on ice for ~1–1.5 h but should not be stored longer as it tends to aggregate over time, and this will compromise polymerization behavior.
14. If your glass is silanized and cleaned well, the initial perfusion may prove to be difficult. One can either pipette the solution a few times and try not to introduce bubbles or use a vacuum line to draw the liquid through.
15. In our experience, having casein in the chamber during imaging tends to compromise image quality.
16. We often check the seed dilution before proceeding to make sure there are a good number of seeds in the field of view and the chamber is not sparse or crowded. It is a good idea to prepare a few chambers at different seed dilutions to determine the correct density before starting assays for the day.
17. Do not perfuse ice-cold solutions into the chamber as this will depolymerize the seeds! You can hold the perfusion in your hand for 1–3 s to warm it up.
18. Drift correction can be performed using spots on the glass as fiducial markers. We almost always have a few bright spots on the glass that we are able to use as landmarks for registration. Fiducial beads can also be immobilized in the chamber with the seeds if needed, but we have not found this to be necessary.

Acknowledgment

A.R.M. is supported by the intramural programs of the National Institute of Neurological Disorders and Stroke (NINDS) and the National, Heart, Lung, and Blood Institute (NHLBI).

References

1. Kuriyama R, Miki-Noumura T (1975) Light-microscopic observations of individual microtubules reconstituted from brain tubulin. J Cell Sci 19(3):607–620
2. Mitchison T, Kirschner M (1984) Dynamic instability of microtubule growth. Nature 312 (5991):237
3. Walker RA, O'Brien ET, Pryer NK, Soboeiro MF, Voter WA, Erickson HP, Salmon ED (1988) Dynamic instability of individual microtubules analyzed by video light microscopy: rate constants and transition frequencies. J Cell Biol 107(4):1437–1448
4. Horio T, Hotani H (1986) Visualization of the dynamic instability of individual microtubules by dark-field microscopy. Nature 321 (6070):605
5. Summers K, Kirschner MW (1979) Characteristics of the polar assembly and disassembly of microtubules observed in vitro by darkfield light microscopy. J Cell Biol 83(1):205–217
6. Akhmanova A, Steinmetz MO (2015) Control of microtubule organization and dynamics: two ends in the limelight. Nat Rev Mol Cell Biol 16(12):711–726. https://doi.org/10.1038/nrm4084
7. Guo H, Xu C, Liu C, Qu E, Yuan M, Li Z, Cheng B, Zhang D (2006) Mechanism and dynamics of breakage of fluorescent microtubules. Biophys J 90(6):2093–2098
8. Aumeier C, Schaedel L, Gaillard J, John K, Blanchoin L, Thery M (2016) Self-repair promotes microtubule rescue. Nat Cell Biol 18 (10):1054–1064. https://doi.org/10.1038/ncb3406
9. Vigers GP, Coue M, McIntosh JR (1988) Fluorescent microtubules break up under illumination. J Cell Biol 107(3):1011–1024
10. Widlund PO, Podolski M, Reber S, Alper J, Storch M, Hyman AA, Howard J, Drechsel DN (2012) One-step purification of assembly-competent tubulin from diverse eukaryotic sources. Mol Biol Cell 23 (22):4393–4401. https://doi.org/10.1091/mbc.E12-06-0444
11. Vemu A, Garnham CP, Lee DY, Roll-Mecak A (2014) Generation of differentially modified microtubules using in vitro enzymatic approaches. Methods Enzymol 540:149–166. https://doi.org/10.1016/B978-0-12-397924-7.00009-1
12. Vemu A, Atherton J, Spector JO, Szyk A, Moores CA, Roll-Mecak A (2016) Structure and dynamics of single-isoform recombinant neuronal human tubulin. J Biol Chem 291 (25):12907–12915. https://doi.org/10.1074/jbc.C116.731133
13. Minoura I, Hachikubo Y, Yamakita Y, Takazaki H, Ayukawa R, Uchimura S, Muto E (2013) Overexpression, purification, and functional analysis of recombinant human tubulin dimer. FEBS Lett 587(21):3450–3455. https://doi.org/10.1016/j.febslet.2013.08.032
14. Ti SC, Pamula MC, Howes SC, Duellberg C, Cade NI, Kleiner RE, Forth S, Surrey T, Nogales E, Kapoor TM (2016) Mutations in human tubulin proximal to the kinesin-binding site alter dynamic instability at microtubule plus- and minus-ends. Dev Cell 37(1):72–84. https://doi.org/10.1016/j.devcel.2016.03.003
15. Johnson V, Ayaz P, Huddleston P, Rice LM (2011) Design, overexpression, and purification of polymerization-blocked yeast alphabeta-tubulin mutants. Biochemistry 50 (40):8636–8644. https://doi.org/10.1021/bi2005174
16. Chaaban S, Jariwala S, Hsu CT, Redemann S, Kollman JM, Muller-Reichert T, Sept D, Bui KH, Brouhard GJ (2018) The structure and dynamics of C. elegans tubulin reveals the mechanistic basis of microtubule growth. Dev Cell 47(2):191–204, e198. https://doi.org/10.1016/j.devcel.2018.08.023
17. Weisenberg RC (1972) Microtubule formation in vitro in solutions containing low calcium concentrations. Science 177 (4054):1104–1105
18. Vemu A, Atherton J, Spector JO, Moores CA, Roll-Mecak A (2017) Tubulin isoform composition tunes microtubule dynamics. Mol Biol Cell 28(25):3564–3572. https://doi.org/10.1091/mbc.E17-02-0124

19. Geyer EA, Burns A, Lalonde BA, Ye X, Piedra FA, Huffaker TC, Rice LM (2015) A mutation uncouples the tubulin conformational and GTPase cycles, revealing allosteric control of microtubule dynamics. Elife 4:e10113. https://doi.org/10.7554/eLife.10113
20. Andrecka J, Arroyo JO, Lewis K, Cross RA, Kukura P (2016) Label-free imaging of microtubules with sub-nm precision using interferometric scattering microscopy. Biophys J 110 (1):214–217
21. Mahamdeh M, Simmert S, Luchniak A, Schaeffer E, Howard J (2018) Label-free high-speed wide-field imaging of single microtubules using interference reflection microscopy. J Microsc 272(1):60–66
22. Simmert S, Abdosamadi MK, Hermsdorf G, Schäffer E (2018) LED-based interference-reflection microscopy combined with optical tweezers for quantitative three-dimensional microtubule imaging. Opt Express 26 (11):14499–14513
23. Katsuki M, Drummond DR, Osei M, Cross RA (2009) Mal3 masks catastrophe events in Schizosaccharomyces pombe microtubules by inhibiting shrinkage and promoting rescue. J Biol Chem 284(43):29246–29250. https://doi.org/10.1074/jbc.C109.052159
24. Kandel ME, Teng KW, Selvin PR, Popescu G (2017) Label-free imaging of single microtubule dynamics using spatial light interference microscopy. ACS Nano 11(1):647–655. https://doi.org/10.1021/acsnano.6b06945
25. Ueno H, Nishikawa S, Iino R, Tabata KV, Sakakihara S, Yanagida T, Noji H (2010) Simple dark-field microscopy with nanometer spatial precision and microsecond temporal resolution. Biophys J 98(9):2014–2023. https://doi.org/10.1016/j.bpj.2010.01.011
26. Edelstein AD, Tsuchida MA, Amodaj N, Pinkard H, Vale RD, Stuurman N (2014) Advanced methods of microscope control using muManager software. J Biol Methods 1 (2):e10. https://doi.org/10.14440/jbm.2014.36
27. Ziółkowska NE, Roll-Mecak A (2013) In vitro microtubule severing assays. Methods Mol Biol 1046:323–334. https://doi.org/10.1007/978-1-62703-538-5_19; pmid: 23868597
28. Schindelin J, Arganda-Carreras I, Frise E, Kaynig V, Longair M, Pietzsch T, Preibisch S, Rueden C, Saalfeld S, Schmid B, Tinevez JY, White DJ, Hartenstein V, Eliceiri K, Tomancak P, Cardona A (2012) Fiji: an open-source platform for biological-image analysis. Nat Methods 9(7):676–682. https://doi.org/10.1038/nmeth.2019
29. Mukherjee A, Jenkins B, Fang C, Radke RJ, Banker G, Roysam B (2011) Automated kymograph analysis for profiling axonal transport of secretory granules. Med Image Anal 15 (3):354–367. https://doi.org/10.1016/j.media.2010.12.005
30. Kapoor V, Hirst WG, Hentschel C, Preibisch S, Reber S (2019) MTrack: automated detection, tracking, and analysis of dynamic microtubules. Sci Rep 9(1):3794. https://doi.org/10.1038/s41598-018-37767-1
31. Ruhnow F, Zwicker D, Diez S (2011) Tracking single particles and elongated filaments with nanometer precision. Biophys J 100 (11):2820–2828. https://doi.org/10.1016/j.bpj.2011.04.023

Chapter 5

In Vitro Reconstitution of Dynamic Co-organization of Microtubules and Actin Filaments in Emulsion Droplets

Kim J. A. Vendel, Celine Alkemade, Nemo Andrea, Gijsje H. Koenderink, and Marileen Dogterom

Abstract

In vitro (or cell-free) reconstitution is a powerful tool to study the physical basis of cytoskeletal organization in eukaryotic cells. Cytoskeletal reconstitution studies have mostly been done for individual cytoskeleton systems in unconfined 3D or quasi-2D geometries, which lack complexity relative to a cellular environment. To increase the level of complexity, we present a method to study co-organization of two cytoskeletal components, namely microtubules and actin filaments, confined in cell-sized water-in-oil emulsion droplets. We show that centrosome-nucleated dynamic microtubules can be made to interact with actin filaments through a tip-tracking complex consisting of microtubule end-binding proteins and an actin-microtubule cytolinker. In addition to the protocols themselves, we discuss the optimization steps required in order to build these more complex in vitro model systems of cytoskeletal interactions.

Key words In vitro reconstitution, Cytoskeleton, Confinement, Self-organization, Microtubule dynamic instability, Cytolinkers

1 Introduction

Microtubules (MTs) and actin filaments (AFs) both play key roles in dynamic cellular processes such as cell migration, polarization and division. They are often studied independently, but it is increasingly recognized that dynamic co-organization is necessary to ensure that these processes proceed correctly [1–3]. A relatively well-studied example is cell migration, which involves interactions between dynamic plus ends of MTs with actin stress fibers and membrane-bound actin arrays at the front and rear end of the cell (Fig. 1a). It has been shown that AFs and MTs influence each

Electronic Supplementary Material: The online version of this chapter (https://doi.org/10.1007/978-1-0716-0219-5_5) contains supplementary material, which is available to authorized users.

Kim J. A. Vendel and Celine Alkemade contributed equally to this work.

Helder Maiato (ed.), *Cytoskeleton Dynamics: Methods and Protocols*, Methods in Molecular Biology, vol. 2101, https://doi.org/10.1007/978-1-0716-0219-5_5,

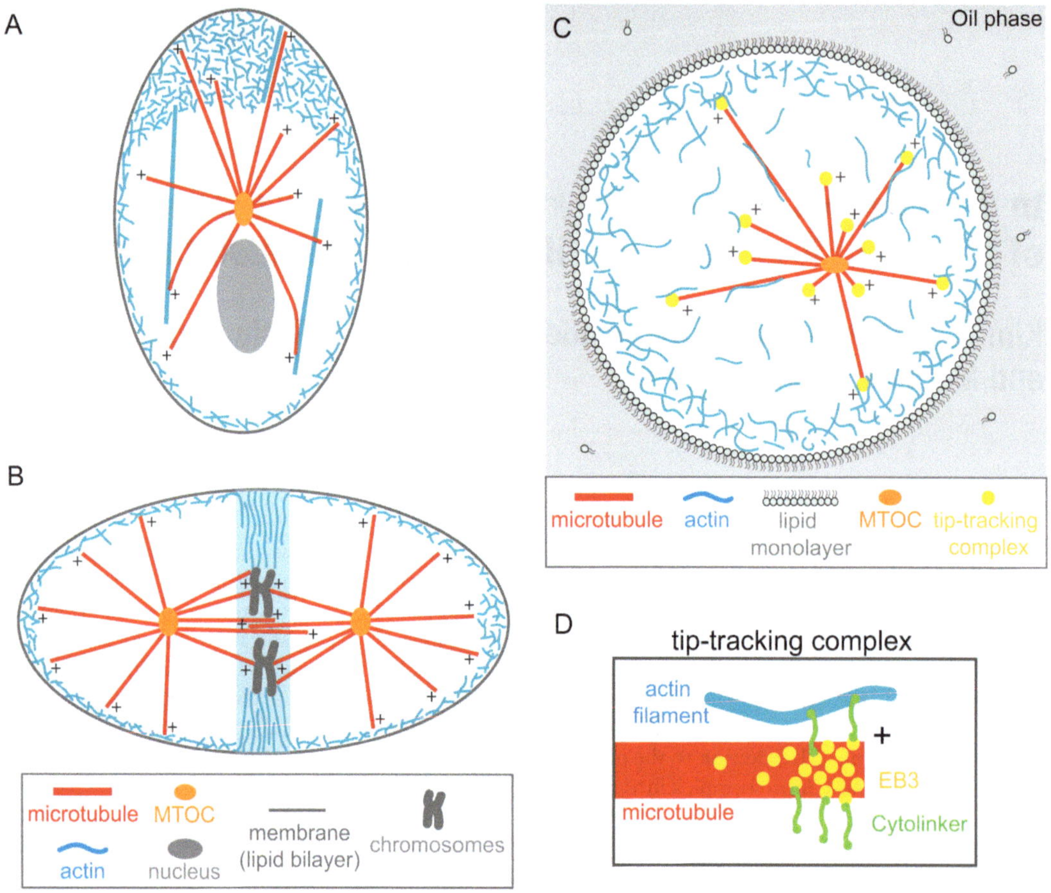

Fig. 1 Typical geometries and organizations of MTs and AFs in cells and in emulsion droplets. (**a**) Schematic interphase cell, where MT plus ends (indicated with '+') interact with different types of actin structures. (**b**) Schematic mitotic cell, where the actin cortex positions the MT spindle and where MTs instruct the formation of the actin cytokinetic ring. (**c**) Schematic of the proposed organization of MTs and AFs in a water-in-oil emulsion droplet for in vitro reconstitution, mimicking the MT and AF organizations of cells. (**d**) Detailed schematic of the tip-tracking complex, consisting of end-binding proteins (EB3) and cytolinkers, that can link AFs to growing MT plus ends

other's organization, and that this is necessary for directional migration. The actin cytoskeleton is thought to direct MT growth towards cortical sites in the vicinity of focal adhesions. This happens by a combination of guiding growing MTs along actin stress fibers and restriction of MT growth by dense actin structures—which act as a physical barrier or sweep back MTs via actin retrograde flows [4]. MT guidance involves cytolinkers: proteins that physically link MTs and AFs [5–7]. Conversely, MTs influence the organization of AFs, for example by locally promoting actin polymerization at a growing MT tip [8] or by activating myosin contractions via Rho-GTPase pathways [9].

Cell division is another process where AFs and MTs are known to influence each other's organization (Fig. 1b). The actin cortex influences MT spindle positioning via interactions with MT plus ends [10, 11]. Conversely, the MT spindle provides cues to position and orient the actin-myosin ring that mediates constriction [12].

The physical mechanisms that underlie actin-MT co-organization remain poorly characterized, even though its importance has been recognized for a long time [13]. In vitro reconstitutions provide a powerful tool to study physical interactions of cytoskeletal components without the full complexity of cells. In vitro experiments have taught us a great deal about the physical properties of actin and MTs in isolation, but only a handful of studies have so far been done on the co-organization of MTs and AFs [5, 8]. These first studies were performed in unconfined environments, using 3D solutions and networks or quasi-2D surface assay geometries. It is known that physical confinement within cell-sized boundaries strongly affects the self-organization of isolated cytoskeletal components, both in case of MTs [14–18] and AFs [19–25]. Hence it is expected that confinement should influence the interaction between MTs and AFs as well. For example, a recent study using *Xenopus* egg extracts showed how different actin network architectures influence the mobility of MT asters in droplets [11].

Here, we present a method to study dynamic co-organization of MTs and AFs in micron-scale water-in-oil emulsion droplets that mimic the physiological geometrical confinement and size of cells, as schematically shown in Fig. 1c. We first describe the protocols to obtain dynamic MT asters nucleating from purified centrosomes encapsulated in emulsion droplets, using small adaptations to previous protocols [26–28]. We then increase the experimental complexity by adding the end-binding protein EB3, which provides the possibility to visualize and quantify MT dynamics inside cell-sized confinement. Subsequently, we describe how to include AFs and establish an interaction between MTs and actin through a tip-tracking complex consisting of EB3 proteins and cytolinkers (Fig. 1d). The cytolinker we use here is TipAct, an engineered protein consisting of the actin- and MT-tip-binding domains from the spectraplakin MACF1 (also called ACF7) that was used previously in TIRF assays [5]. TipAct tracks the growing MT plus tip by binding to EB3.

Simple as it may sound to add extra proteins, these experiments are challenging due to the sensitivity of individual components to experimental conditions. Each new protein may need a different set of optimal experimental conditions, which can interfere with the fragile balance between the already incorporated components. Therefore, new optimization steps are required each time an extra protein is added. Here we not only provide the protocols themselves, but in addition point out which experimental 'knobs' can be tuned during optimization and what changes were needed to get each consecutive experiment working.

2 Materials

2.1 Buffer Conditions

All our experiments are performed in MRB80 buffers (MRB80: 80 mM Pipes, 4 mM $MgCl_2$, 1 mM EGTA, pH 6.8 adjusted with KOH). Monomeric (G-)actin is stored and diluted in G-buffer to prevent polymerization (G-buffer: 2 mM Tris-HCl, 0.2 mM Na_2ATP, 0.2 mM $CaCl_2$, 5 mM dithiothreitol (DTT), pH 7.8 adjusted with HCl).

The components of our final reaction mix are stated in Table 1, including the protein concentration ranges that we generally use. For the pre-polymerized AFs, we use 15% labeled actin and add phalloidin in a 1:1 molar ratio.

2.2 Proteins

1. Tubulin: lyophilized porcine brain tubulin (unlabeled, or labeled with TRITC Rhodamine, HiLyte®488, or HiLyte®647) was obtained from Cytoskeleton (Denver, USA) and resuspended at 50–100 μM in MRB80, snap-frozen and stored at −80°C until use.
2. Actin: G-actin was purified from rabbit skeletal muscle powder [29] and stored in G-buffer at −80 °C. Before use, G-actin was thawed at 4 °C overnight, and clarified by centrifugation for

Table 1
Composition of our final reaction mix, that is MRB80-buffer based

Component	Final concentration
DTT	4 mM
Catalase	0.2 mg/ml
Glucose oxidase	0.4 mg/ml
D-glucose	50 mM
Bovine serum albumin (BSA)	2 mg/ml
κ-casein	0.5 mg/ml
KCl	50 mM
Guanosine 5′-triphosphate (GTP) sodium salt hydrate	3 mM
Tubulin	20–40 μM
Labeled tubulin	0.75–1.5 μM
EB3-GFP	20–150 nM
F-actin + phalloidin	10–1000 nM
Cytolinker	10–50 nM
Centrosomes	10% of total volume

10 min at 149,000×*g*. Alexa Fluor® 647 and 594 succinimidyl esther dyes (Molecular Probes, Life Technologies) were used to produce labeled G-actin [30].

3. EB3: 6xHis-tagged recombinant human EB3, GFP-EB3 and mCherry-EB3 [31], purified from bacteria, were kind gifts from Michel Steinmetz (Laboratory of Molecular Research, Paul Scherrer Institute, Switzerland) and Anna Akhmanova (Department of Biology, Utrecht University, the Netherlands).
4. TipAct: GFP-TipAct is an engineered cytolinking protein [5], that was purified from bacteria.
5. Centrosomes: isolated from human lymphoblastic KE37 cells [32], or a kind gift from the Pierre Gönczy lab (Swiss Institute for Experimental Cancer Research (ISREC), School of Life Sciences, Swiss Federal Institute of Technology (EPFL) Lausanne, Switzerland).

2.3 Lipids, Oils and Surfactants

1. DOPS: 1,2-dioleoyl-sn-glycero-3-phospho-L-serine (sodium salt) in chloroform;
2. Mineral oil;
3. Span80 nonionic surfactant.

2.4 Other Components

1. SU-8 3025 photoresist;
2. Polydimethylsiloxane (PDMS) RVT615;
3. Rain-X rain repellent 200 ml;
4. MicrewLock™ microcentrifuge tubes;
5. Microfluidic MFCS pressure controller;
6. Phalloidin;
7. SiR-tubulin.

3 Methods

3.1 Microtubule Asters in Emulsion Droplets with End-Binding Protein Comets

In this section, we explain in detail the individual steps necessary to study centrosome-nucleated MTs with EB3 comets in confinement. We start by preparing a lipid mix (Subheading 3.1.1) and the final reaction mix (Subheading 3.1.2). Next, we discuss two strategies to create water-in-oil emulsion droplets, namely via microfluidic devices and via pipetting, as well as their (dis-) advantages (Subheading 3.1.3). We then elaborate on the individual steps required to create droplets via the microfluidic method (Subheading 3.1.5). In addition, we explain the preparations for microscopic imaging, such as preparing coverslips (Subheading 3.1.6) and PDMS wells (Subheading 3.1.7). Furthermore, we discuss SiR-tubulin as an alternative method to visualize MTs with reduced background signal and hence better MT visibility (Subheading 3.1.8). Finally, we explain how to image these droplets for long times and analyse the data (Subheading 3.1.9).

3.1.1 Lipid Mix Preparation

The lipid mix consists of mineral oil, phospholipids and a surfactant, where the latter two form a monolayer at the interface between water and oil to stabilize the water-in-oil emulsion droplets. The procedure for making the lipid mix is as described before by Roth et al. [26] and Vleugel et al. [28].

1. Rinse glass syringes three times with chloroform before use;
2. Pipette chloroform-dissolved, 1,2-dioleoyl-sn-glycero-3-phospho-L-serine (DOPS) into a glass vial using a glass syringe (*see* **Note 1**);
3. Dry the lipids into a thin film on the wall of the glass vial by evaporating the chloroform in a gentle nitrogen gas flow (*see* **Note 2**);
4. Evaporate all remaining chloroform in a vacuum dessicator for about 1 h (*see* **Note 3**);
5. Dissolve the lipids in mineral oil and 2.7% (v/v) Span80 to a total lipid concentration of 0.64 mg/ml (*see* **Note 4**) by sonication for 30 min in a bath sonicator (40 kHz);
6. Store the lipid mix at 4 °C until use (*see* **Note 5**).

3.1.2 Reaction Mix Preparation

This reaction mix contains all components required to nucleate MT asters with protein comets in an in vitro environment.

1. Thaw a centrosome aliquot quickly (in your hands) and place in a 37 °C water bath for about 15 min (*see* **Note 6**);
2. In the meantime, prepare a reaction mix on ice with concentrations as in Table 2 but do not add the centrosomes yet. In the end the centrosomes should amount for a maximum of 10% of the total volume, depending on the centrosome stock concentration;
3. Centrifuge the reaction mix 5 min in the airfuge with a cold rotor at 150,000×*g* to remove protein aggregates;
4. Transfer the reaction mix to an 0.5 ml Eppendorf tube, warm up in your hands, add to the centrosomes aliquot (*see* **Note 7**) and mix well but gently (*see* **Note 8**). Keep the reaction mix at room temperature from now on and use to prepare emulsion droplets.

Note that the reaction mix does not contain Tween20, in contrast to previously described experiments [26–28], as we found that this surfactant influences the efficacy of functionalized phospholipids (for example with a biotin or Ni-NTA attached to the lipid head group) in addition to the DOPS lipids. The function

Table 2
Composition of the reaction mix for encapsulation of MT asters with EB3-GFP comets

Component	Final concentration
50× oxygen scavenger mix[a]	1×
D-glucose	50 mM
BSA	2 mg/ml
κ-casein	0.5 mg/ml
KCl	50 mM
GTP	3 mM
Tubulin	38.5 μM
Rhodamine tubulin	1.5 μM
EB3-GFP	20 nM
Centrosomes	10% of total volume

All components are dissolved and diluted in MRB80
[a]200mM DTT, 10mg/ml catalase, 20mg/ml glucose oxidase

of Tween20 was to prevent nonspecific sticking of proteins to the droplet boundary. As a result of leaving it out, we observe an increase in nonspecific cortex boundary sticking over time (*see* **Note 9**).

3.1.3 Emulsion Droplets

We used two different strategies to create water-in-oil emulsion droplets with the previously prepared reaction mix and lipid mix. The first is to create droplets at the T-junction of a microfluidic chip, *see* Fig. 2a and Subheading 3.1.5. The second strategy is to emulsify 0.5 μl of reaction mix into 40 μl of lipid mix by pipetting up and down 40 times with a 10 μl pipette, as schematically depicted in Fig. 2b (*see* **Notes 10** and **11**). Both methods have advantages and disadvantages that we will shortly discuss. Microfluidics produces uniformly sized droplets and in addition it is supposed to be a more gentle way of droplet production, which is advantageous for certain fragile proteins like tubulin or protein complexes like centrosomes. However, proteins tend to stick to PDMS walls of the microfluidic channels, leading to a decrease of protein concentration and functionality in the droplets and possibly also to a loss of centrosomes. Droplet production by pipetting produces a wider variety of droplet sizes—which can be both an advantage and a disadvantage depending on the goal of the experiment—but does not have the problem of proteins sticking to channels. Also, the droplet production is much faster, which is especially convenient for experiments that compare different experimental conditions. Finally, a much smaller volume of the reaction

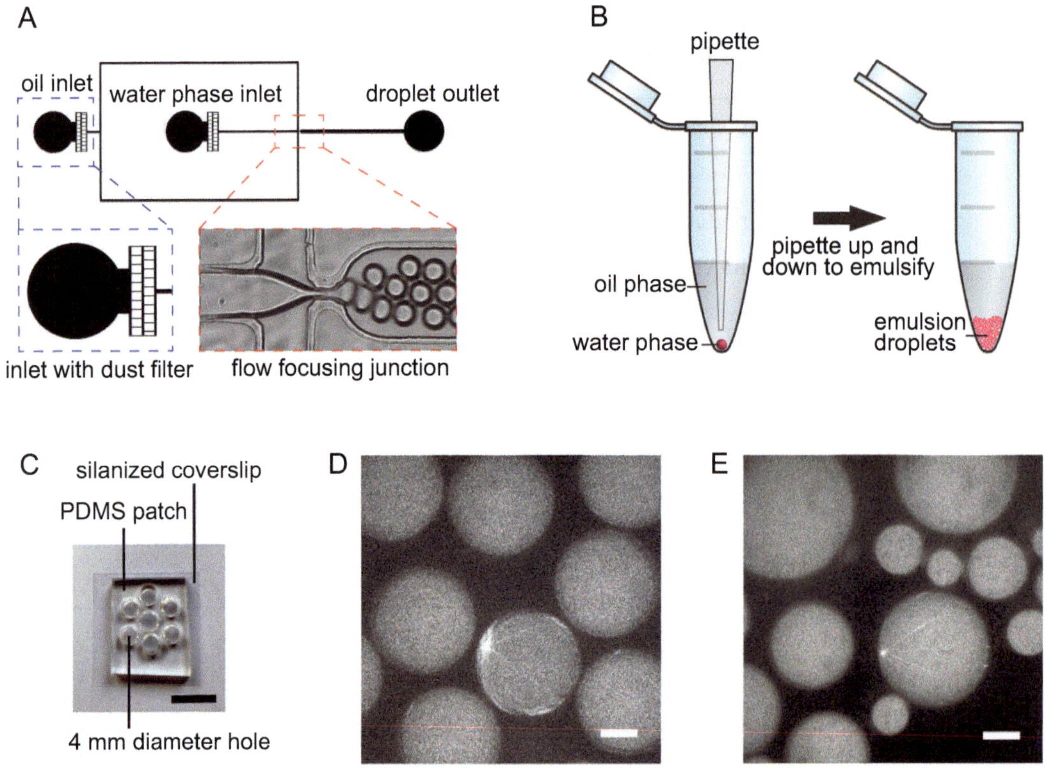

Fig. 2 Water-in-oil emulsion droplets were produced in two ways, as schematically visualized in (**a**) and (**b**). (**a**) Microfluidic chip for emulsion droplet production with one inlet for the oil phase and one for the water phase, which come together at a flow focusing T-junction where droplets are pinched off. Dust filters at the inlets prevent (PDMS) particles from entering the channels and blocking the T-junction. (**b**) Another method for emulsion droplet formation involves pipetting the oil phase and water phase up and down to generate the shear stress that leads to droplet formation. (**c**) For long-term imaging, droplets are transferred into a PDMS well consisting of a PDMS patch with holes on a silanized coverslip. (**d**) and (**e**) are typical fields of view observed when imaging emulsion droplets containing centrosomes and fluorescent rhodamine-tubulin (*see* Table 2) with a confocal spinning disc fluorescence microscope, for droplets produced with microfluidics (**d**) or by pipetting (**e**). Scale bars: 10 mm for (**c**), and 10 μm for (**d**) and (**e**).

mix is needed to produce droplets by pipetting (0.5 μl instead of 10–15 μl for microfluidics). This allows multiple droplet batches of the same reaction mix while optimizing a single component, like actin concentration. We have been able to perform the experiments described in this chapter with both strategies for droplet production, yet the results shown in the figures were obtained in droplets prepared by pipetting unless stated otherwise.

3.1.4 Microfluidic Chips

The procedure for microfluidic chip preparation is exactly as described in previous methods papers [26–28], but for completeness we will repeat the steps. First, an SU-8 mold is prepared in the cleanroom (**steps 1–5**). With this reusable mold, microfluidic PDMS chips can be prepared (**steps 6–13**).

1. Design a photomask with three inlets as in [26] (*see* **Note 12**);
2. Order a low-cost flexible photomask with the desired design at Selba S.A. with a negative polarity, i.e. a dark photomask with transparent structures (*see* **Note 13**);
3. The mold is then produced in the cleanroom according to standard photolithography protocols. Follow the SU-8 manufacturer's instructions for exact times, spinning velocities, temperatures, etc. (*see* **Notes 14** and **15**). Spincoat the SU-8 photoresist onto a 4 in. silicon wafer and perform a soft bake at 95°C;
4. Expose to UV light through the photomask (*see* **Note 16**) and bake again at 95 °C;
5. Develop with the manufacturer's developer for the SU-8 type you used (again follow the manufacturer's instructions on time). Rinse the wafer first with acetone to remove the developer (*see* **Note 17**) and subsequently with isopropanol to remove the acetone. Dry in nitrogen flow and perform a final hard bake at 150 °C to make the structures more robust.
6. To prepare PDMS chips from these SU-8 molds, mix PDMS pre-polymer with the PDMS curing agent in a volume ratio of 9 to 1 (*see* **Note 18**), prepare 40 ml PDMS per SU-8 mold;
7. Centrifuge for 3 min in a 50 ml falcon tube at 300 rcf to remove large air bubbles;
8. Create a cup of aluminum foil around each 4 in. silicon wafer so the wafer is the bottom of the cup and the sides of aluminum foil are at least 2 cm high (*see* **Note 19**);
9. Pour the PDMS into the cup to get a 5 mm thick layer of PDMS on top of the wafer, place in a vacuum dessicator for approximately 1 h to remove remaining air bubbles (*see* **Note 20**) and afterwards cure for 30 min at 100 °C;
10. In the meantime, spincoat the leftover PDMS onto glass slides (5 s at $3.5 \times g$ followed by 30 s at $1400 \times g$) and cure for at least 1 h at 100 °C;
11. Carefully remove the PDMS from the mold, cut out the areas of the chips (one wafer usually contains structures for four microfluidic chips) and punch 0.75 mm inlet and outlet holes (*see* **Note 21**);
12. Apply an oxygen plasma to the PDMS surfaces of glass slide and PDMS chip with a corona discharger for a few seconds. Immediately place the PDMS chip onto the glass slide and bake at 100 °C for at least 1 h to bond the surfaces together;
13. Store in a dust-free place until use (for a few months to a year).

3.1.5 Microfluidic Setup

The flow for the droplet formation is controlled by an MFCS™ pressure controller connected to a Fluiwell accessory. The microfluidic chip, and specifically droplet production at the T-junction (*see* Fig. 2a), are monitored with an inverted bright-field microscope with a 10× magnification air objective.

1. Put water phase and oil phase mixtures into two MicrewLock tubes suitable for the Fluiwell assessory and connect the polyether ether ketone (PEEK) tubing to the microfluidic chip inlets (*see* **Note 22**);
2. Adjust the pressure such that first all channels of the microfluidic chip are filled with oil phase;
3. Introduce the water phase and tune the pressures to obtain a stable production of droplets with the desired diameter at the T-junction (*see* **Note 23**);
4. Wait until a few microliters of droplets have emerged at the outlet (*see* Fig. 2a), harvest 1.5 μl using a pipette with cut-off tip, mix with 8.5 μl lipid mix (*see* **Note 24**) and transfer to the PDMS well (Subheading 3.1.7);
5. Clean the tubing with isopropanol after use. It is also possible to rinse the microfluidic chip with isopropanol and store it until the next use. In that case, always check at the microscope if the channels of the chip are not clogged before using it the next time.

3.1.6 Coverslip Silanization with Rain-X

Emulsion droplets must be placed onto a coverslip for imaging with the confocal spinning disc fluorescence microscope. In principle coverslips are hydrophilic, often resulting in a collapse of the aqueous interior of the droplets onto the coverslip. Coverslips are made hydrophobic to prevent this, for example by spincoating a thin layer of polydimethylsiloxane (PDMS) or by silanization. The advantage of silanization compared to a layer of PDMS is that the imaging quality is much better because the light is not refracted and diffused by the non-transparent PDMS. One can silanize glass in various ways, and we use Rain-X [17] as it contains less harmful chemicals than other silanization solutions and can be purchased at the local car shop.

1. Place 24 × 24 mm^2 coverslips in a holder, bath sonicate the coverslips in ethanol for 20 min and dry in nitrogen air flow;
2. Immerse the coverslips in Rain-X for 5 min; (*see* **Note 25**) and leave to dry in air (*see* **Note 26**);
3. Store between paper tissues until use (*see* **Note 27**).

3.1.7 PDMS Well Preparation

To immobilize the droplets during imaging over longer time periods, a PDMS well is placed onto the Rain-X silanized coverslips (*see* Fig. 2c) to hold the emulsion droplet mix [28]. In general, one can use the PDMS that is left over after cutting out the required areas for microfluidic chips (Subheading 3.1.4). If emulsion droplets are produced with the pipetting method instead of microfluidics, first follow **steps 6–9** from the microfluidic chips protocol in Subheading 3.1.4 with a 4 in. silicon wafer (whether it has structures or not does not matter, we just need a smooth hydrophobic surface to prepare a PDMS slab).

1. Cut the 5 mm thick PDMS slab into 2×2 cm^2 squares;
2. Punch 5 mm diameter holes in each square (*see* **Note 28**) and store them in a dust free box;
3. Before use: press a PDMS square onto a Rain-X-coated coverslip with the smooth side (the side that was in contact with the silicon wafer) facing the coverslip;
4. After use: discard the Rain-X coverslip and carefully rinse the PDMS well with isopropanol to clean it for the next use. You can keep cleaning and re-using PDMS wells for years until they break or get lost.

3.1.8 Rhodamine Tubulin Versus SiR-Tubulin

Multiple options exist to visualize MTs. The conventional method is to add fluorescently labeled tubulin to the reaction mix as is shown in Table 2. Another option is to employ SiR-tubulin, a label based on a far-red fluorophore (silicon rhodamine (SiR)) and the MT binding drug docetaxel [33]. The main advantage of SiR-tubulin is that its fluorescence intensity increases tenfold when bound to a MT compared to being in solution. As a result, the background intensity is lower and the visibility of the MTs is much better than in case of conventional fluorescent probes like rhodamine (compare Fig. 3a left and right). A disadvantage of SiR-tubulin is that, due to the docetaxol, it can have a stabilizing effect on the MTs (*see* **Note 29**).

To use SiR-tubulin instead of rhodamine tubulin, follow the following procedure (*see* **Note 30**):

1. Dissolve SiR-tubulin in dimethyl sulfoxide (DMSO) to a concentration of 500 μM and store at −20 °C (*see* **Note 31**);
2. Mix this solution in the lipid mix (*see* **Note 32**) to a final SiR-tubulin concentration of 1 μM (*see* **Note 33**);
3. Use the lipid mix containing SiR-tubulin to prepare emulsion droplets as described in Subheading 3.1.3 (*see* **Note 34**).

3.1.9 Imaging and Data Analysis

Transfer 10 μl of emulsion droplets, either prepared by microfluidics (Fig. 2a) or by pipetting (Fig. 2b), into the PDMS well (Fig. 2c). We image with a Spinning Disc confocal fluorescence microscope with a Nikon CFI Plan Apo Lambda 100X oil immersion lens (NA 1.45, working distance 0.13 mm). A custom-made temperature control system with a heating ring around the

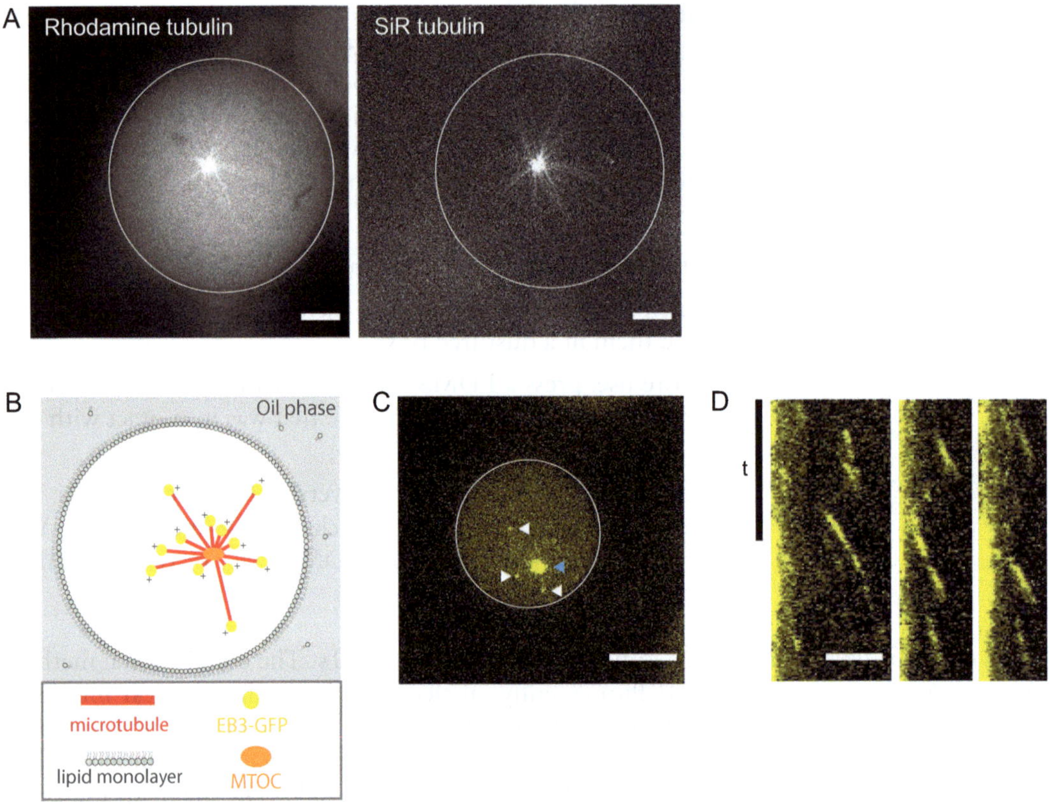

Fig. 3 (**a**) Water-in-oil emulsion droplet containing both rhodamine tubulin (left) and SiR tubulin (right), showing a reduced background and increased visibility of the MT aster inside the droplet for SiR tubulin compared to rhodamine tubulin. (**b**) Schematic of a MT aster containing end-binding (EB) proteins in an emulsion droplet. (**c**) EB3-GFP (yellow) protein comets (indicated with the white arrows) at the tips of MTs (not shown in this image), emerging from a MT aster inside an emulsion droplet. EB3 has an affinity to the centrosome as well, as we can see by the bright yellow signal in the center of the MT aster (blue arrow). (**d**) Tracking these EB3-GFP protein comets in time results in kymographs, showing the typical tip-tracking behaviour of EB3 proteins. White circles indicate droplet edge. Scale bars: 10 μm for (**a**) and (**b**), and 2 min (vertical) and 5 μm (horizontal) for (**c**)

objective lens establishes a sample temperature of 30 °C (*see* **Note 35**), which is required for MTs to polymerize inside emulsion droplets at these tubulin concentrations (20–40 μM). Typical fields of view of an experiment with droplets prepared by microfluidics or by pipetting are shown in Fig. 2d and e, respectively (*see* **Note 36**). Depending on the yield of the centrosome purification, a certain percentage of droplets contains a MT aster (*see* **Note 37**).

Previous studies on emulsion droplets have for example followed MT aster positioning over time, with or without additional MT-associated proteins [14, 26, 28], or MT organization in the presence of crosslinking motors [17, 18]. In those situations, one snapshot every few minutes suffices to track the aster position or MT organization. However, if the objective is to study MT dynamics by tracking end-binding protein comets at the growing MT tips, as is schematically depicted in Fig. 3b, we need to make timelapses with much smaller timesteps, in the order of one per second. To establish this while keeping the signal-to-noise ratio high enough to visualize the protein comets, as is shown in Fig. 3c and Movie S1, we increased laser power and used exposure times of 300–500 ms (*see* **Note 38**). An important parameter for MT dynamics is the MT growth velocity. This can be extracted from the timelapses by making kymographs as depicted in Fig. 3d, and calculating the slope of the protein comet trace.

3.2 Increasing Complexity: Adding Cytolinkers and Actin Filaments

We can make this experimental assay more complex by adding pre-polymerized actin filaments (AFs), as is schematically shown in Fig. 4a (*see* **Note 39**). From these experiments, as is shown in Fig. 4b and in Movie S2, we can study co-organization resulting from steric interactions between MTs and AFs. In addition, we can physically link AFs to MTs by adding a cytolinker, as is schematically shown in Fig. 4c. We observe that AFs co-align with growing MTs (indicated with a blue arrow in Fig. 4d), and that single AFs can be transported by a tip-tracking complex at growing MT ends, as is shown in the kymograph of Fig. 4e and in Movie S3.

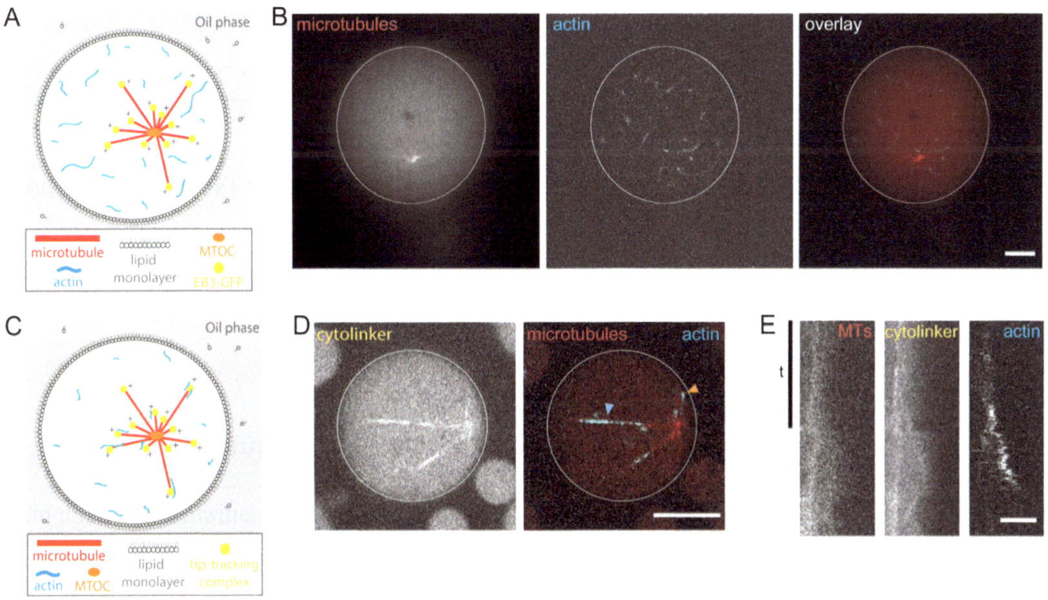

Fig. 4 Co-organization of MTs and AFs in droplets. (**a**) Schematic of an emulsion droplet containing a MT aster, end-binding (EB) proteins, and filamentous actin. (**b**) Dynamic MTs with EB-comets (not shown in this image) in droplets together with pre-polymerized AFs. (**c**) Schematic of MT/actin co-organization in an emulsion droplet with a cytolinker present. (**d**) Co-alignment of AFs to MTs (right) when a cytolinker (left) is added, indicated by blue arrow. (**e**) Kymograph to show a short event of actin transport by the growing MT plus end via the cytolinker. White circles indicating droplet edge; orange arrow (in (**d**)) indicates the AF that is shown in (**e**). Scale bars: 10 μm for (**b**) and (**d**), and 2 min (vertical) and 5 μm (horizontal) for (**e**)

To prepare these assays, one can follow the previously described methods with some adaptations. In this section, we discuss the main differences: an additional preparation step for the polymerization of AFs (Subheading 3.2.1) and their addition to the final reaction mix (Subheading 3.2.2). Furthermore, we discuss the challenges related to imaging the contents of these droplets (Subheading 3.2.3).

3.2.1 Actin Mix

Here, we describe how to prepare pre-polymerized AFs that can be added to the final reaction mix. It is also possible to replace the F-actin mix by a mixture of monomeric (G-)actin, so that AFs will polymerize within confinement, as described earlier [34], but this will not be described here (*see* **Note 40**).

1. Dilute your stock G-actin in G-buffer and pre-mix labeled and unlabeled G-actin (*see* **Note 40**) in the preferred ratios (*see* **Note 42**);
2. To start polymerization, add MRB80 buffer and the other components of the F-actin mix, except for phalloidin, as described in Table 3 (*see* **Note 43**);
3. After about 30 min, add phalloidin to stabilize the polymerized F-actin (*see* **Note 44**). Now, the F-actin mix is ready for use. It can be stored at room temperature for about 1 day.

3.2.2 Final Reaction Mix

The preparation of the final reaction mix is very similar to the one described before (in Subheading 3.1.2), except for the addition of AFs and a cytolinker.

1. Prepare the reaction mix as described in Table 4, but without the centrosomes and the F-actin mix;
2. As described in Subheading 3.1.2, centrifuge this final reaction mix for 5 min in the airfuge at 150,000 × *g* using a cold rotor to remove protein aggregates;

Table 3
Composition of the reaction mix for pre-polymerizing AFs

Component	Final concentration in F-actin mix
Unlabeled actin	6.38 μM
Labeled actin	1.125 μM
DTT	6.8 μM
KCl	50 mM
Mg-ATP	0.2 mM
Phalloidin	1 μM

This mix is based on MRB80 buffer

Table 4
Composition of the final reaction mix for reconstituting co-organization of MTs and AFs in droplets

Component	Final reaction mix concentration
50× oxygen scavenger mix[a]	1×
D-glucose	50 mM
BSA	2 mg/ml
κ-casein	0.5 mg/ml
KCl	50 mM
GTP	3 mM
Tubulin	38.5 μM
Rhodamine tubulin	1.5 μM
EB3	150 nM
Cytolinker	20 nM
F-actin mix	1 μM
Centrosomes	10% of total volume

This mix is based on MRB80 buffer
[a]200mM DTT, 10mg/ml catalase, 20mg/ml glucose oxidase

3. Next, transfer this reaction mix to an 0.5 ml Eppendorf tube and warm up in your hands, before adding to the centrosomes aliquot;
4. Finally, add the desired amount of F-actin mix, and mix well but gently to not break the centrosomes and the AFs (*see* **Note 45**);
5. From this final reaction mix, emulsion droplets can be created using either the microfluidic or pipetting method, as discussed in Subheading 3.1.3 (*see* **Note 46**).

3.2.3 Imaging

To image the dynamic co-organization of MTs and AFs in emulsion droplets there are a few challenges. Firstly, compared to imaging MT asters with EB3-GFP comets, the system has become more complex with more components to consider for imaging. To visualize proteins inside 3D emulsion droplets, the brightness of the fluorescent probe is more important than for quasi-2D experiments on a glass surface, where TIRF microscopy results in good contrast. For confocal Spinning Disc imaging of emulsion droplets, the background intensity is much higher as a result of the larger depth of focus than for TIRF microscopy. Besides, light is refracted by the spherical water-oil interface, which induces aberrations that further decrease the image quality. Hence it is necessary to think carefully about which fluorescent probe to put on which protein. As GFP is the brightest probe we have available, we label the protein of interest with GFP. Depending on the goal of the experiment this is either EB3 or the cytolinker. For the MTs, Rhodamine tubulin works in our hands the best out of the commercially available fluorescently labeled tubulins to visualize MTs in droplets (*see* **Note 47**). Due to the limited amount of available wavelenghts on our microscope, this leaves 647 nm as the imaging wavelength for actin. Otherwise, when using SiR tubulin (visible with 647 nm), actin must be labeled with a 561 nm fluorescent probe.

Secondly, the increase of complexity also comes with an increase of colors to visualize per experiment. Especially when you are interested in relatively fast processes (on the timescale of seconds), the switching between different wavelengths can slow down the acquisition. Therefore, one could consider imaging two wavelengths simultaneously, for example by using dual-imaging on one camera, or using multiple cameras. But as dual-imaging comes at the expense of the intensity of the two wavelengths signals, this can only be used in case of sufficiently high intensities.

Thirdly, another factor to take into account for visualization is the mobility of AFs. As they are more flexible than MTs and not part of a large network or structure (like the MT aster), their diffusion constant is larger. If they move too fast, they are difficult to image. Shorter exposure times and higher laser power are needed to get an image like in Fig. 4b, which can lead to increased photobleaching as a side effect. When AFs and MTs interact and become co-organized, the AFs diffuse less and are easier to visualize (Fig. 4d).

Besides these previously mentioned factors to take into account for optimization of the experiments, there are some additional 'knobs' for fine tuning of experimental conditions, like protein concentrations (*see* **Note 48**), salt concentrations (*see* **Note 49**), the ratio of AFs to cytolinkers (*see* **Note 50**), the density of AFs (*see* **Note 51**) and concentrations of surfactants and excipients (*see* **Note 52**). These experimental conditions can for instance affect the type of predominant interactions inside the droplet (e.g. MT tip vs MT lattice interactions), and hence the MT-actin co-organization.

4 Notes

1. Always use glass objects instead of plastic ones when working with chloroform.
2. This works best by rotating the glass vial while keeping it under a 45° angle in a gentle nitrogen air flow.

3. Previous protocols stated evaporating the chloroform from lipid mix in vacuum for 1 h [28] or only in nitrogen air flow and not in vacuum [26, 27]. The exact timing is not critical but in our hands drying for 1–2 h works well, and even longer is also no problem.
4. Span80 is difficult to pipette due to the high viscosity, use a cut-off pipette tip to get the correct volume.
5. Usually we can store lipids for a couple of weeks in the fridge, although sometimes the lipid mix becomes milky (opaque), which is a sign that it has gone bad and therefore will not produce stable droplets. In addition, when the droplets are more prone to fuse together or splash open on the glass surface, this is another indication that the lipid mix has gone bad.
6. Keeping centrosomes for 10–20 min in a warm (37 °C) water bath was empirically found to increase the nucleation capability of the centrosomes [28].
7. As the centrosomes are stored in a high sucrose solution, they are tricky to pipette. Therefore, it is better to add the reaction mix to the centrosome aliquot than the other way around.
8. Since the sucrose solution is viscous, pipette the mixture up and down at least 15 times to mix well.
9. If the nonspecific cortex boundary sticking becomes a problem, you can add 0.1% (v/v) Tween20 to the reaction mix in Table 2. The data presented in Fig. 2d, e and Fig. 4d contain Tween20.
10. Avoid making air bubbles by going to the second stop of the pipette, therefore take more than 10 μl and then pipette up and down (to the first stop of the pipette) quickly. The extra volume in the pipette tip prevents air bubble formation.
11. Emulsion droplets can also be produced by vortexing: add 2 μl reaction mix to 80 μl lipid mix in an 0.5 ml eppendorf tube and vortex for 5 s with a standard mini vortexer.
12. We only use two inlets (*see* Fig. 2a)—for the oil phase and water phase. The third inlet is for oil, to dilute the droplets on chip. We do not use it as we dilute the droplets later when transferring them into the PDMS imaging well.
13. A flexible photomask will not work on a UV exposure machine in combination with the vacuum used to hold the photomask in place. Therefore we attach it with Scotch tape to a glass mold that fits in the machine.
14. We used the settings for SU-8 3025 from MicroCHEM USA to obtain structures with 40 μm thickness.
15. As SU-8 ages over the years, one might need to adjust the spinning speeds [27].

16. As the exact dose of UV light depends on the machine used, calibration is needed. Make some test exposures with different exposure times and develop to see which exposure times result in robust structures.
17. This is at the same time a way to check if the development is done. If not, a white layer appears due to left over SU-8 on the wafer. In that case, put the wafer back in the developer for a while.
18. Mix the two components really well by stirring the PDMS with e.g. a spatula or pipette tip, otherwise the PDMS might form an inhomogeneous structure. It is no problem if air bubbles are created by stirring as they will be removed during the next steps.
19. Carefully press the aluminum foil around the wafer to make sure the PDMS can not flow underneath the wafer, as in that case the PDMS chip will not have the desired and homogeneous thickness.
20. Never skip this step, even if no air bubbles are visible, because small air bubbles will expand during baking and create undesired holes in your PDMS.
21. Cover the small and fragile PDMS structures that will form the channels with scotch tape to prevent damage. You remove this before bonding the PDMS to the PDMS-coated coverslip.
22. Before use, rinse both tubings first with isopropanol, then with dry them with air (just make sure the end of the tubing inside the MicrewLock tube is just above the isopropanol surface) and finally rinse with MRB80 or lipid mix respectively.
23. As the final reaction mix is only 10–20 μl in volume, there is not much space to fine tune the pressures to get an optimal droplet production. One solution is to make test droplets with just MRB80 as water phase and note the pressures where the droplet production is as desired. Next, take the water phase tubing out of the microfluidic chip and fill it with air (by applying maximum pressure while the other end of the tubing is just above the MRB80 surface). Replace the MRB80-filled MicrewLock tube by one with reaction mix, carefully increase the pressure until the tubing is filled with reaction mix and then stop the pressure so you can re-insert the end into the microfluidic chip. Now set the pressure to the empirically found values to get good droplet production from the start.
24. Usually the droplet density is very high, which gives problems when imaging droplets at the microscope. Therefore we dilute the droplets in the PDMS well.

25. We usually prepare multiple batches of coverslips and immerse them in the same volume of Rain-X for efficiency and to save Rain-X.
26. Sometimes the Rain-X film does not dry completely homogeneously onto the coverslip, visible as some transparant stains at the edges of the coverslip. This has never proven to be a problem with imaging.
27. As a result of the paper tissue, dust particles often stick to the coverslips. Shortly rinse with ethanol and dry in nitrogen flow to get rid of these without damaging the silanization.
28. It is convenient to have as many holes in each square as possible (*see* Fig. 2c), since you need a clean hole each time you prepare new droplets, also within one experimental condition.
29. In our hands, this happended at SiR tubulin concentrations above 200 μM.
30. The SiR-tubulin is added to the lipid mix, while in bulk in vitro reconstitution experiments it can be added to the reaction mix. We found that SiR-tubulin partitions more into the oil phase (lipid mix) than the water phase (reaction mix) and therefore needs to be added to the oil phase, as the volume of the oil phase is about 80 times larger than that of the water phase when making and imaging emulsion droplets. Some SiR-tubulin will partition into the water phase and bind to the MTs.
31. One can take out the desired volume for an experiment and then refreeze the remaining SiR-tubulin in DMSO at −20 °C.
32. We usually pre-dilute the in DMSO dissolved SiR-tubulin in lipid mix before diluting it into the final lipid mix, due to the high stock concentration of SiR-tubulin. Mix these dilutions very well to get a homogeneous distribution of SiR-tubulin in the lipid mix. Also, pay attention to pipette carefully and slowly to make sure that the pipetted volume of lipid mix is consistent over different experiments because otherwise the SiR-tubulin concentration cannot be compared.
33. Usually we get good results at 1 μM, but also 0.5 μM results in nice MT labelling. As only a small fraction of the SiR-tubulin ends up in the water phase, the final concentration there is probably below the 200 nM mentioned in **Note 29**, because we do not see a strong stabilizing effect on the MTs at 0.1–1 μM SiR-tubulin in the oil phase.
34. The distribution of SiR-tubulin throughout the lipid mix is usually not perfectly homogeneous, as droplets containing stabilized MTs are observed as well as droplets with nicely labeled dynamic MTs as in Fig. 3a (right). We do not consider this a problem for our experiments, since we select for the proper droplets.

35. When placing the PDMS well on top of the microscope, it first needs to heat to 30 °C. Due to glass expansion caused by the temperature change, droplets move out of focus. Therefore it is better to wait a few minutes for the system to equilibrate.
36. As water has a higher density than oil, the droplets sink down when they are pipetted into the PDMS well and therefore move out of focus. Thus, as in (*see* **Note 35**) it is better to wait a few minutes until the droplets have reached the coverglass before starting measurements.
37. Finding MT asters inside emulsion droplets can be difficult, especially if the centrosome concentration is low and the MTs are difficult to see. Easiest is to look at live microscope images with 200 ms exposure time in the wavelength channel where MTs are expected. Move in x and y and scan in z through the droplets while looking for some brighter blob in the droplets. To distinguish a protein aggregate from a centrosome, look at the diffusion rate of the blob. Due to the relatively long MTs nucleating from a centrosome (even if you can not visualize the MTs they probably are present), they are not diffusing around a lot, while protein aggregates are very mobile and can move out of your focal plane within seconds.
38. Previous efforts in our lab to visualize protein comets inside emulsion droplets did not work out. One reason is that bleaching was significant, and therefore laser intensity and exposure times were limited. We now work on a different spinning disc microscope than before, which has a different (more sensitive) camera. Since then, bleaching is not observed to be a problem anymore, hence we can image with higher laser power and exposure times. In addition, imaging on silanized coverslips instead of PDMS-coated coverslips has increased the imaging resolution by removing light refractions in the PDMS. This probably also contributed to the observation of protein comets.
39. To check the functionality of all these different proteins, it is advised to first use an assay with higher spatial (and temporal) resolution, such as the quasi-2D TIRF assay that is described before [35]. However, note that protein concentrations for TIRF assays cannot be compared to droplet experiments (see also **Note 48**).
40. G-actin will start polymerizing the moment you add it to the final reaction mix, so be quick in preparing droplets and starting the image acquisition.
41. First mix unlabelled and labelled G-actin well (dilutions in G-buffer), to get homogeneously labelled F-actin.

42. To vary the concentration of F-actin in droplets, either different concentrations of actin can be used in the F-actin mix, or this F-actin mix could be diluted right before addition to the final reaction mix.
43. After adding MRB80 to the actin-mix, polymerization starts and the mix can be kept at room temperature. Store it in the dark, to prevent photo-bleaching.
44. We add phalloidin after some time (e.g. 30 min), so it does not interfere with actin polymerization and distribution of labels. This timing is not very precise.
45. The shear-forces generated by non-gentle-mixing could break the centrosomes and the pre-polymerized AFs. Interestingly, breaking pre-polymerized F-actin (before adding it to the final reaction mix) could also serve as a method to decrease the average length of these filaments.
46. From now on, keep the final reaction mix at room temperature to prevent depolymerisation of AFs (at lower temperatures), and to prevent MTs from starting to grow (at higher temperatures). This means that also the lipid mix should be at room temperature when making droplets.
47. We have also had nicely visible MT asters inside emulsion droplets using HiLyte647-labeled tubulin from Cytoskeleton. However, after switching to a different labeled tubulin batch, the Rhodamine tubulin gave better results.
48. The tubulin concentration for in vitro experiments in water-in-oil emulsion droplets needs to be a factor 2–3 higher than in quasi-2D channel experiments. It is not clear why, but at lower concentrations the MTs do not nucleate in emulsion droplets.
49. Increasing the KCl concentration from 50 mM to 75 mM reduces nonspecific binding of a tip-tracking protein to the MT lattice, leading to a larger contrast between EB3 at the tip compared to the lattice.
50. The cytolinker we use is a dimer that can also bundle AFs. By changing the AF:linker ratio, one can tune the predominant interactions of the cytolinker inside the droplets.
51. The density of actin should be optimized every day, since it depends on the quality of your actin (the amount of functional actin that polymerizes into filaments) and on how much the AFs are sheared by pipetting (*see* **Note 45**).
52. We observe that addition of surfactants like Tween20 or excipients like BSA or κ-casein (*see* Table 1) can influence both aspecific sticking of proteins to the droplet boundary as well as specific protein-lipid interactions (the latter was only observed for Tween20 and Span80), and can even influence the stability of emulsion droplets.

Acknowledgements

We are grateful to Marjolein Vinkenoog for actin purification and Andrea Martorana for TipAct purification. We thank Sophie Roth, Florian Hüber, Mathijs Vleugel, Anne Doerr and Esengül Yildirim for centrosome purification. This work was supported by the European Research Council (Synergy grant 609822 to M.D.) and the Netherlands Organization for Scientific Research (FOM program nr. 151 to M.D.).

References

1. Dogterom M, Koenderink G (2018) Actin-microtubule crosstalk in cell biology. Nat Rev Mol Cell Biol 20:38–54
2. Huda S, Soh S, Pilans D, Byrska-Bishop M, Kim J, Wilk G, Borisy GG, Kandere-Grzybowska K, Grzybowski BA (2012) Microtubule guidance tested through controlled cell geometry. J Cell Sci 125:5790–5799
3. Bazellières E, Massey-Harroche D, Barthélémy-Requin M, Richard F, Arsanto JP, Le Bivic A (2012) Apico-basal elongation requires a drebrin-E-EB3 complex in columnar human epithelial cells. J Cell Sci 125:919–931
4. Zhou F, Waterman-Storer C, Cohan C (2002) Focal loss of actin bundles causes microtubule redistribution and growth cone turning. J Cell Biol 157:839–849
5. López MP, Hubera F, Grigoriev I, Michel O Steinmetz AA, Koenderink GH, Dogterom M (2014) Actin-microtubule coordination at growing microtubule ends. Nat Commun 5:4778
6. Wu X, Kodoma A, Fuchs E (2008) ACF7 regulates cytoskeletal-focal adhesion dynamics and migration and has ATPase activity. Cell 135:137–148
7. Zaoui K, Benseddik K, Daou P, Salaun D, Badache A (2010) Erbb2 receptor controls microtubule capture by recruiting ACF7 to the plasma membrane of migrating cells. Proc Natl Acad Sci U S A 107:18517–18522
8. Henty-Ridilla JL, Rankova A, Eskin JA, Kenny K, Goode BL (2016) Accelerated actin filament polymerization from microtubule plus ends. Science 352:1004–1009
9. Wojnacki J, Quassollo G, Marzolo M, Caceres A (2014) Rho GTPases at the crossroad of signaling networks in mammals: impact of Rho-GTPases on microtubule organization and dynamics. Small GTPases 5:e28430
10. Kunda P, Baum B (2009) The actin cytoskeleton in spindle assembly and positioning. Trends Cell Biol 19:174–179
11. Colin A, Singaravelu P, Théry M, Blanchoin L, Gueroui Z (2018) Actin-network architecture regulates microtubule dynamics. Curr Biol 28:1–10
12. Bringmann H, Hyman A (2005) A cytokinesis furrow is positioned by two consecutive signals. Nature 436:731–734
13. Abercrombie M (1961) The bases of the locomotory behaviour of fibroblasts. Exp Cell Res 8 (Suppl):188–198
14. Laan L, Pavin N, Husson J, Romet-Lemonne-G, van Duijn M, López MP, Vale RD, Jülicher F, Reck-Peterson SL, Dogterom M (2012) Cortical dynein controls microtubule dynamics to generate pulling forces that position microtubule asters. Cell 148:502–514
15. Alvarado J, Mulder B, Koenderink G (2014) Alignment of nematic and bundled semiflexible polymers in cell-sized confinement. Soft Matter 10:2354–2364
16. Tsai F, Koenderink G (2015) Shape control of lipid bilayer membranes by confined actin bundles. Soft Matter 11:8834–8847
17. Baumann H, Surrey T (2014) Motor-mediated cortical versus astral microtubule organization in lipid-monolayered droplets. J Biol Chem 289(32):22524–22535
18. Juniper MP, Weiss M, Platzman I, Spatz JP, Surrey T (2018) Spherical network contraction forms microtubule asters in confinement. Soft Matter 14:901–909
19. e Silva MS, Alvarado J, Nguyen J, Georgoulia N, Mulder BM, Koenderink GH (2011) Self-organized patterns of actin filaments in cell-sized confinement. Soft Matter 7 (22):10631–10641
20. Pontani L-L, Van der Gucht J, Salbreux G, Heuvingh J, Joanny J-F, Sykes C (2009)

Reconstitution of an actin cortex inside a liposome. Biophys J 96(1):192–198
21. Pinot M, Chesnel F, Kubiak J, Arnal I, Nedelec FJ, Gueroui Z (2009) Effects of confinement on the self-organization of microtubules and motors. Curr Biol 19(11):954–960
22. Shah EA, Keren K (2014) Symmetry breaking in reconstituted actin cortices. Elife 3:e01433
23. Takiguchi K, Yamada A, Negishi M, Tanaka-Takiguchi Y, Yoshikawa K (2008) Entrapping desired amounts of actin filaments and molecular motor proteins in giant liposomes. Langmuir 24(20):11323–11326
24. Tsai F-C, Koenderink GH (2015) Shape control of lipid bilayer membranes by confined actin bundles. Soft Matter 11(45):8834–8847
25. Carvalho K, Tsai F-C, Lees E, Voituriez R, Koenderink GH, Sykes C (2013) Cell-sized liposomes reveal how actomyosin cortical tension drives shape change. Proc Natl Acad Sci 110(41):16456–16461
26. Roth S, Laan L, Dogterom M (2014) Reconstitution of cortical dynein function. Methods Enzymol 540:205–230
27. Taberner N, Lof A, Roth S, Lamers D, Zeijlemaker H, Dogterom M (2015) In vitro systems for the study of microtubule-based cell polarity in fission yeast. Methods Cell Biol 128:1–22
28. Vleugel M, Roth S, Groenendijk C, Dogterom M (2016) Reconstitution of basic mitotic spindles in spherical emulsion droplets. J Vis Exp 2016:e54278
29. Pardee JD, Spudich JA (1982) Mechanism of K_+-induced actin assembly. J Cell Biol 93:648–654
30. Gentry BS, van der Meulen S, Noguera P, Alonso-Latorre B, Plastino J, Koenderink GH (2012) Multiple actin binding domains of Ena/VASP proteins determine actin network stiffening. Eur Biophys J 41:979–990
31. Komarova Y, de Groot CO, Grigoriev I, Gouveia SM, Munteanu EL, Schober JM, Honnappa S, Buey RM, Hoogenraad CC, Dogterom M, Borisy GG, Steinmetz MO, Akhmanova A (2009) Mammalian end binding proteins control persistent microtubule growth. J Cell Biol 184:691–706
32. Moudjou M, Bornens M (1998) Method of centrosome isolation from cultured animal cells. Cell Biol 2:111–119
33. Lukinavičius G, Reymond L, D'este E, Masharina A, Göttfert F, Ta H, Güther A, Fournier M, Rizzo S, Waldmann H *et al* (2014) Fluorogenic probes for live-cell imaging of the cytoskeleton. Nat Methods 11 (7):731–733
34. Claessens MM, Bathe M, Frey E, Bausch AR (2006) Actin-binding proteins sensitively mediate F-actin bundle stiffness. Nat Mater 5:748–753
35. López MP, Huber F, Grigoriev I, Steinmetz MO, Akhmanova A, Dogterom M, Koenderink GH (2014) In vitro reconstitution of dynamic microtubules interacting with actin filament networks. Methods Enzymol 540:301–320

Chapter 6

Studying Tau-Microtubule Interaction Using Single-Molecule TIRF Microscopy

Virginie Stoppin-Mellet, Nassiba Bagdadi, Yasmina Saoudi, and Isabelle Arnal

Abstract

Microtubule architecture depends on a complex network of microtubule-associated proteins (MAPs) that act in concert to modulate microtubule assembly/disassembly and spatial arrangement. In vitro reconstitution of cytoskeleton dynamics coupled to single-molecule fluorescence assays has opened new perspectives to quantify the interaction of MAPs with microtubules. Here, we present a Total Internal Reflection Fluorescence (TIRF) microscopy-based assay enabling the characterization of Tau interaction with dynamic microtubules at the single-molecule level. We describe protein sample preparation in flow cells, single-molecule acquisitions by TIRF microscopy, and quantitative analysis of Tau oligomerization states and dwell time on microtubules.

Key words Tau, Microtubules, Single molecule, Total Internal Reflection Fluorescence microscopy, Cell-free assays

1 Introduction

Microtubule-associated proteins (MAPs) function to regulate the assembly, dynamic behavior, and spatial organization of microtubules. MAPs include microtubule-stabilizing and microtubule-destabilizing factors, plus-end tracking proteins, cross-linkers, and motor proteins [1]. For long, the properties of MAPs were inferred from their effects on microtubule dynamics and spatial organization, as studied in cells or in reconstituted cell-free assays coupled to real-time light imaging [2, 3]. These experiments are very informative to specify the activities of MAPs on microtubule regulation, but they fail in providing the parameters that determine the molecular interaction between MAPs and microtubules (e.g., association/dissociation rates, coefficient of diffusion, speed, and directionality

Virginie Stoppin-Mellet and Nassiba Bagdadi contributed equally to this work.

Helder Maiato (ed.), *Cytoskeleton Dynamics: Methods and Protocols*, Methods in Molecular Biology, vol. 2101,
https://doi.org/10.1007/978-1-0716-0219-5_6,

of displacement). Pioneer works combining in vitro assays with numerical computer simulations represent a powerful approach to decipher the key parameters governing microtubule organization by MAPs [4–6]. However, with the notable exception of molecular motors [7, 8], it was not possible for a long time to observe the behavior of MAPs at the single-molecule resolution and in real time. The development of Total Internal Reflection Fluorescence (TIRF) microscopy and labeling procedures of highly pure tubulin and MAPs overcame this barrier [9].

In TIRF microscopy, the illuminated volume is limited to 200 nm above the glass surface, thus greatly improving the signal-to-noise ratio. Combining the high-quantum efficiency, low noise, and high speed of last-generation EMCCD and c-mos cameras with high-numerical aperture 100× objectives, it is now possible to record single fluorescent molecules with great spatial resolution at streaming rate while interacting with dynamic microtubules. By collecting a high number of individual molecule trajectories, one can calculate the kinetic parameters of binding/unbinding events (on-rates, off-rates, dwell times, diffusion rates, etc.), identify deviant subpopulations, and observe rare and transient events along the microtubule length or at its extremities. Single-molecule approaches combined with TIRF microscopy proved very valuable in deciphering molecular details of the interaction of several MAPs with microtubules [10, 11].

Tau is an abundant neuronal MAP identified in the mid-1970s [12] that has been extensively studied for its ability to stabilize microtubules and to promote microtubule bundling [13]. Tau is an intrinsically disordered protein that can form aggregates upon posttranslational modifications, the best known being abnormal phosphorylation. Since many neurodegenerative diseases including Alzheimer's disease and related dementia are characterized by intracellular inclusions of modified Tau [14], a vast majority of studies focus on Tau aggregation. Little is known about the properties of soluble modified forms of Tau and their interactions with microtubules. Furthermore, recent studies have revealed that Tau is not only a potent microtubule stabilizer but can induce different types of microtubule arrays with specific organization and dynamics [15–17]. In this context, it is now essential to understand how Tau induces and regulates these microtubule arrays. Addressing this requires observing single molecules of Tau while interacting with microtubules.

In this chapter, we describe a TIRF microscopy-based assay to study Tau-microtubule interactions at the single-molecule level. We explain how to calibrate the assay in order to calculate the fluorescence intensity of one Tau molecule and to estimate the oligomerization state of Tau. We then describe how to record fluorescent single Tau molecules while interacting with dynamic microtubules, and we give a short description of the analysis procedure allowing calculation of Tau oligomerization state and residency time on microtubules. A list of the main pitfalls that may alter success and reproducibility of the assay is also provided.

2 Materials

2.1 Flow Chambers

2.1.1 Materials and Equipment

1. Pressurized filtered air.
2. Glass staining jar.
3. Polypropylene tray for slides and coverslips (Dutscher, REF 391058, or equivalent).
4. Microscope slides (Menzel-Gläser-superfrost, 76 × 26 mm).
5. Coverslips (Menzel-Gläser, 76 × 26 mm #1).
6. Double-face precut adhesive tape (70 μm thick, 3 mm wide, LIMA Company, France).
7. Plasma system (Femto low-pressure system, version A, Diener electronic).
8. Sonication bath.
9. Mini-centrifuge with two slide-drying cassettes (VWR, Ref VWRI521-2849_P, or equivalent).
10. Orbital shaker.
11. Glass writing pen with a diamond tip (Agar scientific, REF AGT5347).

2.1.2 Glass Cleaning and Functionalization

1. 2% Hellmanex™ III (Hellma Analytics) in water, freshly prepared.
2. 1 mg/mL mPEG-silane, MW 30k (Creative PEGWorks): Dissolve 200 mg of PEG-silane in 200 mL of 96% ethanol, 0.2% (v/v) HCl. Gently stir the solution at 50 °C until complete dissolution. Keep at room temperature safe from light (*see* **Note 1**).
3. 1 mg/mL Biotin-PEG-silane, MW 3400 (Laysan BIO): Dissolve 200 mg of biotin-PEG-silane in 200 mL of 96% ethanol, 0.2% (v/v) HCl. Gently stir the solution until complete dissolution. Keep at room temperature safe from light. Be aware that aliquots of biotin-PEG-silane powder have to be made under argon and stored in safely sealed tubes at −20 °C.

2.2 Proteins and Microtubule Seeds

1. BRB80: 80 mM Pipes pH 6.8, 1 mM EGTA, 1 mM $MgCl_2$. Adjust pH with KOH.
2. K-BRB80: 80 mM Pipes pH 6.8, 1 mM EGTA, 1 mM $MgCl_2$, 50 mM KCl. Adjust pH with KOH.
3. GMPCPP (Jena Bioscience) 10 mM in H_2O. Store as small aliquots at −20 °C.

4. Purified labeled tubulin: Mammalian tubulin is purified from bovine brains by two polymerization-depolymerization cycles, followed by a cation-exchange chromatography to remove MAP [18]. Tubulin is labeled with biotin or ATTO fluorophores (ATTO-TEC Gmbh) according to [19] and stored in liquid nitrogen. Details of the procedures we are using in the lab can be found in [20].
5. Tau-sfGFP proteins: Human recombinant His-tagged Tau (1N4R isoform) and His-tagged ΔN-Tau (deleted from Tau projection domain [16]) fused to sfGFP [21] at their C-terminal extremities are expressed in bacteria and purified with IMAC chromatography followed by gel filtration according to [20] with few modifications. Tau-sfGFP and ΔN-Tau-sfGFP were cloned in pET28a and expressed in BL21(DE3) *E. coli*. Expression is induced with 0.5 mM IPTG overnight at 18 °C. After lysis and centrifugation, the supernatant containing Tau-sfGFP or ΔN-Tau-sfGFP is not heated at 75 °C as commonly done for Tau purification but is directly incubated with IMAC (cobalt) resin. Tau-sfGFP and ΔN-Tau-sfGFP are stored in K-BRB80 buffer in liquid nitrogen (*see* **Note 2**).
6. Fluorescent microtubule seeds: Biotinylated fluorescent microtubule seeds are obtained by polymerizing a mix of 5 μM of biotin-labeled tubulin and 5 μM of ATTO-565-labeled tubulin for 1 h at 35 °C in the presence of 0.5 mM GMPCPP and 1 mM DTT in BRB80. After centrifugation (5 min at 110,000 × *g* and 25 °C in a TLA-100 rotor, Beckman), the pellet is washed twice with warm BRB80 and resuspended in BRB80 supplemented with 0.5 mM GMPCPP and 1 mM DTT. Small aliquots (2 μL) of microtubule seeds are flash frozen and stored in liquid nitrogen.

2.3 TIRF Assay

2.3.1 Microscope

We are using an inverted microscope (Eclipse Ti, Nikon) equipped with a Perfect Focus System and a CFI Apochromat TIRF 100×/1.49 NA Oil objective (Nikon). The temperature in the flow chamber is maintained constant by a warm stage controller (Linkam Scientific) and an objective heater (OkoLab). The microscope is enclosed in a homemade Plexiglas chamber to limit the temperature variation and airflow. The microscope is associated to an ilas^2 TIRF system (Roper Scientific) and a cooled charged-coupled EMCCD camera (Evolve 512, Photometrics) controlled by MetaMorph software (version 7.7.5, Molecular Devices). Illumination is provided by 488- and 561-nm lasers (150 and 50 mW, respectively) coupled to a Z405/491/561 triple band mirror and ET525/50 and ET605/52 emission filters (Chroma). We recommend a regular check for a perfect alignment of the lasers and homogenous illumination of the imaging field. For simultaneous dual color imaging, we use a dual-view system (DV2) from Roper Scientific equipped with ET525/50 and ET630/75 emission filters (Chroma).

2.3.2 Stock Solutions

For convenience, most of our stock solutions are stored as small aliquots at −20 or −80 °C. The day of the experiments, aliquots are thawed and used to prepare working solutions. Except for BRB80, thawed aliquots are not refrozen and are discarded at the end of the day. Solutions are filtered through a 0.22 μm filter.

1. 10% (w/v) BSA (Sigma-Aldrich Ref A7030) in phosphate buffer saline: Clarify by filtration and store at −20 °C.
2. 5× BRB80: 400 mM Pipes pH 6.8, 5 mM EGTA, 5 mM $MgCl_2$. Adjust pH with KOH. Filter and store at −20 °C.
3. 1 M Hepes pH 7.4: Filter and store at −20 °C.
4. Neutravidin: Make 1 mg/mL stock solution in H_2O and store at −20 °C.
5. PLL-PEG (Jenkem, PLL20K-G35-PEG2K) (*see* **Note 2**): 1 mg/mL in a freshly prepared and filtered 10 mM Hepes pH 7.4 solution. Store as aliquots at −20 °C.
6. 0.2 M DTT in H_2O: Store as small aliquots at −20 °C.
7. 0.5 M KCl in H_2O: Filter and store at −20 °C.
8. 0.1 M GTP in H_2O: Filter and store at −20 °C.
9. Glucose 450 mg/mL in BRB80: Filter and store at −80 °C.
10. Anti-photobleaching cocktail (GODCAT cocktail) (*see* **Note 3**): Prepare a mix containing 3.5 mg/mL catalase (Sigma-Aldrich C9322) and 25 mg/mL glucose oxidase (Sigma-Aldrich G6766) in BRB80. Do not vortex but gently mix the solution until complete dissolution. Filter, snap freeze as small aliquots in liquid nitrogen, and store at −80 °C.
11. Mineral grease for flow chamber sealing (Dow Corning high-vacuum grease, Sigma-Aldrich Ref Z273554).

2.3.3 Working Solutions

1. Methylcellulose (1500 cP, Sigma-Aldrich M-0555): Heat up deionized and filtered water to 60 °C, and add methylcellulose to obtain a 1% (w/v) solution. Gently shake until the solution becomes clear (requires 20–30 min at least). Keep in the dark at 4 °C. This solution can be used for up to 10 days.

   The following solutions should be prepared extemporaneously. Except for BRB-BSA, all working solutions are kept on ice:
2. BRB-BSA: 1% BSA in BRB80.
3. Neutravidin: 25 μg/mL in BRB-BSA.
4. PLL-PEG: 0.1 mg/mL in freshly prepared 10 mM Hepes pH 7.4.
5. 20 mM GTP in BRB80.
6. Glucose 45 mg/mL in BRB80.

2.4 Software

ImageJ is used to process movies and draw kymographs using a homemade plugin available upon request.

3 Methods

3.1 Flow Chamber Preparation

3.1.1 Glass Cleaning (See Note 4)

Dispose slides and coverslips in the propylene tray and incubate in the following solutions under gentle agitation (80 rpm on a horizontal shaker). Rinse **steps 3**, **5**, **7**, and **9** are performed in a 2-L beaker.

1. 20 min in acetone 100%.
2. 10 min in ethanol 96%.
3. Carefully rinse in a large volume of filtered deionized water (*see* **Note 5**).
4. 20 min in 2% Hellmanex preheated at 60 °C (*see* **Note 6**).
5. Carefully rinse in a large volume of filtered deionized water.
6. 15-min sonication in freshly prepared 1 M NaOH in a bath sonicator (sweep mode).
7. Carefully rinse in a large volume of filtered deionized water.
8. 15-min sonication in ethanol in a bath sonicator (sweep mode).
9. Carefully rinse in a large volume of filtered deionized water.
10. Dry slides and coverslips. Coverslips can be dried in a mini-centrifuge. For coverslips, we recommend drying by filtered air or argon flow. At this step, clean slides and coverslips can be kept in the fridge, in a clean and sealed box to avoid dust and moisture. Extensive storage (several weeks) is not recommended.

3.1.2 Glass Functionalization

1. Activate slides and coverslips in a plasma cleaner (3 min, 50 Hz, 80% full power of 100 V) (*see* **Note 7**).
2. Immediately deep slides and coverslips (disposed in propylene trays) in the mPEG-silane solution and biotin-PEG-silane solutions, respectively. Incubate overnight (less than 18 h) in the dark with moderate horizontal shaking (80 rpm).
3. Rinse slides and coverslips by dipping them one by one ten times in ethanol and then in filtered deionized water. If the functionalization is successful, water will homogenously run over the surface of the glass.
4. Dry as in Subheading 3.1.1, **step 10**. Keep at 4 °C in the dark, in a clean and sealed box. Use within a week.

3.1.3 Flow Chamber Assembly

Using the diamond pen, cut a 76 × 26 mm coverslip in 8 pieces of 19 × 13 mm. Note that from this step, only the upper face of the coverslip is now clean.

Place two bands of double-face tape on a slide, about 0.5 cm apart, and fix a coverslip piece with the clean side facing the interior of the chamber. You will obtain a flow chamber of approximately 10 μL. Press over the tape surface so that it is perfectly flat. This is important to ensure that the coverslip surface will be parallel to the microscope stage. Several chambers can be mounted on the same slide.

3.2 Single-Molecule Photobleaching Experiments (SMPB)

The purpose of this experiment is to test the solubility of Tau proteins and find the concentration at which Tau proteins appear as single molecules. It also allows analysis of the oligomerization state of the proteins.

3.2.1 Sample Preparation

Tau proteins are diluted in cold K-BRB80 to a concentration of a few tenths of micromolar and centrifuged to remove aggregates (5 min/4 °C/215,000 × *g* in TLA-100 rotor (Beckman)). The concentration of soluble Tau proteins is checked using a Bradford assay with BSA as a standard. Proteins are kept on ice until use.

Just before use, Tau proteins are further serial diluted to a final concentration of few pM (10–50 pM). All intermediate dilutions are discarded.

3.2.2 Photobleaching of Single Molecules of Tau

1. Perfuse 50 μL of the diluted Tau-sfGFP or ΔN-Tau-sfGFP in the flow chamber for 5 min at room temperature.
2. Wash three times with 100 μL K-BRB80 to remove unbound proteins.
3. Seal the chamber with mineral grease.
4. Image samples with the TIRF microscope. Fluorescent Tau proteins should appear as dispersed isolated particles. If not, adjust the dilution of your sample.
5. Record images in continuous streaming mode with the shortest exposure time (we use 50 ms) and sufficient laser power to photobleach the isolated molecules. In our case, photobleaching is very fast and 2-min movies are sufficient (Fig. 1).

3.2.3 Analysis of Fluorescence Intensity Traces

The accuracy of the analysis implies that all molecules included in the analysis have been illuminated the same way. So first check for the illumination of the imaging field by applying the Fire lookup table of ImageJ to your movie and select a region of interest (ROI) with a uniform illumination. Crop the stack to the ROI and reactivate the gray scale. The next step consists in collecting the fluorescence intensity fluctuations over time of each fluorescent Tau molecule. To do so, we use the Thunderstorm plug-in [22] in ImageJ that detects true-signal spots on each frame. Thunderstorm output is a data file containing the coordinates and the intensity of each spots, from which we reconstitute individual traces using a

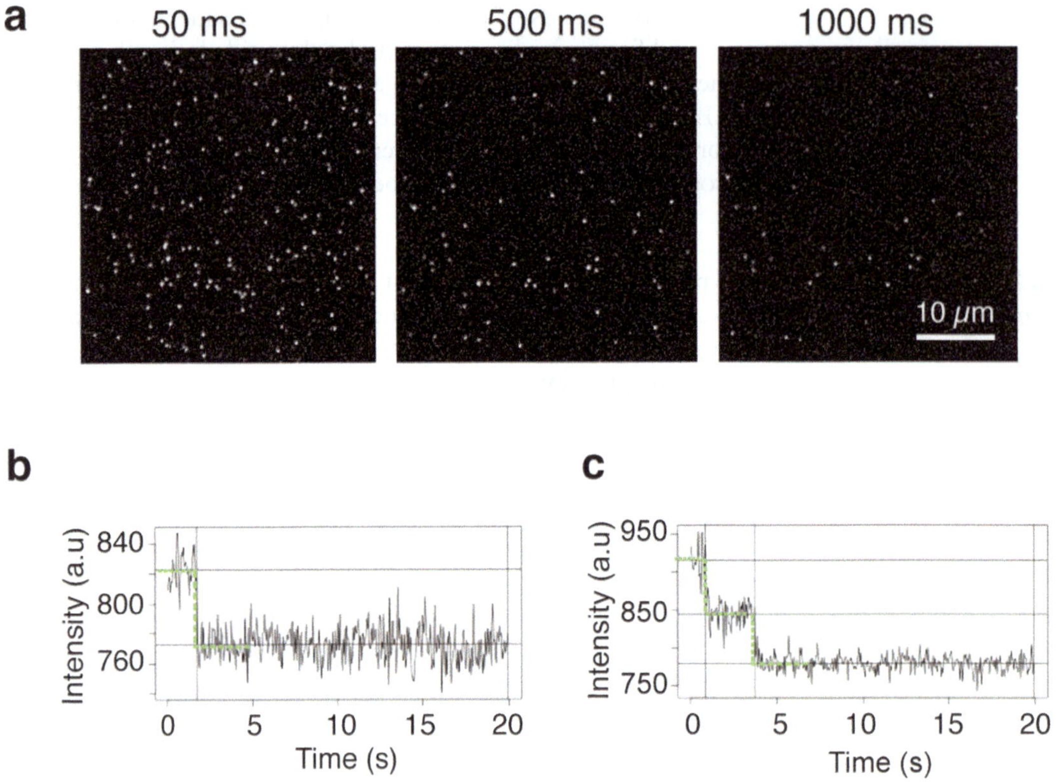

Fig. 1 Single-molecule photobleaching experiments. (**a**) Snapshots of Tau-sfGFP during photobleaching. Each spot represents one single molecule of Tau-sfGFP. Molecules are continuously irradiated with a 488 nm laser and recorded in the streaming mode with 50 ms exposure time. Individual spot successively disappears over time due to photobleaching. The contrast of images has been enhanced for better visualization of spots. (**b**, **c**) Fluctuations of fluorescence intensity of Tau-sfGFP over time. One-step (**b**) and two-step (**c**) bleaching indicate the presence of monomeric and dimeric proteins, respectively

homemade R script in Rstudio [23]. Other automated methods such as SONIC [24] are available.

Ideally, you should obtain steplike traces where an abrupt decrease in intensity corresponds to the bleaching of one sfGFP, and the height of a step corresponds to the intensity of one sfGFP (only suitable for the specific imaging conditions used during the experiment). Thus, the number of steps gives the number of Tau monomers in a single molecule. For example, the proportion of one-step (monomeric proteins) and two-step (dimeric proteins) traces provides information about the presence of subpopulations in your sample (Fig. 1) (*see* **Note 8**). Aggregates generate high-intensity traces that gradually decrease without any detectable individual bleaching events.

3.3 Interaction of Single Molecules of Tau with Microtubules

3.3.1 Preparation of Proteins and TIRF Assay Mix

1. Tau-sfGFP and ΔN-Tau-sfGFP are prepared as for the control SMPB experiment. Diluted samples should be prepared just before use.
2. Stock solution of concentrated tubulin samples are prepared by mixing ATT0-565 tubulin (10–30% of the final concentration) and unlabeled tubulin. Aggregates are removed by centrifugation at 4 °C for 5 min at 215,000 × *g* in a TLA-100 rotor (Beckman), and concentration of the soluble tubulin is estimated by absorbance at 280 nm (we routinely use the approximation of 1 $OD_{280} = 10$ μM). Store at small aliquots (1–2 μL) in liquid nitrogen until use. As an example, if the tubulin final concentration for the assay is 15 μM, we prepare a stock solution of tubulin of about 300 μM, and add 1.5 μL of this sample to the 30 μL TIRF assay mix.
3. Few minutes before use, fluorescent microtubule seeds are rapidly thawed, diluted in warm BRB80, and immediately perfused in the flow chamber.
4. The TIRF assay mix (30 μL total volume) is prepared just before use by mixing the following reagents in order and on ice:
   (a) Cold H_2O qsp 30 μL.
   (b) 3 μL 5× BRB80 (final 1×).
   (c) 3 μL 0.5 M KCl (final 50 mM).
   (d) 3 μL 10% BSA (final 1%).
   (e) 1.5 μL 20 mM GTP (final 1 mM).
   (f) 0.7 μL 0.2 M DTT (final 4 mM).
   (g) 0.7 μL 45 mg/mL glucose (1 mg/mL final).
   (h) 0.7 μL Anti-photobleaching cocktail.
   (i) 5 μL 1% Methylcellulose (0.15% final).
   (j) 10–15 μM Tubulin mix.
   (k) 50 pM sfGFP-Tau or ΔN-Tau-sfGFP.

   Homogenize carefully avoiding bubbles.

3.3.2 TIRF Assay and Imaging

Perfusions are performed by introducing solutions with a 200 μL pipette from one side of the flow chamber and absorbing the liquid with absorbent paper from the opposite site, in order:

1. 30 μL Neutravidin for 5 min.
2. 50 μL PLL-PEG for 30 s to 1 min.
3. Three washes with 100 μL BRB-BSA.
4. Fluorescent microtubule seeds for 5 min. During this time, prepare the TIRF assay mix.
5. Three washes with 100 μL BRB-BSA.
6. 30 μL TIRF assay mix.

Seal the chamber with mineral grease and rapidly place the slide on the warm microscope stage for imaging. Record fluorescent Tau proteins in the exact same conditions as for the control experiments. Engage the dual-view system if you want to simultaneously record assembly of microtubules (labeled with ATTO-565 tubulin) and Tau proteins. You can also take two pictures of the microtubules at the beginning and at the end of Tau recording to evaluate the position and the length of microtubules.

3.3.3 Measurements of the Fluorescence Intensity and the Bleaching of Single Tau Molecules

To obtain specific parameters of the Tau-microtubule interaction such as the oligomerization state of Tau proteins on microtubule or the number of molecules that accumulate at a microtubule binding site for instance due to a cooperative binding, you need to count the number of Tau molecules. To do so, you need to estimate the fluorescence intensity of one single molecule of Tau-sfGFP (or ΔN-Tau-sfGFP) the same day and in the exact same imaging conditions than those used to record Tau binding to microtubules (Subheading 3.3.2).

1. Prepare a TIRF assay mix as before (Subheading 3.3.1, **step 4**) but without tubulin.
2. Perfuse 50 μL of the mix in a flow chamber for 5 min at room temperature.
3. Wash three times with 100 μL K-BRB80 to remove unbound proteins.
4. Seal the chamber with mineral grease.
5. Image samples with the TIRF microscope in the same conditions as you used before (laser power, exposure time, recording mode, and duration).
6. Measure the fluorescence intensity of each single spot as explained in Subheading 3.2.3.

Average the intensity of each spot over several successive frames (around 10 frames), and plot the distribution of all spot intensities. If your protein sample is composed of a unique population of oligomers, you will obtain a unimodal Gaussian distribution whose peak value is the average intensity of a single molecule. By comparing the mean spot intensity with those of a standard protein (sfGFP alone or EB3-sfGFP), one can estimate the oligomerization state of the protein [25]. Multimodal intensity distributions indicate either the presence of subpopulations of proteins with different oligomerization states or the presence of aggregates. Note that the mean fluorescence intensity you measure with this approach should be close to the amplitude of the full drop of intensity observed in SMPB experiments (Subheading 3.2).

To estimate the bleaching, plot the number of fluorescent spots over time. Fitting the data with a mono-exponential curve gives the mean bleaching time. The bleaching time should be at least one order of magnitude larger than the dwell time measured as explained below. In this case, photobleaching does not interfere with measurements of single-molecule dwell times (*see* ref. 25 for an example).

3.3.4 Analysis of Tau-Microtubule Interaction with Kymographs

In ImageJ, do a maximum-intensity Z-projection of the movie of fluorescent Tau proteins (Fig. 2a). Tau-sfGFP (or ΔN-Tau-sfGFP) molecules that bind to microtubules will appear as dotted straight lines. Background corresponding to Tau molecules adsorbed to the glass should be minimal. Draw a selection line along a microtubule, report this selection to the movie of fluorescent Tau proteins and draw a kymograph. To generate kymographs (Fig. 2b), we are using a homemade toolset (available upon request; *see* ref. 20 for details), but other tools such as Multi Kymograph in ImageJ or the kymograph tool of Metamorph are also suitable. Analysis of the fluorescent Tau traces on kymographs provides information about the kinetic of the Tau-microtubule interaction:

1. The vertical length of Tau traces is the duration of a binding event of one Tau molecule to the microtubule. Fitting the distribution of these data with a mono-exponential function gives the mean dwell time (in s), inverse of this value being the dissociation rate constant (in s^{-1}) (Fig. 2b, c) (*see* **Note 9**).
2. The total number of events is used to calculate the binding frequency (in number of events per M of Tau/min/μm).
3. The shape of the traces reveals potential displacement of Tau molecules along the microtubules. Vertical traces represent immobile molecules, whereas oblique traces indicate diffusion along the microtubules, the slope, and the orientation of the traces indicating the rate (in μm/s) and the direction of the movement (toward the (+) or the (−) microtubule end).
4. Dividing the intensity of one trace with the intensity of one single Tau molecule obtained from the calibration experiment (Subheading 3.3.3) gives the number of Tau molecules bound to the microtubule at this location, providing that calibration experiments and Tau-microtubule experiments have been performed the same day, with the same microscope settings and that their respective movies have been processed similarly.

4 Notes

1. mPEG-silane and biotin-PEG-silane solutions should stay anhydrous and be kept in the dark. Change for fresh solutions as soon as you observe unusual noise on your coverslips during TIRF imaging. In our lab, we prepare new solutions every two

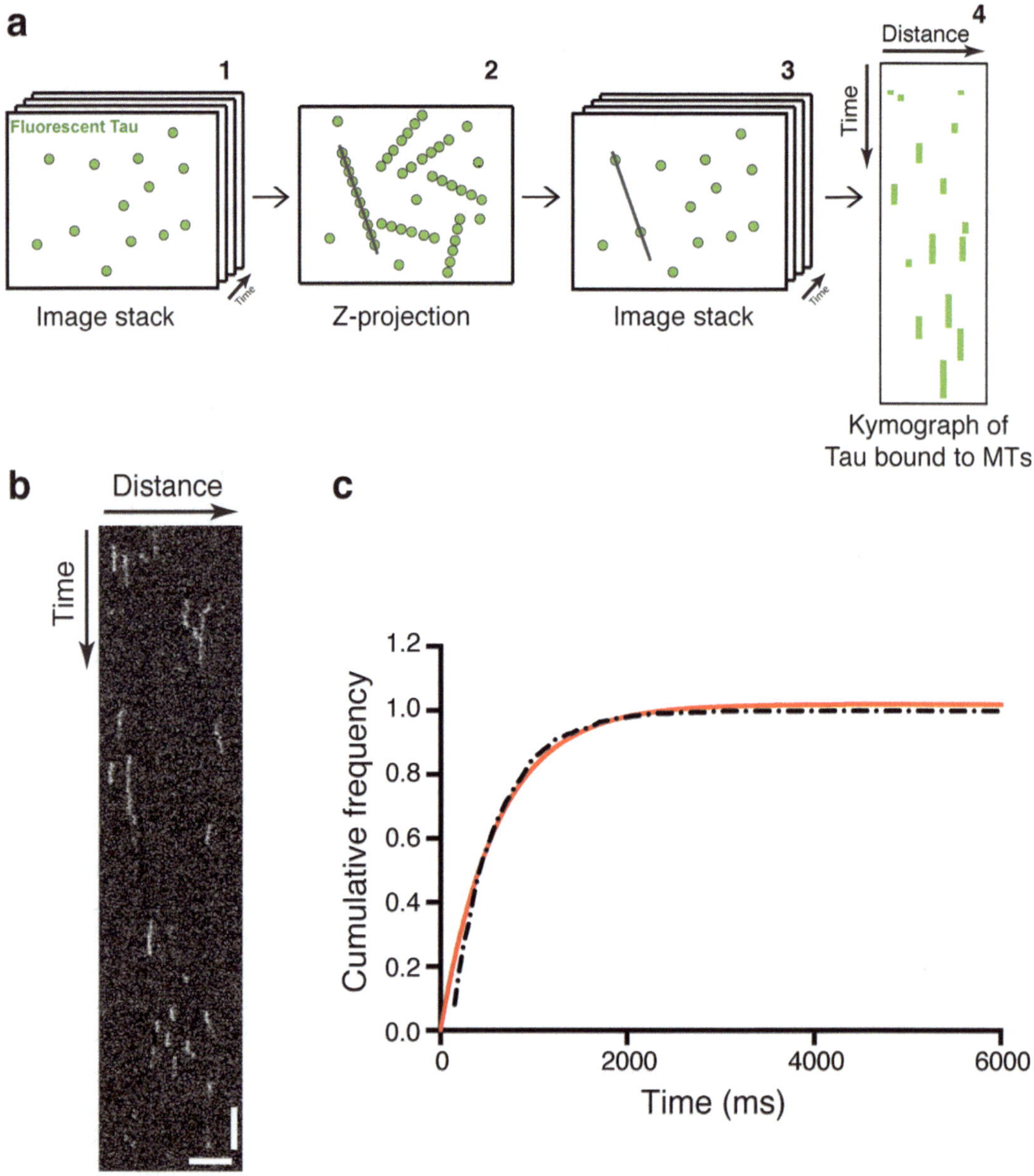

Fig. 2 Interaction of single Tau molecules with microtubules. (**a**) Scheme showing the different steps to generate a kymograph of single Tau molecules interacting with a microtubule. In Image J, a maximum-intensity Z-projection of the movie of fluorescent Tau is used to localize the molecules of Tau bound to microtubules (steps 1 and 2). Selection lines corresponding to Tau-bound microtubules (step 3, black line) are then transferred to the initial image stack to generate kymographs (step 4). MTs, microtubules. (**b**) Representative kymograph of single molecules of ΔN-Tau-sfGFP bound to a microtubule. The contrast of the kymograph has been enhanced for better visualization of the traces. (**c**) Cumulative frequency of the dwell times measured for ΔN-Tau-sfGFP (red curve). The mean dwell time ($\tau = 607$ ms) is obtained by fitting the curve with a mono-exponential function (black dotted line). Scale bars: horizontal, 5 μm; vertical, 1.5 s

months. Be aware that aliquots of mPEG-silane, biotin-PEG-silane, and PLL-PEG powders have to be prepared under argon and stored in safely sealed tubes at −20 °C.

2. We choose to label Tau and ΔN-Tau with sfGFP to obtain a 1 to 1 stoichiometry between Tau proteins and the fluorophore, but other labeling strategies are possible. *See* ref. 26 for a complete discussion about protein labeling for single molecule experiments. During purification of Tau proteins and further handling during TIRF experiments, take care of potential aggregation of Tau as aggregates can greatly interfere with the results of single-molecule experiments [27].
3. When thawed, anti-photobleaching cocktail should be kept on ice and changed for a new aliquot if efficiency loss is observed (1 or 2 h after being thawed). We obtained good results with GODCAT cocktail [28], but other anti-fading systems for single-molecule experiments are available [29] (*see* ref. 26 for a complete discussion on this point).
4. This cleaning step should be done carefully to obtain ultraclean slides and coverslips. We recommend wearing gloves during the whole procedure. Use clean vessels and dedicated jars. Check your solutions regularly and change for new ones after several uses. Use high-quality filtered air or argon to dry coverlips. Protect slides and coverslips from dust. We prepare slides and coverslips as close as possible before use and use them within a week.
5. Rinsing steps with high-quality water are important and should be done with high care so as to remove any trace of chemicals that might interfere with the following steps.
6. Hellmanex solutions can be used at different temperatures, providing the incubation time is adapted according to the manufacturer recommendations.
7. As an alternative, slides and coverslips can be activated by sonication in KOH (*see* ref. 30 for details). In our experience, plasma cleaner activation gives better results.
8. As control experiments, you might perform SMPB experiments with highly soluble proteins of known oligomerization state. We use sfGFP and EB3-sfGFP as standard for monomeric and dimeric proteins, respectively. Standards have to be imaged the same day with the exact same imaging settings than the SMPB experiments of Tau proteins.
9. If the dwell time is shorter or close to the exposure time, then all events that occur between the beginning of the recording and the first image are not observed, leading to an incorrect estimate of the mean dwell time. This can be compensated by collecting a large number of data [31].

Acknowledgments

We thank Eric Denarier (Grenoble Institute of Neurosciences), Clarissa Liesche and Joël Baudoin (Institute of Structural Biology, Grenoble, France), and Florence Forbes and Théo Moins (INRIA, Grenoble) for helpful discussions regarding data analysis. This work was supported by grants from the French National Research Agency (ANR-17-CE11-0026-01) and from the Union France Alzheimer et Maladies Apparentées and the Fédération pour la Recherche sur le Cerveau (FRC) (AAP SM 2015). This work is developed in the framework of Grenoble Alpes Data Institute (ANR-15-IDEX-02). NB is supported by a PhD fellowship from the French National Research Agency (ANR-17-CE11-0026-01). Image acquisition was performed on the Photonic Imaging Centre at Grenoble Institute Neuroscience (Univ Grenoble Alpes—INSERM U1216), which is part of the IBiSA-certified ISdV core facility.

References

1. Goodson HV, Jonasson EM (2018) Microtubules and microtubule-associated proteins. Cold Spring Harb Perspect Biol 10(6): a022608
2. Zanic M (2016) Measuring the effects of microtubule-associated proteins on microtubule dynamics in vitro. Methods Mol Biol 1413:47–61
3. Zwetsloot AJ, Tut G, Straube A (2018) Measuring microtubule dynamics. Essays Biochem 62(6):725–735
4. Verde F, Dogterom M, Stelzer E, Karsenti E, Leibler S (1992) Control of microtubule dynamics and length by cyclin A- and cyclin B-dependent kinases in Xenopus egg extracts. J Cell Biol 118(5):1097–1108
5. Surrey T, Nedelec F, Leibler S, Karsenti E (2001) Physical properties determining self-organization of motors and microtubules. Science 292(5519):1167–1171
6. Janson ME, Loughlin R, Loïodice I, Fu C, Brunner D, Nédélec FJ, Tran PT (2007) Crosslinkers and motors organize dynamic microtubules to form stable bipolar arrays in fission yeast. Cell 128(2):357–368
7. Howard J, Hudspeth AJ, Vale RD (1989) Movement of microtubules by single kinesin molecules. Nature 342(6246):154–158
8. Vale RD, Funatsu T, Pierce DW, Romberg L, Harada Y, Yanagida T (1996) Direct observation of single kinesin molecules moving along microtubules. Nature 80(6573):451–453
9. Axelrod D (2001) Total internal reflection fluorescence microscopy in cell biology. Traffic 2:764–774
10. Maurer SP, Fourniol FJ, Bohner G, Moores CA, Surrey T (2012) EBs recognize a nucleotide-dependent structural cap at growing microtubule ends. Cell 149(2):371–382
11. Subramanian R, Wilson-Kubalek EM, Arthur CP, Bick MJ, Campbell EA, Darst SA, Milligan RA, Kapoor TM (2010) Insights into antiparallel microtubule crosslinking by PRC1, a conserved nonmotor microtubule binding protein. Cell 142:433–443
12. Weingarten MD, Lockwood AH, Hwo SY, Kirschner MW (1975) A protein factor essential for microtubule assembly. Proc Natl Acad Sci U S A 72(5):1858–1862
13. Morris M, Maeda S, Vossel K, Mucke L (2011) The many faces of tau. Neuron 70(3):410–426
14. Arendt T, Stieler JT, Holzer M (2016) Tau and tauopathies. Brain Res Bull 126(Pt 3):238–292
15. Qiang L, Sun X, Austin TO, Muralidharan H, Jean DC, Liu M, Yu W, Baas PW (2018) Tau does not stabilize axonal microtubules but rather enables them to have long labile domains. Curr Biol 28(13):2181–2189
16. Prezel E, Elie A, Delaroche J, Stoppin-Mellet V, Bosc C, Serre L, Fourest-Lieuvin A, Andrieux A, Vantard M, Arnal I (2018) Tau can switch microtubule network organizations: from random networks to dynamic and stable bundles. Mol Biol Cell 29(2):154–165

17. Janning D, Igaev M, Sündermann F, Brühmann J, Beutel O, Heinisch JJ, Bakota L, Piehler J, Junge W, Brandt R (2014) Single-molecule tracking of tau reveals fast kiss-and-hop interaction with microtubules in living neurons. Mol Biol Cell 25(22):3541–3551
18. Shelanski ML, Gaskin F, Cantor CR (1973) Microtubule assembly in the absence of added nucleotides. Proc Natl Acad Sci U S A 70 (3):765–768
19. Hyman A, Drechsel D, Kellogg D, Salser S, Sawin K, Steffen P, Wordeman L, Mitchison T (1991) Preparation of modified tubulins. Methods Enzymol 196:478–485
20. Ramirez-Rios S, Serre L, Stoppin-Mellet V, Prezel E, Vinit A, Courriol E, Fourest-Lieuvin-A, Delaroche J, Denarier E, Arnal I (2017) A TIRF microscopy assay to decode how tau regulates EB's tracking at microtubule ends. Methods Cell Biol 141:179–197
21. Pédelacq JD, Cabantous S, Tran T, Terwilliger TC, Waldo GS (2006) Engineering and characterization of a superfolder green fluorescent protein. Nat Biotechnol 24(1):79–88
22. Ovesný M, Křížek P, Borkovec J, Švindrych Z, Hagen GM (2014) ThunderSTORM: a comprehensive ImageJ plugin for PALM and STORM data analysis and super-resolution imaging. Bioinformatics 30(16):2389–2390
23. R Core Team (2012) R: a language and environment for statistical computing. R Foundation for Statistical Computing, Vienna. ISBN: 3-900051-07-0. http://www.R-project.org/
24. Liesche C, Grussmayer KS, Ludwig M, Wörz S, Rohr K, Herten DP, Beaudouin J, Eils R (2015) Automated analysis of single-molecule photobleaching data by statistical modeling of spot populations. Biophys J 109 (11):2352–2362
25. Bieling P, Kandels-Lewis S, Telley IA, van Dijk J, Janke C, Surrey T (2008) CLIP-170 tracks growing microtubule ends by dynamically recognizing composite EB1/tubulin-binding sites. J Cell Biol 183(7):1223–1233
26. Gell C, Bormuth V, Brouhard GJ, Cohen DN, Diez S, Friel CT, Helenius J, Nitzsche B, Petzold H, Ribbe J, Schäffer E, Stear JH, Trushko A, Varga V, Widlund PO, Zanic M, Howard J (2010) Microtubule dynamics reconstituted in vitro and imaged by single-molecule fluorescence microscopy. Methods Cell Biol 95:221–245
27. Brouhard GJ (2010) Quality control in single-molecule studies of kinesins and microtubule-associated proteins. Methods Cell Biol 97:497–506
28. Roy R, Hohng S, Ha T (2008) A practical guide to single-molecule FRET. Nat Methods 5(6):507–516
29. Aitken CE, Marshall RA, Puglisi JD (2008) An oxygen scavenging system for improvement of dye stability in single-molecule fluorescence experiments. Biophys J 94(5):1826–1835
30. Prezel E, Stoppin-Mellet V, Elie A, Zala N, Denarier E, Serre L, Arnal I (2017) TIRF assays for real-time observation of microtubules and actin coassembly: deciphering tau effects on microtubule/actin interplay. Methods Cell Biol 141:199–214
31. Telley IA, Bieling P, Surrey T (2009) Obstacles on the microtubule reduce the processivity of Kinesin-1 in a minimal in vitro system and in cell extract. Biophys J 96(8):3341–3353

Chapter 7

In Vitro FRET- and Fluorescence-Based Assays to Study Protein Conformation and Protein-Protein Interactions in Mitosis

Stephanie C. Ems-McClung and Claire E. Walczak

Abstract

Proper cell division and the equal segregation of genetic material are essential for life. Cell division is mediated by the mitotic spindle, which is composed of microtubules (MTs) and MT-associated proteins that help align and segregate the chromosomes. The localization and characterization of many spindle proteins have been greatly aided by using GFP-tagged proteins in vivo, but these tools typically do not allow for understanding how their activity is regulated biochemically. With the recent explosion of the pallet of GFP-derived fluorescent proteins, fluorescence-based biosensors are becoming useful tools for the quantitative analysis of protein activity and protein-protein interactions. Here, we describe solution-based Förster resonance energy transfer (FRET) and fluorescence assays that can be used to quantify protein-protein interactions and to characterize protein conformations of MT-associated proteins involved in mitosis.

Key words FRET, Fluorescence, Protein-protein interactions, Protein conformation, Microtubule affinity, Kinesin, Mitosis

1 Introduction

MTs and their associated proteins, including the kinesin superfamily, play fundamental roles during cell division [1, 2]. Kinesins are key players in the assembly of the spindle and the focusing of spindle poles, as well as in the alignment and segregation of chromosomes. The roles of individual proteins in mitosis are well characterized by molecular genetics, RNAi-mediated knockdown, transient transfection of GFP-tagged proteins, and, more recently, CRISPR-genome editing. How these individual proteins are regulated biochemically and how they interact with other proteins in the spindle have been challenging to dissect. Thus, we need novel methods to address these questions. Fluorescence-based approaches, like FRET, offer unique advantages to the characterization of how and where mitotic proteins function during cell division because they can be both qualitative and quantitative;

Helder Maiato (ed.), *Cytoskeleton Dynamics: Methods and Protocols*, Methods in Molecular Biology, vol. 2101,
https://doi.org/10.1007/978-1-0716-0219-5_7,

thus, they can bridge the gap between physiological function and biochemical activity if used in tandem. The challenge with FRET assays is to choose the correct pair of fluorescent proteins for your question and detection system and then to carefully design and carry out the assay to be able to properly interpret the results. Here, we will discuss important considerations for choosing a FRET pair for solution-based biochemical assays and detail the needed controls for careful analysis and interpretation of the results.

FRET is a powerful tool that can measure distances between parts of proteins (intramolecular) or between individual proteins (intermolecular) on the order of nanometers, as well as measure the percentage of those interactions [3–7]. When two fluorochromes (a donor and an acceptor) are attached to an individual protein or to separate proteins, are in close enough proximity, and have enough spectral overlap between the donor emission and the acceptor excitation, the donor can donate energy to the acceptor. This energy transfer to the acceptor will cause the donor emission to decrease and its decay rate or lifetime to get faster, inverse to the distance between the fluorochromes [8]. For fluorescence-based acceptors, this transfer of energy causes the acceptor to have increased fluorescence as the distance between the fluorochromes becomes smaller [4, 9]. Thus, both donor- and acceptor-based FRET detection methods can measure distance and protein-protein interactions quantitatively.

There are different modalities for performing FRET that can be used to study protein conformation and protein-protein interaction. For example, FRET can occur between fluorescently labeled compounds and intrinsic tryptophan residues in proteins and has been used to identify and characterize allosteric inhibitors of Eg5 and kinesin-1 [10–12] and the nucleotide binding kinetics of ATP in rice kinesin-7 motors [13]. This modality is dependent on a tryptophan being in or near the binding site of the ligand, which is not guaranteed for all proteins. Dye-labeling of proteins provides an advantage over intrinsic FRET of tryptophan residues in that a label can be put almost anywhere on a protein, given the presence of a cysteine residue for chemical coupling [14]. This technique became common with the increased ease of mutagenesis, recombinant protein expression, and purification and has been used extensively to study kinesin stepping along MTs [15–19]. Dye labeling requires some knowledge of the protein structure to be able to generate "cysteine light" proteins that have all but one surface cysteine mutated so that labeling can be precise.

More recently, fluorescent proteins are being used in FRET assays (FP-FRET) when co-expressed as a fusion protein due to the wide spectrum of fluorescent proteins available [14, 20]. FP-FRET offers some important advantages over dye-labeled proteins. First, full-length fluorescent fusion proteins can be readily expressed and purified without the need to make "cysteine light" versions that could be potentially deleterious to protein structure and function. Second, fluorescent proteins allow the protein(s) of interest to be visualized in physiological contexts like cells and cellular extracts as well as studied in vitro biochemically. For example, FP-FRET has been used to visualize kinesin-interacting partners in vivo [21–23] as well as to understand the intricate functions of the different kinesin domains in vitro [24, 25]. Finally, FP-FRET has the power to directly correlate in vitro mechanism with in vivo function.

There are two categories for detecting FRET, visual and solution based. Visual detection of FRET is spatial but can also be used temporally for asking when and where an interaction occurs. Common methods of visual FP-FRET include sensitized emission, acceptor photobleaching, and FLIM-FRET [3, 26]. These methods require high-cost microscopy equipment and complicated analysis. Solution-based FP-FRET has several advantages compared to visual-based FP-FRET detection. Solution-based FP-FRET is easy to set up, uses lower cost equipment like filter or monochromatic spectrophotometers, can be done under physiological buffer conditions, and has simple calculations for cross talk between excitation and emission channels, making it ideal for biochemical characterizations of protein-protein interactions and protein conformation. For example, FP-FRET and fluorescence-based detection of protein-protein interactions in solution are more powerful than typical gel-based binding assays because detection is independent of protein molecular weight, the need for antibody detection, and methods to separate bound from unbound protein. Furthermore, fluorescence-based solution assays were historically done in fluorometers, but these instruments require high concentrations and large volumes of protein, making some experiments prohibitive. The recent technical advances in plate readers that use 96- and 384-well formats have made FP-FRET experiments more amenable by allowing the use of less protein and smaller volumes. Acquisition of solution-based FP-FRET is also fast and is thus effective for determining the kinetics of interactions, as well as allowing for higher throughput when used in conjunction with multi-well plates.

The key to developing an effective FP-FRET biochemical assay is to pick a FRET pair with minimal cross talk. There are two types of cross talk. First, the donor emission can bleed through into the acceptor emission spectrum, which could mask the detection of FRET. For example, GFP is much brighter than RFP/mCherry, and its emission bleeds into the RFP emission spectrum, making this a poor pair for acceptor-based detection of FRET [14, 20]. Second, the acceptor should have minimal excitation by the donor excitation wavelength to prevent spurious acceptor emission not caused by FRET. To measure the relative FRET using acceptor fluorescence, one can calculate the FRET ratio, I_{AD}/I_{DA}, where

I_{AD} is the acceptor fluorescence in the presence of the donor, and I_{DA} is the fluorescence of the donor in the presence of the acceptor [9, 27]. If the acceptor exhibits cross talk, this must be subtracted when calculating FRET by including appropriate acceptor-only controls. For the assays we describe here, we use CFP/YFP-derived FRET pairs because of their reduced cross talk and brighter acceptor fluorescence compared to the classical GFP/RFP FRET pair.

In this chapter, we will describe the development of in vitro FP-FRET and fluorescence-based strategies to study the regulation of mitotic kinesins using multi-well plate formats for high-throughput analysis. We will first describe how intermolecular FP-FRET can be used to study the regulation of known protein-protein interactions, using the Kinesin-14 XCTK2 and its binding partner importin α as an example. Next, we will describe how FP-FRET can be used to study the intramolecular conformation of proteins using the Kinesin-13 MCAK as an example. Lastly, we will describe an enhanced method to measure the MT affinity of MT-associated proteins using fluorescence that increases the throughput of the assay.

2 Materials

2.1 Solutions

1. FPLC buffer: 20 mM PIPES pH 6.8, 1 mM $MgCl_2$, 1 mM EGTA, 0.1 mM EDTA stored at 4 °C. Buffer the pH with KOH and store at 4 °C (*see* **Note 1**).
2. 1 M KCl FPLC buffer: 1 M KCl, 20 mM PIPES pH 6.8, 1 mM $MgCl_2$, 1 mM EGTA, 0.1 mM EDTA stored at 4 °C. Buffer the pH with KOH and store at 4 °C (*see* **Note 1**).
3. XB dialysis buffer: 10 mM HEPES pH 7.7, 100 mM KCl, 25 mM NaCl, 50 mM sucrose, 0.1 mM EDTA, 0.1 mM EGTA stored at 4 °C.
4. 10 mg/mL Casein: Made in water and stored in aliquots at −20 °C.
5. 1 M DTT: Made in water and stored in aliquots at −20 °C.
6. 5× BRB49: 245 mM PIPES pH 6.8, 5 mM $MgCl_2$, 5 mM EGTA. Buffer the pH with KOH, sterile filter, and store at 4 °C (*see* **Note 1**).
7. BRB49/DTT: 49 mM PIPES pH 6.8, 1 mM $MgCl_2$, 1 mM EGTA, 1 mM DTT. Made fresh from 5× BRB49 stock and 1 M DTT.
8. 3 M KCl: Made in water, autoclaved, and stored at room temperature (RT).
9. 0.5 M MgATP: Made with equimolar amounts of ATP and $MgCl_2$. Measure concentration precisely (*see* **Note 2**). The solution is stored in aliquots at −20 °C.

10. 5× BRB80: 400 mM PIPES pH 6.8, 5 mM $MgCl_2$, 5 mM EGTA. Buffer with KOH, sterile filter, and store at 4 °C (*see* **Note 1**).
11. BRB80/DTT: 80 mM PIPES pH 6.8, 1 mM $MgCl_2$, 1 mM EGTA, 1 mM DTT. Made fresh from 5× BRB80 stock and 1 M DTT stock.
12. BRB80/$CaCl_2$: 5 mM $CaCl_2$, 80 mM PIPES pH 6.8, 1 mM $MgCl_2$, 1 mM EGTA. Made fresh and stored on ice.
13. 10 mM Taxol: Made in DMSO and stored as 20 μL aliquots at −20 °C in a container with desiccant.
14. 200 μM Taxol: Diluted to 200 μM in DMSO from the 10 mM stock, aliquoted, and stored at −20 °C.
15. BRB80/DTT/Taxol: 80 mM PIPES pH 6.8, 1 mM $MgCl_2$, 1 mM EGTA, 1 mM DTT, 10 μM Taxol. Made fresh from 5× BRB80 stock, 1 M DTT, and 200 μM Taxol and stored at RT.
16. 10 mM GMPCPP (Jena Biosciences #NU-405L): Aliquoted and stored at −20 °C.

2.2 Proteins

Indicated protein concentrations are suggestions based on the purification yield, interaction affinities, and the salt concentrations of the protein preps. For example, bacterially expressed proteins generally yield higher concentrations of proteins than baculovirus-expressed proteins, and more concentrated proteins may be needed for some assays when affinities are weaker or when proteins require larger dilutions because higher salt concentrations are needed for protein stability during purification. All proteins were flash frozen in liquid nitrogen in small single-use aliquots and stored at −80 °C.

1. ≥50 μM Donor control proteins: 6His-CyPet and 6His-mCereulean bacterially expressed, NiNTA (Qiagen) purified, and dialyzed into XB dialysis buffer [28].
2. ≥50 μM Acceptor control proteins: 6His-YPet and 6His-mCitrine bacterially expressed, NiNTA purified, and dialyzed into XB dialysis buffer [28].
3. ≥50 μM FRET control proteins: 6His-CyPet-YPet fusion protein (FCP1) and 6His-mCitrine-mCerulean fusion protein (FCP2) bacterially expressed, NiNTA purified, and dialyzed into XB dialysis buffer [28].
4. ≥5 μM CyPet-XCTK2: Baculoviral expressed and purified by traditional chromatography with a final buffer composition of 300 mM KCl, FPLC buffer, 1 mM DTT, 10 μM MgATP, 0.1 μg/mL leupeptin, pepstatin, chymostatin (LPC), and 10% sucrose [29].
5. ≥20 μM 6His-importin α-YPet: Bacterially expressed, purified, and dialyzed into XB dialysis buffer [30].

6. ≥5 μM FMCAK: FRET biosensor of MCAK expressed in insect cells with baculovirus and purified by traditional chromatography [28] with a final buffer composition of 37 mM KCl, FPLC buffer, 1 mM DTT, 10 μM MgATP, 0.1 μg/mL LPC, and 10% sucrose.
7. ≥5 μM FMΔNT: FRET biosensor of MCAK with its N-terminal domain deleted, consisting of amino acids 187–730. Expressed and purified as FMCAK [28]. Final buffer composition of 50 mM KCl, FPLC buffer, 1 mM DTT, 10 μM MgATP, 0.1 μg/mL LPC, and 10% sucrose.
8. ≥5 μM FMΔCT: FRET biosensor of MCAK with its C-terminal domain deleted, consisting of amino acids 2–592. Expressed and purified as FMCAK [28]. Final buffer composition of 40 mM KCl, FPLC buffer, 1 mM DTT, 10 μM MgATP, 0.1 μg/mL LPC, and 10% sucrose.
9. ≥50 μM YPet-XTail: Bacterially expressed, NiNTA purified, and dialyzed into XB250 dialysis buffer (XB dialysis buffer with 250 mM KCl instead of 100 mM KCl).
10. ≥50 μM Cycled tubulin: Cycled phosphocellulose-purified bovine brain tubulin in BRB80.

2.3 Equipment

1. Plates: Costar #3964 half-area 96-well non-treated black plate and Nunc #264556 384-well non-treated black plate (*see* **Notes 3** and **4**).
2. Plate readers: Synergy H1 (BioTek) with monochromator-based optics, Appliskan (Thermo Scientific) with filter-based optics, or equivalent instrument from another vendor.
3. Excitation filters (center wavelength/bandwidth): 430/10 and 485/10 nm.
4. Emission filters (center wavelength/bandwidth): 480/10 and 535/20 nm.
5. Plate centrifuge: Eppendorf 5810R centrifuge with A-4-81 MTP rotor and 96-well plate adapters.
6. Ultracentrifuge: Beckman TLX with TLA100 rotor.

3 Methods

3.1 Design and Production of FRET Biosensors for Solution-Based FRET

In their guide to fluorescent FRET pairs, Bajar et al. [14] highlight the common types of FRET biosensors and methodologies, as well as the various categories of FP-FRET pairs available. While we provide a description for designing FRET biosensors below, Kaláb and Soderholm [26] provide additional details, especially in regard to the design of intramolecular FRET biosensors. Since fluorescent proteins are an active area of research with continual enhancement to the fluorescence properties of the proteins, we strongly encourage one to research the current literature prior to choosing a FP-FRET pair.

There are also several online resources that are extremely helpful for choosing a FP-FRET pair because they are interactive and have a vast amount of aggregated information in one place. Nikon offers a thorough introduction to the basic principles of FP-FRET, fluorescent proteins, and the use of various FP-FRET methods in their MicroscopyU webpage (https://www.microscopyu.com/applications/fret/basics-of-fret-microscopy) [31]. FPbase (https://www.fpbase.org) is an open-source searchable fluorescent protein database that provides many specifications for most fluorescent proteins, including, but not limited to, interactive "genetic" trees, graphical and downloadable excitation and emission profiles, as well as their biochemical and fluorescence attributes (excitation/emission maxima, quantum yield, brightness, photostability, etc.) [32]. Thermo Fisher offers an interactive spectral viewer (https://www.thermofisher.com/us/en/home/life-science/cell-analysis/labeling-chemistry/fluorescence-spectraviewer.html) that allows one to visualize and compare the emission/excitation spectra of a wide variety of fluorescent reagents from common dye-labeled secondary antibodies to fluorescent proteins [33]. One can also apply different light sources for excitation (lasers or lamps), as well as tunable excitation/emission filters to determine whether your detection system is competent for excitation and detection of the chosen fluorescent proteins.

In this section, we provide a pipeline to design, produce, quantify, test, and interpret your control proteins and FRET biosensors for solution-based FRET assays to study either intermolecular or intramolecular protein interactions (Fig. 1a).

3.1.1 Design: FRET Pair, Control Proteins, and FRET Biosensors (Fig. 1ai)

Choose a FRET pair based on your experimental and detection systems using the literature and the references mentioned above.

1. Consider what types of FRET assays you would like to do with the FRET biosensors. Are they just biochemical or would you also like to use them in visual assays? For example, GFP/RFP pairs do not work well for measuring acceptor-based FRET but work well for donor-based FRET like acceptor photobleaching and FLIM-FRET, whereas CFP/YFP pairs are well suited for solution-based and all visual-based assays, including sensitized emission FRET [14, 20, 34]. If you plan to pair solution-based biochemical assays with visual assays, be sure to choose a FRET pair that will work well in both types of assays so that results can be correlated appropriately.
2. Consider the detection method and attributes of the plate reader and/or visual system. Is it filter- or monochromator-based? Does the plate reader have the right filter sets for

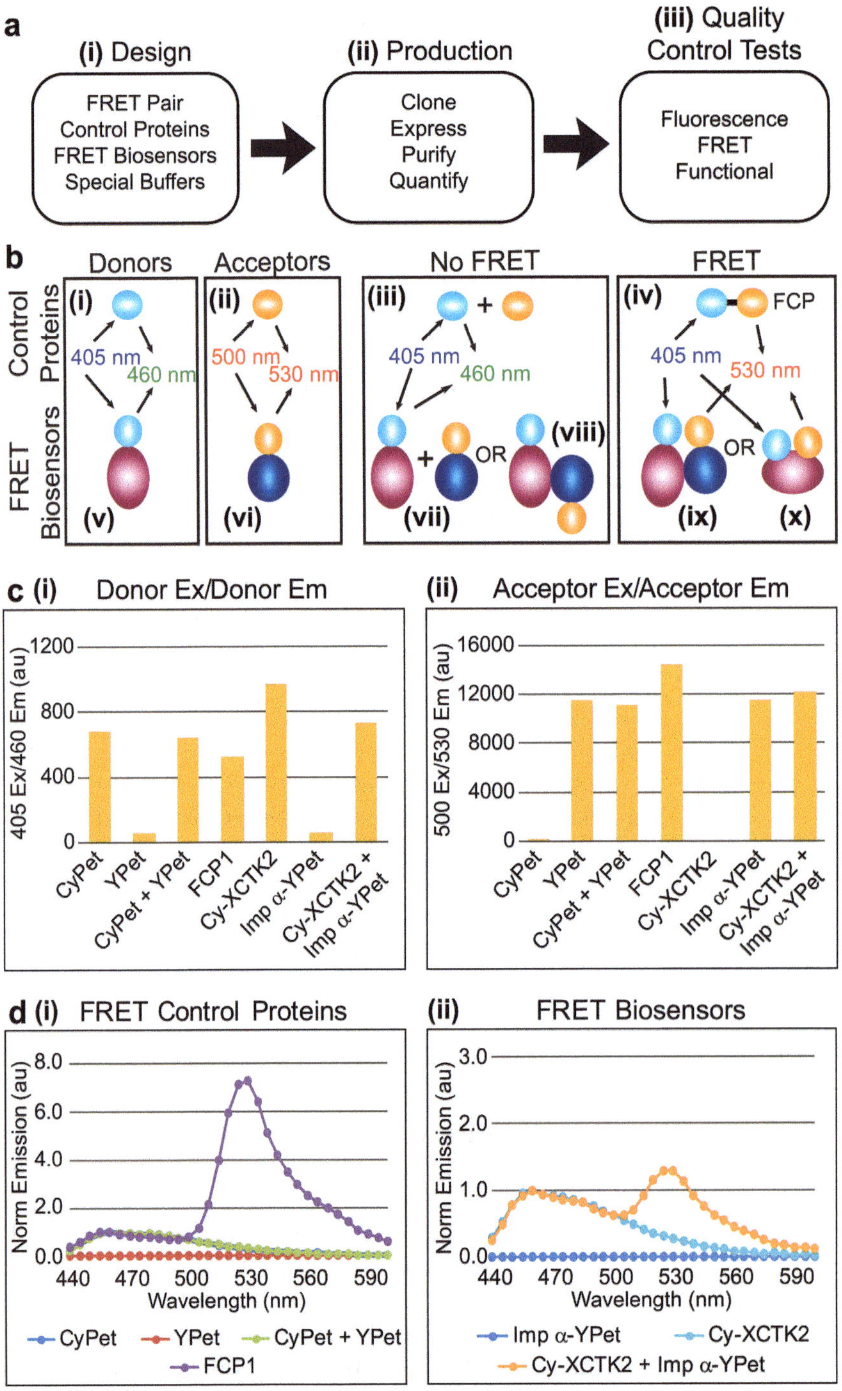

Fig. 1 (**a**) Pipeline for the design, production, and validation of FRET biosensors. (**b**) Schematics of control proteins and FRET biosensors for intermolecular and intramolecular FRET. (i) Donor control protein, e.g., CyPet. (ii) Acceptor control protein, e.g., YPet. (iv) FRET control protein (FCP). (v) Donor FRET biosensor, e.g., CyPet-XCTK2. (vi) Acceptor FRET biosensor, e.g., importin α-YPet. (x) Intramolecular FRET biosensor, e.g., FMCAK. The positive FRET control protein (FCP) is the donor and acceptor separated by a short amino acid

detection of the fluorescent proteins? Can the plate reader perform a spectral scan? A filter-based plate reader may lead to more cross talk than a monochromatic reader and thus be less sensitive. Filter- or monochromator-based optics are optimal for kinetic assays for fast reads, whereas steady-state assays can be done with spectral scans to survey the full-emission spectrum, which can be helpful for quality control of your experiments.

3. Consider the stability (folding properties, pH stability, optimal expression temperature), brightness, and amount of cross talk of a FRET pair [8, 20]. For example, CyPet and YPet produce a seven-fold higher FRET signal than their CFP and YFP parents, which can be helpful in some visual experimental systems [26, 35]. Some fluorophores do not fold well at 37 °C, which can hinder cellular studies and protein expression (*see* **Note 5**) [20].
4. Consider previous work with your proteins of interest. Have they been successfully tagged with fluorophores without hindering function or activity? If they have been successfully expressed previously, consider on which end of the protein the fluorophore was attached, as this can affect function and activity. What buffer compositions (buffer, salt concentration, nucleotide, reducing agents) were used in the previous studies? Will downstream experiment conditions impose limitations upon the protein-protein interaction you are studying?
5. Design control proteins. Control proteins will serve as negative and positive controls for your experimental conditions, e.g., protein concentration, buffer, salt, and pH, which can affect fluorescence and thus FRET. Negative control proteins include the individual donor (Fig. 1bi) and acceptor (Fig. 1bii) fluorescent proteins, which will not display FRET when mixed together (Fig. 1biii). A positive FRET control protein (FCP) should be made as a fusion protein of the donor and acceptor connected by a short amino acid linker that will display FRET [36, 37] (Fig. 1biv).
6. Intermolecular FRET biosensors by design have the donor and acceptor fluorophores on separate proteins (Fig. 1bv, vi) and are most often used to study protein-protein interactions but can also be used to study the effects of posttranslational

Fig. 1 (continued) linker. (**c**) Examples of donor excitation and donor emission (i) and acceptor excitation and acceptor emission (ii) of CyPet, YPet, FCP1 control proteins, and the FRET biosensor proteins CyPet-XCTK2 (Cy-XCTK2) and importin α-YPet (Imp α-YPet). (**d**) Example spectral scan FRET assay of control proteins (i) and FRET biosensors (ii) excited at 405 nm with emission collection from 440 to 600 nm. The CyPet donor emission spectrum in (i) is masked behind the CyPet + YPet spectrum

modifications [38]. Intermolecular FRET biosensors will not display FRET if they do not interact stably (Fig. 1bvii) or if the donor and acceptor fluorophores are not in close enough proximity for energy transfer (Fig. 1bviii). FRET will occur only if the two proteins interact and juxtapose the donor and acceptor close enough for energy transfer (Fig. 1bix). In multi-protein complexes, consider previous work on which protein might interact with whom and how that interaction takes place. For example, importin α tagged with an N-terminal fluorescent protein cannot interact effectively with importin β to form an inhibitory complex (unpublished results).

7. Intramolecular FRET biosensors will typically have the donor and acceptor on either end of the polypeptide chain and will display FRET if the donor and acceptor are near one another based on the secondary or tertiary structure of the protein (Fig. 1bx). These types of biosensors can be used to understand protein conformation through deletion analysis [28], to study protein-protein interactions [23, 27], to characterize post-translational modifications that affect protein conformation [28, 38], or to visualize cellular gradients by using a short segment of protein that can act as a docking site for other proteins or be modified posttranslationally [39, 40].

3.1.2 Production: Expression, Purification, and Quantification of FRET Biosensors and Control Proteins (Fig. 1aii)

1. Depending on your protein(s) of interest, clone and express your FRET biosensors in the appropriate expression system to preserve structure and functionality. For example, most full-length kinesins require expression in eukaryotic systems like insect cells, but smaller truncations and many binding partners can often be expressed in bacteria (*see* **Note 5**).
2. The purified proteins should be devoid of degradation products because they interfere with analysis by contributing background fluorescence that may or may not contribute to FRET. Consider downstream purification schemes like gel-filtration or alternate cloning or expression strategies if this cannot be avoided.
3. When assaying the quality of the proteins, keep in mind that Bradford assays quantify total protein, which will include contaminating host proteins as well as degradation products of the expressed proteins. It is imperative to check the protein prep quality and concentration using a gel-based assay and the extinction coefficient of the fluorescent protein. We follow the general guidelines of determining the total protein concentration with a Bradford assay using BSA as a standard followed by measuring the expressed protein concentration by absorbance (*see* **Note 6**) and/or by densitometry using a gel-based assay (*see* **Note 7**). For a clean protein, the total protein concentration should be the same as the concentration based on

the absorbance and/or densitometry. If there is a disparity between the Bradford and absorbance assays, the prep may still be usable if the gel-based concentration matches the absorbance assay, indicating that there is minimal contamination of free fluorophore. For an intramolecular FRET biosensor, the concentration of both the donor and acceptor fluorophores should be the same, since they are inherently 1:1.

3.1.3 Quality Control Tests: Assess the Precision of Your Quantification, the FRET of Your Control Proteins, and the Activity of Your FRET Biosensors (Fig. 1aiii)

1. Measure the fluorescence of your purified fluorescent control proteins and FRET biosensors side by side at defined concentrations to assess the precision of quantification, the buffer conditions, and the settings of the plate reader for the specific fluorophore (i.e., gain, step size) (*see* **Note 8**) (Fig. 1c). Reactions containing the same donor proteins, e.g., negative control donor proteins (CyPet and CyPet + YPet), should have similar fluorescence when excited at the donor wavelength (*see* **Note 8**) (Fig. 1ci). Similarly, the reactions containing the negative control acceptor proteins (YPet and CyPet + YPet) or the acceptor-tagged FRET biosensors (importin α-YPet and CyPet-XCTK2 + importin α-YPet) should have similar fluorescence when excited at the acceptor excitation wavelength (*see* **Note 9**) (Fig. 1cii).
2. Measure the FRET of the control proteins and the FRET biosensors by measuring three fluorescence components: donor excitation and donor emission (donor Ex/donor Em), donor excitation and acceptor emission (donor Ex/acceptor Em), and acceptor excitation and acceptor emission (acceptor Ex/acceptor Em) (*see* **Note 10**) (Fig. 1d). Here, we measured equivalent amounts of donor (CyPet), acceptor (YPet), or FRET (FCP1) control proteins (Fig. 1di) and the FRET biosensors (CyPet-XCTK2 and importin α-YPet) (Fig. 1dii) in a FRET assay like what will be used in the intermolecular FRET assay described in Subheading 3.2.
3. Assess the donor Ex/donor Em measurement. In reactions containing donor fluorophore and no FRET, the maximum emission of a spectral read will be the donor emission with very little or no emission at the maximum emission of the acceptor. For example, the negative control protein reactions, CyPet and CyPet + YPet (Fig. 1di), and the donor FRET biosensor CyPet-XCTK2 (Fig. 1dii) show typical emission spectra of CyPet with maximal emission at 460 nm and very little at 530 nm, the maximum emission of the acceptor. Note that the CyPet emission profile is behind the CyPet + YPet profile because their emission spectra are essentially identical.
4. Assess the donor Ex/acceptor Em measurement. In reactions containing acceptor fluorophore that do not exhibit FRET, the emission at the acceptor maximum should be minimal to reduce

cross talk. For example, the control protein reactions YPet and CyPet + YPet (Fig. 1di) and importin α-YPet (Fig. 1dii) have minimal emission at 530 nm. In reactions containing donor and acceptor fluorophores that do undergo FRET, the maximum emission should be at the acceptor maximum emission, indicating a FRET signal. For example, the FRET control protein, FCP1 (Fig. 1di), and CyPet-XCTK2 + importin α-YPet (Fig. 1-dii) have substantial emission at the acceptor emission wavelength of 530 nm, which demonstrates FRET. The strength of the emission at the acceptor emission wavelength is proportional to the efficiency of energy transfer; thus, the FCP1 protein has higher FRET than CyPet-XCTK2 + importin α-YPet.

(a) Assess the acceptor Ex/acceptor Em measurement. The acceptor Ex/acceptor Em (Fig. 1cii) and the donor Ex/donor Em (Fig. 1ci) intensities are assessed for the precision of protein addition as outlined in Subheading 3.1.3, **step 1**.

5. Carry out functional tests, if possible, to determine whether the fluorophores interfere with basic biochemical activities or physiological functions. For example, enzymatic assays and alternate binding studies for biochemical activity and/or depletion add-back in extracts or cellular expression for localization/function in physiological settings.
6. If FRET was not detected with your FRET biosensors, consider what is known about your proteins of interest based on previous work. Was a direct interaction demonstrated? Were there other proteins involved in the interaction? What is known structurally about the proteins? One may need to redesign the FRET biosensors taking these findings into account. For example, one may need to put one of the fluorophores on the other end of the protein or tag another protein involved in the interaction. Otherwise, move forward with characterizing the protein-protein interaction in question.

3.2 Intermolecular FRET Assay to Study Protein-Protein Interactions

To study the regulation of protein-protein interactions by FRET, one must first establish the relative stoichiometry of binding between the proteins of interest by measuring FRET with constant donor concentration and increasing acceptor concentrations. To demonstrate this methodology, we will use the Kinesin-14 XCTK2, which binds to importin α through a nuclear localization signal sequence [30]. We will describe how to determine the stoichiometry of binding between CyPet-XCTK2 (donor) and importin α-YPet (acceptor) followed by example adaptations to this method to characterize the regulation of protein-protein interactions.

3.2.1 Design the Experiment to Determine the Stoichiometry of Donor and Acceptor Binding (Fig. 2ai)

The stoichiometry of binding can be done with a small set of reactions by doing a series of acceptor protein concentrations ranging from zero to less than and above the concentration of the donor. Alternatively, a more extensive series of concentrations of the acceptor can be used to establish apparent affinities by determining a K_d. Note that the buffer composition used in this example was optimized for the protein-protein interaction in other assays and chosen for the planned downstream applications; thus, the negative control proteins CyPet and YPet and the positive FRET control protein, FCP1, are not included in the assay but were tested in pilot assays like those described in Subheading 3.1.3, **step 2**.

1. Determine the order of reactions to be added to the multi-well plate, being sure to include the appropriate control reactions. In this example, wells A1-8 of a 384-well black plate are used (*see* **Notes 3** and **4**).
2. Control reactions include the buffer-only control (donor buffer + acceptor buffer, well A1) and the acceptor controls in acceptor buffer (importin α-YPet, wells A2–4), prepared as a set of serial dilutions (*see* **Note 11**).
3. Experimental reactions include the donor only in donor buffer (CyPet-XCTK2, well A5) as well as the donor + acceptor proteins (CyPet-XCTK2 + importin α-YPet, wells A6–8).

3.2.2 Set Up and Perform the Stoichiometry of Donor and Acceptor Binding Experiment (Fig. 2aii)

1. Make master mixes on ice (*see* **Note 12**).
   (a) Donor buffer (125 mM KCl FPLC buffer, 0.2 mg/mL casein) (*see* **Note 13**).
   (b) Acceptor buffer (XB dialysis buffer, 0.2 mg/mL casein) (*see* **Note 13**).
   (c) Donor protein: 2× donor protein (100 nM CyPet-XCTK2) in donor buffer (*see* **Note 14**).
   (d) Acceptor protein: 2× serial dilution of acceptor protein (0–2 μM importin α-YPet) in acceptor buffer (*see* **Note 14**).
2. Add 10 μL acceptor buffer to wells A1 and A5 of a 384-well black plate and importin α-YPet acceptor protein to wells A2–4 and A6–8.
3. Add 10 μL donor buffer to wells A1–4 (control reactions) and 10 μL CyPet-XCTK2 donor protein to wells A5–8 (experimental reactions) for a final well volume of 20 μL. Incubate 10 min at RT to equilibrate.
4. Collect contents to the bottom of the plate by centrifuging in an Eppendorf 5810R centrifuge equipped with A-4-81 MTP rotors for 30 s at 1000 rpm (180 × *g*).

5. Perform spectral FRET scans by exciting the donor and collecting over a range of wavelengths that include the donor and acceptor emission profiles (*see* **Note 10**). For this example, we used 405 nm excitation and 440–600 nm emission scan with 5 nm steps, which is specific to CyPet and YPet (*see* **Note 15**).
6. Perform acceptor excitation with acceptor emission reads (acceptor Ex/acceptor Em) as quality control for the equivalent addition of the acceptor protein to the control and experimental reactions. For this example, YPet was excited at 500 nm and the emission read at 530 nm, which was optimized for this instrument (*see* **Note 15**).
7. Export the data to Excel.

3.2.3 Analyze the Stoichiometry of Donor and Acceptor Binding Experiment Results (Fig. 2aiii)

1. Copy the original raw data into a new Excel sheet called "Spectra" with the wavelengths in column A, the reaction headers in row 1, and the emission values in columns B–I below the headers.
2. Plot the control (Fig. 2bi) and experimental data (Fig. 2bii) in separate graphs in the Excel "Spectra" sheet with emission intensity on the *y*-axis and emission wavelength on the *x*-axis.
3. Inspect the data. The emission spectra of the acceptor-only reactions should be minimal as to not contribute to cross talk or bleed through (Fig. 2bi). Note how 0.01 and 0.1 μM importin α-YPet spectra have minimal emission at 530 nm. In contrast, 1 μM importin α-YPet has higher emission at 530 nm. This cross talk will be corrected for in Subheading 3.2.3, **step 6**.
4. If FRET occurs or if there are changes in FRET, this should be visible in the raw data in which the acceptor emission increases or changes in intensity. In this example, increasing the acceptor concentration of importin α-YPet in the presence of the donor CyPet-XCTK2 caused a stepwise increase in acceptor emission at 530 nm (Fig. 2bii).
5. The donor emission should also decrease with FRET, demonstrating the quenching of the donor fluorescence due to energy transfer. In this example, CyPet-XCTK2 + 0.1 μM or 1 μM importin α-YPet emissions at 460 nm are less than the emission of CyPet-XCTK2 alone and CyPet-XCTK2 + 0.01 μM importin α-YPet (Fig. 2cii).
6. Correct for cross talk of the acceptor bleed through emission due to the excitation of the acceptor at the donor wavelength (*see* **Note 16**). Copy the "Spectra" sheet headers and wavelength values into a new sheet in the same Excel Workbook and label it "Corr Spectra" (*see* **Note 17**). For each concentration of acceptor, create a separate equation in the "Corr Spectra" sheet that subtracts the acceptor fluorescence at each

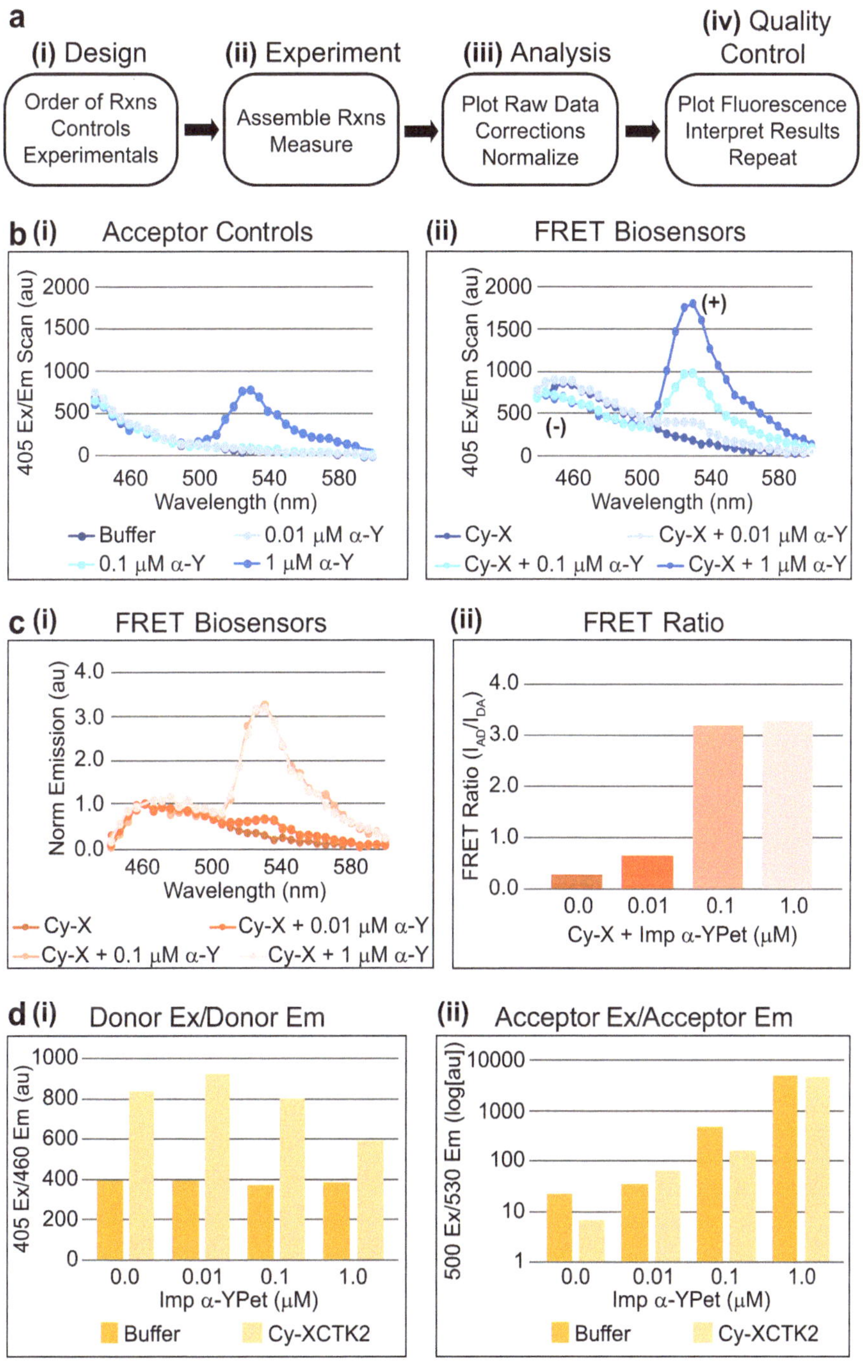

Fig. 2 (**a**) Pipeline for the design, execution, analysis, and interpretation of intermolecular FRET assays to study protein-protein interactions. (**b**) Example of raw spectral data of the control (i) and experimental (ii) intermolecular FRET assays in which the donor CyPet-XCTK2 (Cy-X) was excited at 405 nm, and the emission was measured from 440 to 600 nm. The acceptor importin α-YPet (α-Y) was added at increasing concentrations. When FRET occurs, the acceptor emission will increase (+) and the donor emission will decrease (−). (**c**) The YPet-corrected and YPet-normalized FRET spectral emissions (i) and FRET ratios (I_{AD}/I_{DA}) (ii) of the data in (**b**). (**d**) Quality control fluorescence of the donors (i) and acceptors (ii) in the reactions presented in (**b**)

wavelength in the "Spectra" sheet from the acceptor fluorescence at the same wavelength, e.g., ='Spectra'!B2 – 'Spectra'!$B2 (*see* **Note 18**). The returned values for the acceptor-only reactions should be zero. Repeat for each acceptor concentration, and then copy the concentration-specific acceptor equations to the appropriate reaction columns containing donor + acceptor. The graphs below the data table will be automatically updated. Adjust scale of the *y*-axis if needed.

7. Inspect your data. The acceptor-only spectra should be zero, and the donor + acceptor spectra should decrease relative to the raw data, representing the subtraction of the acceptor fluorescence from the raw data.
8. Normalize the spectra to the donor maximum emission (Fig. 2c). Copy the "Spectra" sheet headers and wavelength values into a new sheet in the same Excel Workbook, and label the new sheet "Norm Corr Spectra." Create an equation at each wavelength in the "Norm Corr Spectra" sheet by dividing the YPet-corrected emission value in the "Corr Spectra" sheet by the maximum donor emission value. For example, divide the CyPet-XCTK2 reaction 440 nm emission value (row 2, column F) by the CyPet-XCTK2 reaction 460 nm emission value (row 6, column F) by creating the equation ='Spectra'!F2 – 'Spectra'!F$6 (*see* **Note 18**). Copy this equation to the remaining cells in the reaction column, e.g., CyPet-XCTK2, and in all the other reaction columns (*see* **Note 19**). The graphs below the data table will update automatically; adjust the scale of the *y*-axis if necessary (Fig. 2ci). This normalized acceptor-corrected value at the acceptor maximum wavelength (530 nm in this example) is effectively the FRET ratio (I_{AD}/I_{DA}).
9. Alternatively, graph the FRET ratio (I_{AD}/I_{DA}) versus concentration of acceptor as a bar graph (Fig. 2cii).

3.2.4 Quality Control of the Stoichiometry of Donor and Acceptor Binding Experiment (Fig. 2aiv)

1. Plot the maximum donor emission from the excitation of the donor in the spectral scan (Subheading 3.2.2, **step 5**) and the acceptor emission excited at the wavelength for the acceptor in the single-point read (Subheading 3.2.2, **step 6**) to assess the signal strength and quality of the experiment (Fig. 2d). Large deviations in the fluorescence or low fluorescence may indicate a reaction with improper addition of components, which may need to be excluded from the final analysis or be repeated. Copy the maximum donor emission from the donor-excited spectral scan (donor Ex/donor Em) in the "Spectra" sheet into a new Excel sheet containing the same headers and call it "Fluorescence." In this example, CyPet emits maximally at 460 nm. Copy the acceptor emission from the acceptor end-point read (acceptor Ex/acceptor Em) into the "Fluorescence" sheet under the same headers as the donor. In this example,

YPet was excited at 500 nm and read at 530 nm for emission. Create bar graphs of the donor Ex/donor Em (Fig. 2di) and acceptor Ex/acceptor Em (Fig. 2dii) intensities relative to acceptor concentration.

2. The donor Ex/donor Em intensities for the buffer- and acceptor-only controls should be similar, representing background fluorescence (Fig. 2di, dark orange).
3. The donor Ex/donor Em intensities in the presence of donor should be highest without any added acceptor or with very low levels of acceptor and should decrease as the acceptor concentration increases, indicative of being quenched due to FRET (Fig. 2di, light orange).
4. The acceptor Ex/acceptor Em intensities should be equivalent in reactions containing the same concentration of acceptor. In this example, the acceptor emissions for both the acceptor-only control reactions and the reactions containing donor are similar and increase logarithmically with increasing acceptor concentration (Fig. 2dii).
5. Interpret your results. Did the FRET ratio plateau at the higher concentrations of acceptor, indicating saturation of binding (Fig. 2c)? Is the FRET ratio dramatic enough to tell differences if you wanted to test the regulation of the interaction? If the FRET ratio did not plateau at the highest concentrations of acceptor tested, repeat with higher concentrations of acceptor. Alternatively, consider the buffer composition if a plateau was not attained or if the FRET ratio was low. High-salt concentrations or high-ionic-strength buffers like BRB80 can interfere with protein-protein interactions and the fluorescence of the fluorophores.
6. Repeat the experiment at least three times for reproducibility once conditions are optimized for protein concentration and buffer composition.

3.2.5 Adaptations for Characterizing the Regulation of Protein-Protein Interactions

Once a stoichiometry of binding is established for your FRET biosensors, one can now characterize contributions of other protein interactions or protein modifications on that interaction. For example, importin β binds to importin α and promotes its binding to proteins containing nuclear localization signal sequences. Thus, one can ask how much importin β is needed to fully stimulate importin α binding to XCTK2 by fixing the CyPet-XCTK2 and importin α-YPet concentrations and doing a serial dilution of importin β in the steps above.

3.3 Intramolecular FRET Assay to Study Protein Conformation

Here, we describe how FRET can be used to assess contributions of different protein domains to conformation. For intramolecular FRET, the donor and acceptor fluorophores are intrinsically present at 1:1 stoichiometries, so it is not necessary to first determine the stoichiometry of binding or perform corrections for acceptor bleed through. We will first describe an example that involves looking at how the N- and C-terminal domains of MCAK contribute to the closed conformation of the full-length protein and then will describe adaptations to the assay that can be done to characterize the effects of posttranslational modifications on protein conformation [28]. While this is a specific example, the methods described should be applicable to many different proteins.

3.3.1 Design the Intramolecular FRET Assay to Study Protein Conformation (Fig. 3ai)

MCAK in solution exists in a compact conformation in which its N- and C-termini are close together [28, 41, 42]. This closed/compact conformation can be demonstrated by FRET when the N- and C-termini are tagged with FRET pairs [28]. To understand how the N- and C-termini are involved in MCAK conformation, we generated FRET biosensors of N- and C-terminal domain truncation mutants of MCAK that we call FMΔNT and FMΔCT and compared them to full-length FRET MCAK, called FMCAK. In this example, the FRET pair is mCerulean (donor) and mCitrine (acceptor), which are derivatives of CFP and YFP, respectively. The BRB49 buffer composition used in this example was chosen to optimize FRET for downstream applications involving MT binding, which prefers high-ionic-strength solutions like PIPES.

1. Determine order and number of reactions to be added to the multi-well plate, being sure to include the appropriate controls. In this example, wells A1-6 of a 96-well half area black plate are used (*see* **Notes 3** and **4**).
2. Control reactions include (Fig. 3bi):
   (a) Buffer control (reaction control buffer, well A1).
   (b) Negative control donor + acceptor control proteins in reaction buffer (mCerulean + mCitrine, well A2).
   (c) Positive control protein in reaction buffer (FCP2, well A3).
3. Experimental reactions include (Fig. 3bii):
   (a) Full-length protein: FMCAK in reaction buffer (well A4).
   (b) C-terminal domain (CT) deletion: FMΔCT in reaction buffer (well A5).
   (c) N-terminal domain (NT) deletion: FMΔNT in reaction buffer (well A6).

3.3.2 Set Up and Perform the Intramolecular FRET Assay to Study Protein Conformation (Fig. 3aii)

1. Make master mixes on ice (*see* **Note 12**).
   (a) Reaction buffer (24 mM KCl, 2 mM MgATP, 0.2 mg/mL casein, BRB49/DTT) (*see* **Notes 20** and **21**).
   (b) Negative control proteins: 100 nM mCerulean +100 nM mCitrine in reaction buffer.
   (c) Positive control protein: 100 nM FCP2 in reaction buffer.

(d) Experimental proteins: 100 nM FMCAK in reaction buffer, 100 nM FMΔNT in reaction buffer, and 100 nM FMΔCT in reaction buffer.

2. Aliquot 50 μL control buffer, negative control proteins, positive control protein, and FMCAK proteins to 96-well half-area black plate, and incubate for 10 min at RT to equilibrate.

3. Measure the following FRET components and the fluorescence of the acceptor protein sequentially per well in a plate reader based on the spectral properties of your donor and acceptor fluorophores. In this example, the fluorophores are mCerulean and mCitrine, and the reactions were read in a filter-based plate reader (*see* **Note 10**).
   (a) Donor excitation and donor emission read (donor Ex/donor Em) with 430/10 nm excitation and 480/10 nm emission filters.
   (b) Donor excitation and acceptor emission read (donor Ex/acceptor Em) with 430/10 excitation and 535/20 nm emission filters.
   (c) Acceptor excitation and acceptor emission read (acceptor Ex/acceptor Em) with 485/10 nm excitation and 535/20 nm emission filters.

4. Export the data to Excel.

3.3.3 Analyze the Intramolecular FRET Assay to Study Protein Conformation (Fig. 3aiii)

1. Plot the raw emission of the donor Ex/donor Em and donor Ex/acceptor Em reactions (Fig. 3c). Copy the donor Ex/donor Em and donor Ex/acceptor Em read data into separate tables in Excel with appropriate column and row headers (*see* **Note 17**). Graph the donor Ex/donor Em (Fig. 3ci) and donor Ex/acceptor Em (Fig. 3cii) emission intensities in separate bar graphs, and ensure that the donor emission values are three- to fourfold higher than the buffer background levels.

2. Correct the donor Ex/donor Em and donor Ex/acceptor Em intensity values for background fluorescence. Copy the column and row headers into a new table called "Corr." Correct the control and experimental emission intensities by subtracting the appropriate buffer-only donor Ex/donor Em or donor Ex/acceptor Em fluorescence values to account for the buffer and well background fluorescence (*see* **Note 18**). Plot corrected values in separate bar graphs and compare to the raw data. Note that the buffer emission should be zero, and the other reactions should have reduced levels from the background subtraction.

3. Calculate the FRET ratio (I_{AD}/I_{DA}) (Fig. 3d). Copy the column and row headers into a new table called "FRET Ratio." Calculate the FRET ratio by dividing the corrected donor Ex/acceptor Em values (I_{AD}) by the corrected donor

Ex/donor Em values (I_{DA}) in the "FRET Ratio" table. Graph the FRET ratios (I_{AD}/I_{DA}) according to reaction condition in a bar graph (Fig. 3d).

3.3.4 Quality Control of the Intramolecular FRET Assay to Study Protein Conformation (Fig. 3aiv)

1. Plot the acceptor Ex/acceptor Em values into a separate graph (Fig. 3e). Copy the raw acceptor Ex/acceptor Em read values into a new table in Excel with appropriate column and row headers labeled "Acc/Acc." Plot the acceptor emission values in a bar graph (Fig. 3e).
2. The donor emission intensity in the absence of FRET should be similar for equivalent concentrations of proteins, whereas in the presence of FRET, the donor emission intensity will be less and the degree dependent on the amount of energy transfer (Fig. 3ci). For example, mCer + mCit has higher donor fluorescence than FCP2, and FMΔCT has higher fluorescence than F2MCAK and FMΔNT, which have higher FRET.
3. The acceptor emission values should also be similar for equivalent concentrations of proteins (Fig. 3e). For example, mCer + mCit and FCP2 have similar acceptor emission intensities, and FMCAK, FMΔCT, and FMΔNT have similar acceptor emission intensities.
4. Interpret results based on your quality control analysis, your predictions, or hypotheses, and then repeat for reproducibility. In this example, deleting the CT of MCAK reduced FRET relative to FMCAK, suggesting that in FMCAK the CT is near the NT in space [28]. Removing the NT increased FRET relative to FMCAK, consistent with the CT being important for the closed conformation of MCAK.

3.3.5 Adaptations to the Intramolecular FRET Assay to Study Protein Conformation

This method can be readily expanded to include the regulation of protein conformation through phosphorylation by incubating with kinase or through protein-protein interactions by performing FRET in the presence of serial dilutions of protein as described in the adaptations for intermolecular protein-protein interactions (Subheading 3.2.5). For example, we showed using FRET that phosphorylation of MCAK by Aurora B opens the conformation of MCAK [28]. Because FRET detection by monochromator- or filter-based reads is fast, temporal effects of phosphorylation or protein-protein interaction can also be determined.

3.4 Fluorescence-Based Determination of Protein Affinity

With the sensitivity and ease of fluorescence measurement, we sought to develop an MT-binding experiment using a fluorescence microtiter plate assay because traditional MT affinity assays involve laborious gel electrophoresis followed by Coomassie staining or Western blotting. Coomassie staining requires relatively high concentrations of proteins for detection of partial binding, and Western blotting, while more sensitive, adds additional variables due to

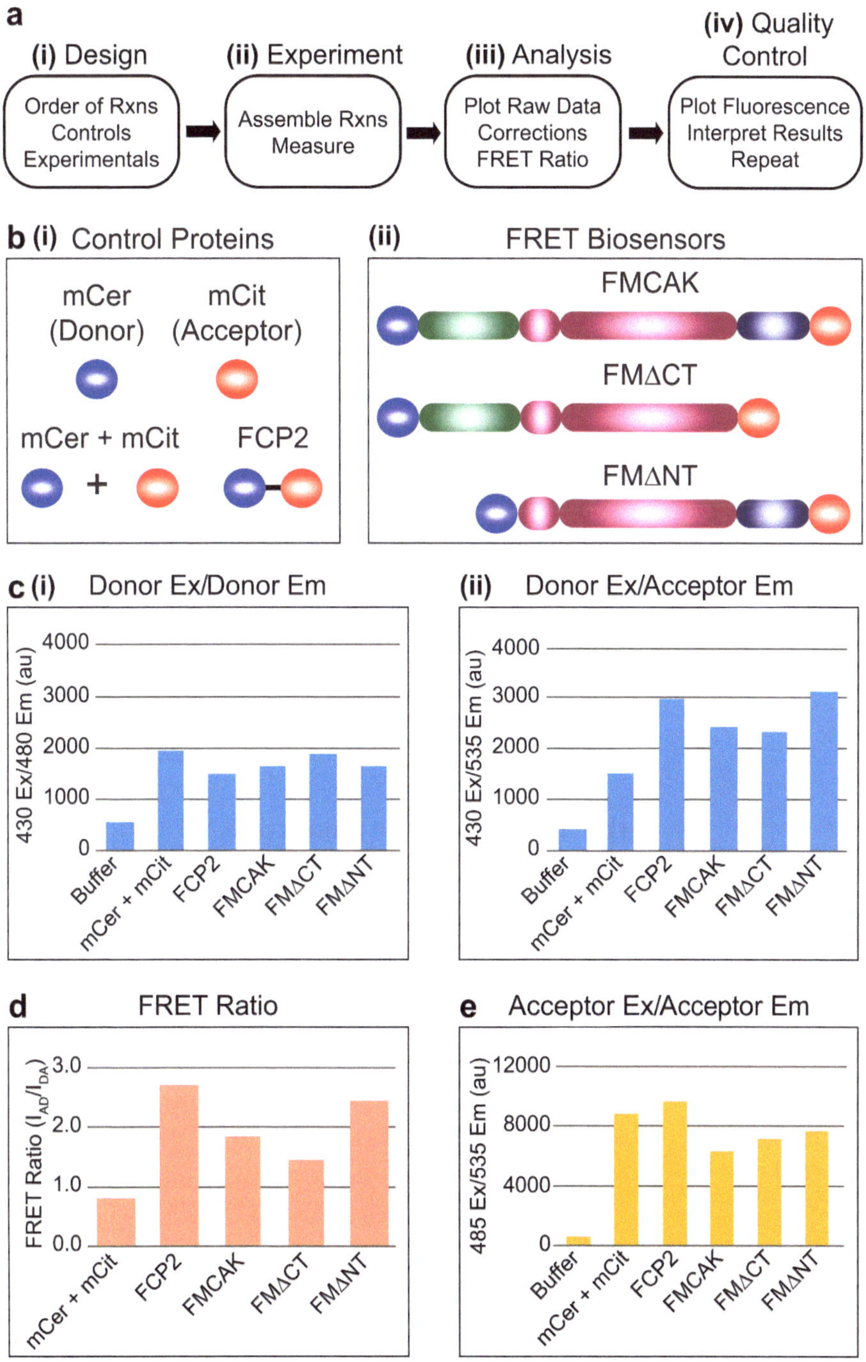

Fig. 3 (**a**) Pipeline for the design, execution, analysis, and interpretation of intramolecular FRET assays to study protein conformation. (**b**) Schematics of negative mCerulean donor (mCer) and mCitrine acceptor (mCit) control proteins, and positive FCP2 FRET control protein (i). Schematics of FMCAK, FMΔCT, and FMΔNT FRET biosensors for full-length, CT-deleted, and NT-deleted MCAK, respectively (ii). MCAK NT (amino acids 2–186) is in green, MCAK neck and catalytic domain (amino acids 187–592) is in magenta, and MCAK CT (amino acids 593–730) is in purple. (**c**) Example of raw filter-based emission data of the intramolecular FRET assay in which

blotting and antibody detection. Here, we describe the use of fluorescence to determine the affinity of the XCTK2 tail domain for MTs and then describe possible adaptations to this assay. The development of this assay provides faster and higher throughput of samples and easier quantification.

3.4.1 Design of the Experiment for Fluorescence-Based Determination of MT Affinity

MT affinity assays consist of combining an MT-binding protein with a range of MT concentrations, incubating for approximately 15 min to reach steady state, sedimentation of the MTs at high *g*-forces to separate the MT-bound and not unbound proteins, and then analyzing the supernatant and pellet fractions. In this example, we will use the XCTK2 tail domain tagged with YPet, YPet-XTail.

1. Determine order and number of reactions, being sure to include appropriate control reactions. The rotor we use in this assay has 20 positions for tubes, and we typically use 10 positions per protein for each concentration range of MTs. Thus, one rotor can accommodate two sets of MT concentrations if desired. In this example, we will only do one concentration range.
2. Control reactions include:
   (a) YPet-XTail without MTs (tube #1). YPet-XTail without MTs will be used to background subtract the amount of YPet-XTail that sediments nonspecifically in the absence of MTs.
   (b) Input reaction of protein plus MT buffer that will **not** be sedimented in the centrifuge (tube #11). The Input reaction will serve as a control for the reaction components sticking to the centrifuge tubes or loss due to sedimentation and resuspension of the pellet.
3. Experimental reactions include YPet-XTail mixed with a concentration series of MTs (tubes #2–10).

3.4.2 Set Up and Perform the Fluorescence-Based Determination of MT Affinity Experiment (Fig. 4aii)

1. Make fresh: BRB80/DTT and BRB80/$CaCl_2$, place on ice; and BRB80/DTT/Taxol, place at RT (*see* **Note 12**).
2. Polymerize MTs [43] by diluting cycled tubulin to 10 μM in BRB80/DTT and then adding GMPCPP to 0.5 mM. Incubate on ice for 5 min. Polymerize MTs for 30 min at 37 °C, supplementing with 200 μM Taxol to a final concentration of 20 μM at 20 min and continuing incubation for an additional 10 min.
3. Sediment MTs from non-polymerized tubulin at 45,000 rpm (90,000 × *g*) for 10 min in an Optima TLX ultracentrifuge at

Fig. 3 (continued) the donor was excited at 430 nm, and the emission was measured for the donor at 480 nm (i) and for the acceptor at 535 nm (ii). (**d**) The background-corrected FRET ratios (I_{AD}/I_{DA}) of the data presented in (**c**). (**e**) Quality control fluorescence of the acceptor excitation and emission

35 °C, remove the supernatant, and resuspend the pelleted MTs in a small volume of BRB80/DTT/Taxol. Store at RT (*see* **Note 22**).

4. Determine the MT concentration in terms of tubulin concentration by diluting 1 μL of MTs in 59 μL of cold BRB80/$CaCl_2$ and incubating on ice for 10–15 min to depolymerize the MTs. Measure the absorbance of the depolymerized MTs at 280 nm, A_{280}, using BRB80/$CaCl_2$ as a blank. Calculate the concentration of tubulin using Beer's law with 115,000 M^{-1} cm^{-1} as the extinction coefficient of tubulin.
5. Serially dilute MTs into nine concentrations using BRB80/DTT/Taxol, and store at RT (*see* **Note 22**).
6. Make master mixes (*see* **Note 12**): 1 μM YPet-XTail in 100 mM KCl, 0.4 mg/mL casein, FPLC buffer (*see* **Notes 13** and **14**), store on ice; and Resuspension Buffer (5% FPLC Buffer, 0.5× BRB80/DTT/Taxol, 0.2 mg/mL casein), place at RT (*see* **Note 22**).
7. Move YPet-XTail master mix to RT for 3 min to equilibrate (*see* **Note 22**).
8. Aliquot 10 μL of BRB80/Taxol to the Input tube at RT (tube #11). This reaction will not be sedimented.
9. Aliquot 10 μL of BRB80/Taxol to TLA100 tube #1 at RT as a control for YPet-XTail that sediments in the absence of MTs. Aliquot 10 μL of each MT concentration in the serial dilution to the additional TLA100 tubes #2–10 at RT.
10. Add 10 μL of 1 μM YPet-XTail protein to the reaction tubes #1–10 and the Input tube #11 to start the reaction. Incubate at RT for 15 min.
11. Pellet the samples in tubes #1–10 at 45,000 rpm (90,000 × *g*) in TLA100 rotor in an ultracentrifuge centrifuge for 10 min at 22 °C.
12. Save the supernatant in a fresh 1.5 mL tube being careful to avoid the pellet (*see* **Note 23**). Store at RT.
13. Resuspend the pellets in 20 μL resuspension buffer.
14. Move 15 μL of the supernatant and pellet samples to a 384-well black plate (*see* **Notes 3** and **4**), alternating the supernatant and pellet fractions into wells A1–A20. Ten reactions become 20 supernatant and pellet fractions. Be careful not to introduce bubbles (*see* **Note 24**).
15. Add 15 μL of the Input reaction (tube #11) to well A21 (*see* **Note 24**).
16. Spin the plate at 1000 rpm (180 × *g*) for 30 s in a plate centrifuge (e.g., Eppendorf 5810R) to collect the sample to the bottom of the plate.

17. Measure the acceptor fluorescence with plate reader. For YPet in this example, YPet was excited at 500 nm, and emission was read at 530 nm with a gain of 75 in a Synergy H1 plate reader (*see* **Note 15**).
18. Export data to Excel file.

3.4.3 Analyze the Experiment for Fluorescence-Based Determination of MT Affinity (Fig. 4aiii)

1. Calculate the fraction bound. Copy data from the acceptor fluorescence read into a separate sheet in Excel with appropriate column and row headers. Calculate the total protein by adding the values for corresponding supernatant and pellet fractions in an adjacent column labeled "S + P." Determine fraction bound in a third column labeled "Fract Bound" by dividing pellet fraction by total protein.
2. Normalize the concentration of bound protein in a fourth column labeled "μM Bound" by multiplying the fraction bound by the concentration added. In this example, the concentration of protein added was 0.05 μM.
3. Correct the concentration bound for nonspecific pelleting. Copy the concentration bound to a separate table in the Excel sheet labeled with the corresponding MT concentration. Subtract the pelleted protein in the reaction without MTs in an adjacent column (*see* **Note 18**).
4. Copy the background-subtracted data to a program that can fit the data to a quadratic binding curve like GraphPad Prism (Fig. 4b).

3.4.4 Quality Control of the Fluorescence-Based Determination of MT Affinity Experiment (Fig. 4aiv)

1. Graph the total protein fluorescence values from the "S + P" column in Subheading 3.4.3, **step 1**, and the Input fluorescence against the MT concentration for equivalent addition of protein (Fig. 4c).
2. If the total protein of a reaction is drastically awry from the other reactions or from the Input, consider the validity of the data point and whether it should be included in the analysis or whether the experiment needs to be repeated (*see* **Note 25**). Also, if there is a dramatic increase in total protein fluorescence with increasing protein concentration indicating considerable protein loss at the lower concentrations, consider using higher amounts of protein or different blocking agents (e.g., casein or BSA) in the reactions (*see* **Note 13**).
3. Fit the data to a nonlinear regression model for binding affinity in software like GraphPad Prism. An appropriate binding curve will appear hyperbolic when plotted on a linear scale (Fig. 4b).
4. Interpret your results. Ideally, points below and above the K_d are needed to assure an accurate determination of the affinity. If the full binding curve was not represented with the concentrations of MTs used, repeat experiment with adjusted MT

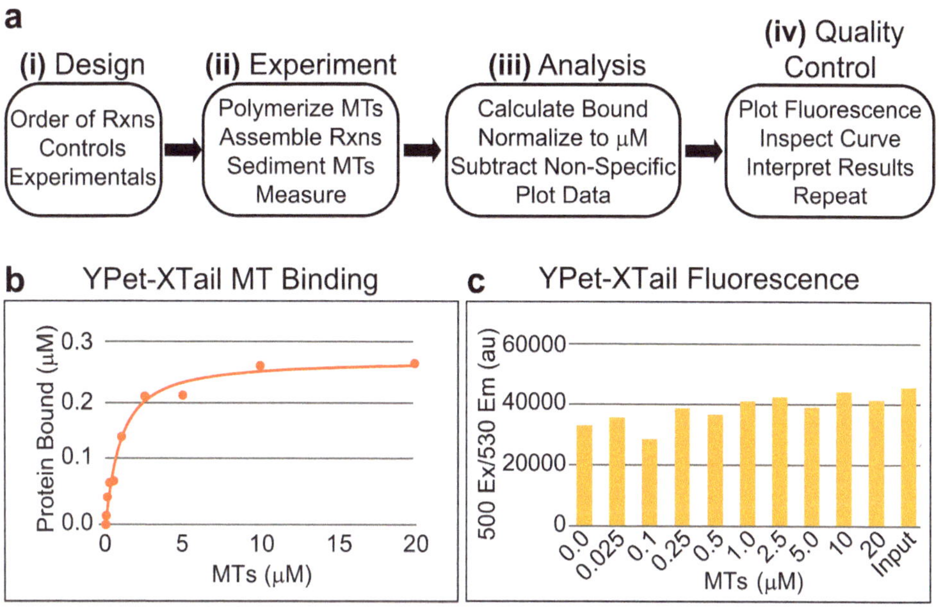

Fig. 4 (**a**) Pipeline for the design, execution, analysis, and interpretation of a fluorescence-based assay to determine MT-binding affinity. (**b**) Amount of YPet-XTail bound to increasing concentrations of MTs that was corrected for nonspecific pelleting in the absence of MTs and then plotted and fit to the quadratic nonlinear regression model for total protein binding in GraphPad Prism. (**c**) Quality control fluorescence of the total YPet-XTail (S + P) and Input in each reaction from (**b**)

concentrations that would be predicted to approximate the full binding curve. Typically, concentrations range from tenfold below and above the K_d.

5. Repeat at least three times for reproducibility once a set of optimal concentrations are determined.

3.4.5 Adaptations

This assay can be easily adapted to measuring the effects of inhibitors on MT binding by adding increasing concentrations of the inhibitors as described for the intermolecular protein-protein interaction assay (Subheading 3.2.5). For example, importin α/β inhibit the binding of the XCTK2 tail domain to MTs [30] and could be added at increasing concentrations to measure inhibition quantitatively using fluorescence of the YPet-XTail domain. Furthermore, for non-sedimentation-based assays, FRET can be used as a read out for the fraction of protein binding by using the maximum FRET ratio at saturating ligand as 100% bound.

4 Notes

1. When preparing GMPCPP-stabilized MTs, it is essential that all solutions be made as potassium salts because a combination of glycerol and sodium can induce hydrolysis of GMPCPP

[44], which decreases the stability of MTs stabilized with GMPCPP.

2. Measure the exact concentration of MgATP by absorbance at 259 nm using 15,400 M^{-1} cm^{-1} as the extinction coefficient for ATP.
3. Black or white plates are ideal for fluorescence to prevent bleed through from adjacent wells.
4. Half area 96-well plates and 384-well plates are very convenient for smaller volumes when samples are precious or in low supply. Minimum recommended volume for half area 96-well plates is 50 and 20 μL for 384-well plates.
5. Solubility of fluorescent protein fusion proteins in bacteria can be improved by induction at a lower temperature, 16–20 °C, overnight or up to 24 h.
6. Determine the concentration of the fluorophore in the sample by measuring the absorbance at the appropriate wavelength for the fluorophore using Beer's law and the extinction coefficient according to the equation $A = l * c * \varepsilon$, where A is the absorbance, l is the pathlength of the light source in cm, c is the concentration in molarity (M), and ε is the extinction coefficient in M^{-1} cm^{-1}.
7. Determine the concentration of your protein of interest by densitometry using a gel-based assay. Run 0.5 μg of protein, based on the Bradford and absorbance assays, on an SDS-PAGE gel with a serial dilution of BSA from 0 to 0.8 μg as a standard followed by Coomassie staining and densitometry. Run more protein on the gel if the total protein and absorbance concentrations are quite different. This may indicate a protein prep that has considerable contamination by host proteins or degradation products.
8. Typically, 50–200 nM protein is needed to get a good fluorescence signal that is at least two- to four-fold higher than the background of a well containing the same buffer as the proteins. Be aware that the donor fluorescence of the positive FRET control protein (FCP1) or the combination of the donor and acceptor FRET biosensors (CyPet-XCTK2 + importin α-YPet) is more difficult to assess because of the effects of FRET on the donor emission.
9. Note that the FRET control protein FCP1 consistently had higher acceptor fluorescence despite being added at an equal molar concentration to CyPet and YPet.
10. The three components of FRET can be measured spectrally or individually with single measurements from monochromator- or filter-based spectrophotometers. We find spectral

measurements of FRET more helpful in assessing the quality of an experiment.

11. The acceptor controls are separate reactions containing the indicated acceptor protein concentrations performed in the absence of the donor. These reactions are necessary to calculate FRET in intermolecular FRET assays, because the bleed through of acceptor emission upon the excitation of the donor at each acceptor concentration must be considered in the final calculations.
12. Be sure to make enough master mix for all your control and experimental reactions plus extra to account for pipetting error.
13. We find for many proteins that adding casein or BSA stabilizes the protein by preventing loss due to nonspecific binding of the proteins to surfaces of the tubes and plates.
14. XCTK2 is purified by conventional chromatography and has a final KCl concentration of 300 mM KCl, which is nonphysiological and often too high for many downstream applications. It is important to have high-protein-concentration preps so that the final salt concentration in the reaction can be diluted to physiological levels. Adjust KCl concentration with 1 M KCl FPLC Buffer.
15. We found that performing a spectral read with excitation at 405 nm in the Synergy H1 with reactions containing CyPet and YPet worked better than using the optimal excitation wavelength of 435 nm for two reasons. First, spectrophotometers have a limit as to how close the emission can be read from the excitation wavelength to prevent damage to the instrument. Second, exciting at 405 nm reduced acceptor cross talk by minimizing the YPet excitation that causes bleed through to the YPet emission wavelength.
16. Correction of the donor bleed through into the acceptor emission is not necessary because this is minimal for this FRET pair and constant for each reaction since the same concentration of donor was added to each reaction.
17. For multistep analyses, it is a good idea to make clear stepwise calculations in the form of different tables within an Excel Sheet or with separate sheets within the Excel Workbook to be able to clearly follow the transformations of the data and prevent data loss. The original data file is always best left unmodified, with the raw/original data copied or saved as a new file on which calculations are done.
18. Use the "$" in front of the column letter or cell number to make the subtracted term static for the column or cell for ease of copying to multiple columns. For example, ='Spectra'!B2 – 'Spectra'!$B2 will make the column letter static, and ='Spectra'!F2 – 'Spectra'!F$6 will make the cell number static.

19. Note that the acceptor-only controls will have the returned value of #DIV/0! because the values are zero in these cells.
20. FMCAK is purified by conventional chromatography with a final concentration of 370 mM KCl. This is not physiological and can interfere with downstream applications. Use a concentrated stock of KCl, e.g., 3 M KCl, to adjust KCl concentration depending on the stock protein concentrations.
21. Some proteins are more stable in the specific nucleotide states. MCAK is most stable in the ATP-bound state; thus, MgATP is added to the reaction.
22. It is important not to induce MT depolymerization by incubating MTs on ice or adding cold buffers to solutions containing MTs. Store MTs at RT.
23. It is convenient to place a mark on the rim of the TLA100 tube, and then place this mark to the outside of the rotor to indicate where the pellet will be as the pellets are small and hard to see.
24. Avoid introducing bubbles by consistently and gently pipetting the sample three times and then adding the sample into the plate by dispensing to the first stop of the Pipetman.
25. It is not uncommon for the total protein fluorescence to increase with increasing concentrations of MTs due to stabilization of the protein by binding to the MTs or by the presence of additional protein in the reaction.

Acknowledgment

We thank Benjamin Walker and Ahmed Ghobashi for insightful comments and editing of the manuscript. C.E.W. is supported by NIH R35GM122482.

References

1. Prosser SL, Pelletier L (2017) Mitotic spindle assembly in animal cells: a fine balancing act. Nat Rev Mol Cell Biol 18:187–201
2. Yount AL, Zong H, Walczak CE (2015) Regulatory mechanisms that control mitotic kinesins. Exp Cell Res 334:70–77
3. Sahoo H (2011) Förster resonance energy transfer—a spectroscopic nanoruler: principle and applications. J Photochem Photobiol C 12:20–30
4. Clegg RM (2009) Förster resonance energy transfer—FRET what is it, why do it, and how it's done. In: Gadella TWJ (ed) FRET and FLIM techniques, 1st edn. Elsevier, Boston, MA, pp 1–57
5. Selvin PR (1995) Fluorescence resonance energy transfer. Methods Enzymol 246:300–334
6. Stryer L, Haugland RP (1967) Energy transfer: a spectroscopic ruler. Proc Natl Acad Sci U S A 58:719–726
7. Förster T (1948) Zwischenmolekulare energiewanderung und fluoreszenz. Ann Phys 437:55–75
8. Lakowicz JR (2006) Principles of fluorescence spectroscopy, 3rd edn. Springer Science, New York
9. Kaláb P, Pralle A (2008) Quantitative fluorescence lifetime imaging in cells as a tool to design computational models of

Ran-regulated reaction networks. Methods Cell Biol 89:541–568
10. Raghav D, Sebastian J, Rathinasamy K (2018) Biochemical and Biophysical characterization of curcumin binding to human mitotic kinesin Eg5: Insights into the inhibitory mechanism of curcumin on Eg5. Int J Biol Macromol 109:1189–1208
11. Kaan HY, Major J, Tkocz K et al (2013) "Snapshots" of ispinesib-induced conformational changes in the mitotic kinesin Eg5. J Biol Chem 288:18588–18598
12. Hopkins SC, Vale RD, Kuntz ID (2000) Inhibitors of kinesin activity from structure-based computer screening. Biochemistry 39:2805–2814
13. Umezu N, Umeki N, Mitsui T et al (2011) Characterization of a novel rice kinesin O12 with a calponin homology domain. J Biochem 149:91–101
14. Bajar BT, Wang ES, Zhang S et al (2016) A guide to fluorescent protein FRET pairs. Sensors (Basel) 16:1–24
15. Verbrugge S, Lansky Z, Peterman EJ (2009) Kinesin's step dissected with single-motor FRET. Proc Natl Acad Sci U S A 106:17741–17746
16. Martin DS, Fathi R, Mitchison TJ et al (2010) FRET measurements of kinesin neck orientation reveal a structural basis for processivity and asymmetry. Proc Natl Acad Sci U S A 107:5453–5458
17. Hajdo L, Skowronek K, Kasprzak AA (2004) Spatial relationship between heads of dimeric Ncd in the presence of nucleotides and microtubules. Arch Biochem Biophys 421:217–226
18. Aoki T, Tomishige M, Ariga T (2013) Single molecule FRET observation of Kinesin-1's head-tail interaction on microtubule. Biophysics (Nagoya-shi) 9:149–159
19. Mori T, Vale RD, Tomishige M (2007) How kinesin waits between steps. Nature 450:750–754
20. Shaner NC, Steinbach PA, Tsien RY (2005) A guide to choosing fluorescent proteins. Nat Methods 2:905–909
21. Wagner OI, Esposito A, Kohler B et al (2009) Synaptic scaffolding protein SYD-2 clusters and activates kinesin-3 UNC-104 in *C. elegans*. Proc Natl Acad Sci U S A 106:19605–19610
22. Cai D, Hoppe AD, Swanson JA et al (2007) Kinesin-1 structural organization and conformational changes revealed by FRET stoichiometry in live cells. J Cell Biol 176:51–63
23. Espenel C, Acharya BR, Kreitzer G (2013) A biosensor of local kinesin activity reveals roles of PKC and EB1 in KIF17 activation. J Cell Biol 203:445–455
24. Hackney DD, Baek N, Snyder AC (2009) Half-site inhibition of dimeric kinesin head domains by monomeric tail domains. Biochemistry 48:3448–3456
25. Hallen MA, Liang ZY, Endow SA (2011) Two-state displacement by the Kinesin-14 Ncd stalk. Biophys Chem 154:56–65
26. Kaláb P, Soderholm J (2010) The design of Förster (fluorescence) resonance energy transfer (FRET)-based molecular sensors for Ran GTPase. Methods 51:220–232
27. Hao Y, Macara IG (2008) Regulation of chromatin binding by a conformational switch in the tail of the Ran exchange factor RCC1. J Cell Biol 182:827–836
28. Ems-McClung SC, Hainline SG, Devare J et al (2013) Aurora B inhibits MCAK activity through a phosphoconformational switch that reduces microtubule association. Curr Biol 23:2491–2499
29. Walczak CE, Verma S, Mitchison TJ (1997) XCTK2: A kinesin-related protein that promotes mitotic spindle assembly in *Xenopus laevis* egg extracts. J Cell Biol 136:859–870
30. Ems-McClung SC, Zheng Y, Walczak CE (2004) Importin α/β and Ran-GTP regulate XCTK2 microtubule binding through a bipartite nuclear localization signal. Mol Biol Cell 15:46–57
31. Basics of FRET Microscopy (2019) MicroscopyU: the source for microscopy education. https://www.microscopyu.com/applications/fret/basics-of-fret-microscopy. Accessed 16 Apr 2019
32. Lambert TJ (2019) FPbase: a community-editable fluorescent protein database. Nat Methods 16:277–278
33. Fluorescence SpectraViewer (2019) https://www.thermofisher.com/us/en/home/life-science/cell-analysis/labeling-chemistry/fluorescence-spectraviewer.html. Accessed 16 Apr 2019
34. Shimozono S, Miyawaki A (2008) Engineering FRET constructs using CFP and YFP. Methods Cell Biol 85:381–393
35. Nguyen AW, Daugherty PS (2005) Evolutionary optimization of fluorescent proteins for intracellular FRET. Nat Biotechnol 23:355–360
36. Thaler C, Koushik SV, Blank PS et al (2005) Quantitative multiphoton spectral imaging and its use for measuring resonance energy transfer. Biophys J 89:2736–2749

37. Koushik SV, Chen H, Thaler C et al (2006) Cerulean, Venus, and VenusY67C FRET reference standards. Biophys J 91:L99–L101
38. Guillaud L, Wong R, Hirokawa N (2008) Disruption of KIF17-Mint1 interaction by CaMKII-dependent phosphorylation: a molecular model of kinesin-cargo release. Nat Cell Biol 10:19–29
39. Kaláb P, Weis K, Heald R (2002) Visualization of a Ran-GTP gradient in interphase and mitotic *Xenopus* egg extracts. Science 295:2452–2456
40. Wang E, Ballister ER, Lampson MA (2011) Aurora B dynamics at centromeres create a diffusion-based phosphorylation gradient. J Cell Biol 194:539–549
41. McHugh T, Zou J, Volkov VA et al (2019) The depolymerase activity of MCAK shows a graded response to Aurora B kinase phosphorylation through allosteric regulation. J Cell Sci 132:1–8
42. Talapatra SK, Harker B, Welburn JP (2015) The C-terminal region of the motor protein MCAK controls its structure and activity through a conformational switch. Elife 4:1–21
43. Desai A, Walczak CE (2001) Assays for microtubule destabilizing kinesins. In: Vernos I (ed) Kinesin protocols, vol 164. Humana Press, Totowa, NJ, pp 109–121
44. Caplow M, Ruhlen RL, Shanks J (1994) The free energy for hydrolysis of a microtubule-bound nucleotide triphosphate is near zero: all of the free energy for hydrolysis is stored in the microtubule lattice. J Cell Biol 127:779–788

Chapter 8

Methods for Studying Membrane-Associated Bacterial Cytoskeleton Proteins In Vivo by TIRF Microscopy

Charlène Cornilleau, Arnaud Chastanet, Cyrille Billaudeau, and Rut Carballido-López

Abstract

MreB proteins are actin homologs present in nonspherical bacteria. They assemble into membrane-associated discrete filamentous structures that exhibit different dynamic behaviors along the bacterial sidewalls. Total internal reflection fluorescence (TIRF) microscopy, a sensitive method for studying molecular events at cell surfaces with high contrast and temporal resolution, is a method of choice to characterize the localization and dynamics of cortical MreB assemblies in vivo. This chapter describes the methods for visualizing fluorescently tagged MreB proteins in live *Bacillus subtilis* cells. We detail how to (1) grow *B. subtilis* strains for reproducible TIRF observations, (2) immobilize cells on agarose pads and (3) in CellASIC® microfluidic plates, and (4) acquire TIRF images and time lapses.

Key words TIRF microscopy, Actin-like MreB proteins, Bacterial cytoskeleton, Agarose pad, CellASIC® microfluidic system, *Bacillus subtilis*

1 Introduction

MreB proteins are structural homologs of eukaryotic actin present in nonspherical bacteria and central to rod-shape determination. Over the last two decades, the use of fluorescent proteins and developments in imaging approaches have revealed that dynamic polymers of MreB associated with the cytoplasmic membrane localize along the cell cylinder, where they are believed to spatiotemporally coordinate enzymes involved in sidewall elongation. Among the fluorescent microscopy methods used to visualize the localization and dynamics of proteins in live cells, total internal reflection fluorescence (TIRF) microscopy stands as a method of choice for studying membrane-associated events in bacterial cells. In conventional wide-field microscopy setups, the incident light is parallel to the optical axis (i.e., the coverslip/sample surface normal) and thus travels across the entire sample, simultaneously exciting all fluorophores in the cell. Unfocused fluorescence from above and below

Helder Maiato (ed.), *Cytoskeleton Dynamics: Methods and Protocols*, Methods in Molecular Biology, vol. 2101,
https://doi.org/10.1007/978-1-0716-0219-5_8, © Springer Science+Business Media, LLC, part of Springer Nature 2020

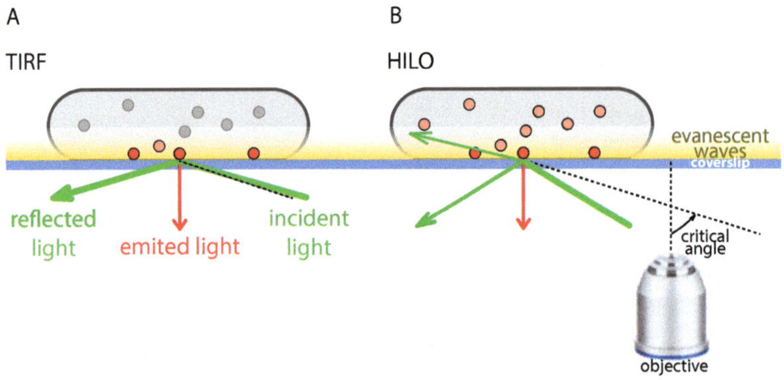

Fig. 1 Principle of TIRF (**A**) and HILO (**B**) illuminations. In a TIRF setup, the incident light hits the coverslip with a high angle relative to the optical axis (while in epifluorescence the light beam is in the axis of the objective; hence the angle is 0°). When the beam hits the coverslip/substrate interface with an angle higher than the critical angle (left panel, TIRF), all photons are reflected and the fluorophores (dots) are not directly excited. Only the evanescent waves generated at the interface and propagating in its close vicinity (yellow area) can excite the neighboring fluorescent particles (red dots). When the angle between the objective and the incident light is reduced below the critical angle (right panel, HILO), a fraction of the photons enters the substrate and a mixed illumination occurs ("dirty TIRF"). The proportion of reflecting photons decreases with the angle, to be minimum when the beam is in the axis of the objective (epifluorescence)

the focal plane is detected generating a high background scatter and thus a low signal-to-noise ratio (contrast). Two related techniques compatible with wide-field setups, TIRF and highly inclined laminated optical sheet (HILO), have been successfully applied to bacterial cells. In TIRF microscopy, the laser beam arrives on the sample with a high incident angle (relative to the optical axis) in which light is totally reflected (critical angle) (Fig. 1A). An evanescent wave is generated at the interface between the coverslip (glass) and the sample/water (aqueous medium), which have different refractive indices. The energy of the evanescent wave decays exponentially into the sample, selectively exciting the fluorophores at the proximity (100–200 nm) of the coverslip [1]. This enables a high contrast and, as a consequence, a high temporal resolution of molecular events at the cell surface. In HILO, the laser incident angle is a bit lower than the critical angle (subcritical angle) and a fraction of the photons penetrates the sample with a highly inclined angle, creating a mixed regime (Fig. 1B). HILO is therefore a middle ground between wide-field and TIRF illumination and can be useful for observations requiring a deeper penetration than TIRF microscopy, but it generates larger amounts of out-of-focus blur and leads to faster bleaching of the fluorophores [2]. When imaged by TIRF microscopy, *B. subtilis* MreB proteins appear as spherical or elliptical assemblies close to the diffraction limit (light microscopy lateral resolution ~250–300 nm) that are evenly distributed along the bacterial cell cylinder [3, 4]. Growing cells contain subpopulations of both nonmobile (constrained) and mobile MreB assemblies, the latter exhibiting a variety of dynamic behaviors that require a high temporal resolution to be imaged: directed processive movement around the cell circumference, random diffusion, and mixed patterns of motions [4, 5].

In this chapter, we describe the methods to grow and immobilize *B. subtilis* cells expressing fluorescently labeled proteins for TIRF acquisitions. Although growing *B. subtilis* in a rich medium such as LB is easy, it is important to take into account a number of factors in order to achieve quality imaging experiments and good reproducibility, and to avoid generating artifacts. The bacterial cytoskeleton is very sensitive to the cell environment and growth conditions. The localization and dynamic properties of actin-like MreB proteins are quickly modified by stresses and are very sensitive to moderate changes (e.g., the speed of MreB assemblies exhibiting directed movement drops in a few seconds when cells are exposed to certain drugs or if the temperature is reduced) [3, 6, 7]. Thus, it is important to control carefully the growth and image acquisition conditions to minimize perturbations that may affect protein localization and/or dynamics.

Because bacteria do not directly adhere to glass and many, like *B. subtilis*, are motile, it is also important to immobilize them to allow quality microscopy acquisitions. Immobilization on agarose pads is minimally perturbative for bacterial cell physiology and therefore the main method used. Furthermore, it allows the growth of bacteria over extended periods under the microscope. Glass functionalization by adhesive molecules is a potential alternative to immobilize bacteria but does not reproducibly support growth over long periods and can deeply affect cell physiology. Cationic polymers such as poly-L-lysine, for example, have been shown to perturb the bacterial cell membrane and the proton motive force, which can affect the localization and/or dynamics of membrane-associated processes [7–9], and thus poly-L-lysine-coated slides are not recommended. Slide and agarose pad mounting can be advantageously replaced by a microfluidic chamber for long time-lapse experiments. In addition to immobilizing cells and supporting growth under constant controlled conditions, microfluidic systems allow the analysis of dynamic perturbations (e.g., change of growth medium, addition of drugs, etc.). Homemade microfluidic devices that allow growing bacterial cells in dead-ended channels (e.g., the "mother machines") or traps are ideal because they are customizable and cheap but require access to facilities and expertise. Here, we describe the use of the commercial CellASIC® ONIX Microfluidic System from Merck, which is a user-friendly and convenient alternative.

The methods described have been optimized for *B. subtilis* but should be generalizable, with some small adaptations, to TIRF imaging of other aerobic bacterial species.

2 Materials

2.1 Bacterial Cell Growth

Bacterial growth requires basic sterile microbiology vessels (test tubes, flasks, or microplates) in glass or plastic, with a cap or lid (to avoid contamination). The only restriction for aerobic bacteria such as *B. subtilis* [10] is, in addition to continuous strong agitation of the culture, to use a recipient large enough to allow a maximum aeration, i.e., the medium should not occupy more than 1/5th of the total volume. A variety of media with different formulations can be used depending on the bacterial species and specific application (transformation, protein expression, microscopy, etc.). Commercial LB medium (lysogeny broth, sometimes referred to as Luria broth or Luria-Bertani medium) is the most widely used rich medium and is routinely used for the study of the *B. subtilis* cytoskeleton [3–5]. However, cell suspensions in LB medium display background autofluorescence, which limits overall sensitivity and thus can be an issue for the observation of weak fluorescent signals. Non-autofluorescent media should be considered instead when needed. Cell growth is monitored by following the optical density of the culture at the single 600 nm wavelength ($OD_{600\ nm}$) using a spectrophotometer.

2.2 Agarose Solution for Pad Preparation

1. Ultrapure agarose (molecular grade).
2. LB (or alternative) medium or sterile pure water.
3. Microwave.
4. Standard microscopy glass slides 26 × 76 mm.
5. Coverslips 24 × 60 and 22 × 22 mm (thickness 170 μm, No. 1.5).
6. Lab tape or commercial frame.

2.3 Microfluidic Device

1. CellASIC® ONIX (Merck) Microfluidic System for live cell imaging.
2. Optically clear CellASIC® ONIX plates for bacterial cells (Ref. B04A).

2.4 TIRF Imaging

Use an inverted microscope equipped with an environmental chamber for temperature control (unless experiments are to be performed at room temperature), a laser, a motorized TIRF arm for incident laser angle control, a high-quantum-yield ultrasensitive camera (e.g., up to 95% quantum efficiency for back-illuminated EMCCD cameras, or >82% for sCMOS), and a 60× or 100× TIRF oil-immersion objective with high numerical aperture (NA > 1.45). Immersion oil permits to use the full numerical aperture of the objective by removing reflection in the light path at the glass slide interface when refractive indexes of the immersion oil and the glass match. Refractive index depends on temperature (manufacturers usually provide its value at 23 °C) and should be adapted depending on specific acquisition conditions (e.g., 37 °C). Immersion oils must have extremely low or no fluorescence to avoid signal degradation due to undesired background. They can be purchased from most microscopy manufacturers (e.g., Immersol 518F, $n = 1.518$ @ 23 °C, Zeiss). The microscope stage must accept the 96-well plate format if microfluidic CellASIC® plates are to be used.

3 Methods

3.1 Growth of *Bacillus subtilis* Strains for TIRF Microscopy

1. On day 1, prepare a pre-culture by inoculating 2 mL of medium (e.g., LB or the medium to be used for the experiment, supplemented with antibiotics and/or with other supplements when needed) from a frozen glycerol stock of the strain of interest and incubate overnight. Alternatively, inoculate the pre-culture with a single colony from a freshly streaked plate (*see* **Note 1**). Adjust the duration of the pre-culture and the temperature of incubation in order to minimize entering into stationary phase. Typically, start the pre-culture as late as possible, grow it at 25–30 °C with strong aeration, and dilute it the next morning.
2. On the day of the experiment, measure the optical density ($OD_{600\ nm}$; *see* **Note 2**) of the pre-culture and make an appropriate dilution into fresh medium to a theoretical OD_{600} of 0.005 (*see* **Note 3**). If comparing several strains, it is important to ensure an identical starting cell density so that all the strains will be observed at an identical OD_{600} after an identical number of generations. Do not add antibiotics to the culture since growth in the presence of antibiotics can affect growth rate and/or shape. If maintaining the selective pressure is absolutely required, include a control strain containing the same resistance marker.
3. Grow the cultures at constant temperature (e.g., 30 or 37 °C) and maximum agitation (no less than 200 rpm) for optimum aeration [10], and monitor bacterial growth by following the OD_{600}.
4. When the desired cell density is reached, add 1–2 μL of the-bacterial culture on a pre-warmed 2% agarose pad, prepared as described below. To increase cell density on the agarose pad (especially if working at low cell density), the culture can be concentrated by gentle centrifugation (e.g., 1–3 min

at 1000 × *g*). If taking several samples throughout growth, be careful to minimize perturbations (e.g., do not leave the cultures standing on your bench for minutes while taking a sample; cells will detect both the aerobic and temperature downshift and reduce their growth accordingly).

3.2 Agarose Pad Preparation and Slide Mounting

1. Prepare a solution of 2% agarose: typically, microwave-melt 100 mg of ultrapure agarose in 5 mL of LB (or the growth medium used for the experiment or water; *see* **Note 4**) in a 50 mL conical tube. It is recommended to prepare the agarose solution daily (*see* **Note 5**).
2. To obtain agarose pads of the desired thickness, create spacers by sticking together two layers of lab tape and placing two of these on a microscope slide as shown in Fig. 2A (*see* **Note 6**).
3. Pipet approximately 250 μL of the melted 2% agarose solution between the two spacers, immediately cover the drop with a 24 × 60 mm coverslip (Fig. 2A, B), and place a weight on top to pressure the solution and create a uniformly flat surface. An aluminum-heating block makes a very good job as it is heavy and efficiently dissipates the heat, speeding up the agarose solidification process.

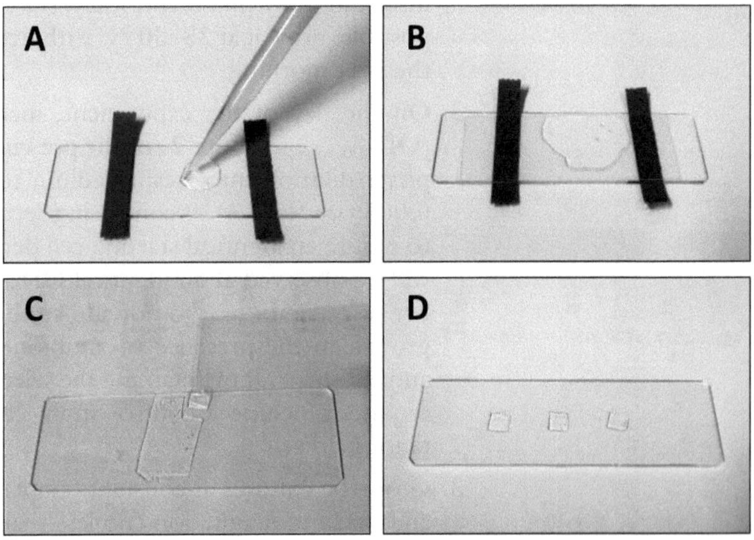

Fig. 2 Agarose pad preparation. (**A**) Stick two pieces of double-layered lab tape on a microscope slide to obtain spacers of the desired thickness. Pipet approximately 250 μL of the melted 2% agarose solution between the two spacers. (**B**) Immediately cover the drop with a 24 × 60 mm coverslip. Cover with a weight (not shown) and allow to set for a few minutes. (**C**) Use the edge of a clean coverslip to cut ~5 × 5 mm mini pads. (**D**) Example of three mini pads placed on the same slide

4. After a few minutes, carefully remove the weight, the coverslip, and the spacers, and use the edge of a clean coverslip to cut ~5 × 5 mm mini pads (Fig. 2C). Carefully lift these mini pads with an edge of the coverslip, and place them on a clean microscope slide. Several mini pads can be placed on the same slide (Fig. 2D) (*see* **Note 7**).
5. Add 1–2 μL of the bacterial culture at the center of the mini pad (ideally pre-warmed), and allow the pad to adsorb the suspension before placing a 22 × 22 mm coverslip on top. Gently press the coverslip to allow complete adherence to the pad. If multiple pads have been disposed on the glass slide, use a large 24 × 60 mm coverslip.

3.3 Microfluidic Setup (CellASIC® ONIX)

The CellASIC® device and plates should be used according to the manufacturer's recommendations. Here, we indicate specific tips based on our own experience for the preparation, loading, and mounting of the plates. We describe the simple case where bacteria are continuously perfused with a single growth medium, but changes of media can be achieved by using the extra wells (#2–5; Fig. 3).

1. Plate preparation: Completely pipet out the PBS buffer from the wells that will contain the cells (#9) and the medium for perfusion (#1) and for loading (#6), and wash once with fresh medium. Also, empty half of the waste (#8) to prevent overflow (Fig. 3). Then, fill in wells #1 and #6 with 1–200 μL of medium and #9 with 1–200 μL of bacterial cell culture.
2. Turn on the CellASIC® device, and seal the plate according to the manufacturer guidelines. Position the plate on the stage so

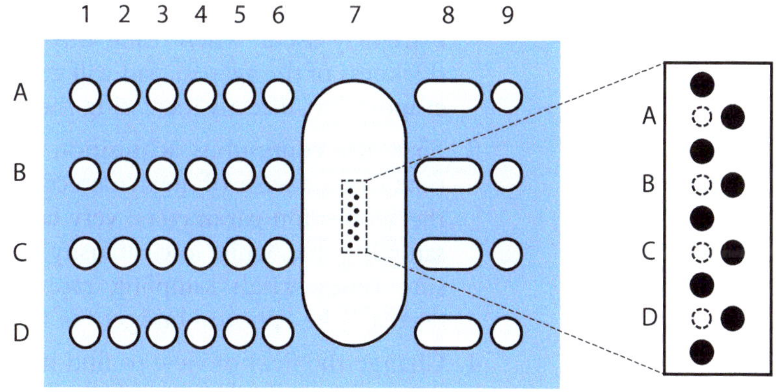

Fig. 3 CellASIC® ONIX B04A microfluidic plate. Schematics of the plate. The plate is organized in four rows (*A–D*) allowing four independent experiments, and nine columns corresponding to: (*1–5*), wells for media/solutions; (*6*), wells for medium for cell loading; (*7*), area containing the four culture micro-chambers (*A–D*); (*8*), trashes, and (*9*), loading wells for the cells

that the growth chambers are aligned with the objective (Fig. 3; *see* **Note 8**). The chambers should be visible in the field of view.

3. Load bacteria into the chamber using a 4 psi pressure (pressure determines the flow rate of medium) for 6 s. *B. subtilis* is typically trapped in the fourth area of the micro-chamber (this depends on cell diameter and can vary with species; *see* **Note 9**). Finally, apply a continuous (5 psi) flow of fresh medium (from well #1) to allow efficient bacterial growth. We recommend to check after a few minutes that the cells are indeed growing and to control frequently the absence of communication errors between the CellASIC® controller and the plate, which lead to a flow arrest.

3.4 TIRF Imaging

1. Put a small drop of microscopy oil on top of the coverslip (or on the objective), and place the slide or microfluidic plate on the temperature-controlled microscope stage. When possible, use white light illumination (bright-field, phase contrast, DIC, etc.) rather than fluorescence to find the focal plane containing the cells to prevent useless bleaching of the fluorophores.
2. Set up the TIRF settings so that the laser power is minimal and the laser/objective angle maximal, ensuring that there is no direct penetration of photons in the sample. From this starting point, carefully reduce the laser angle while adjusting the focus until the fluorescence appears. Stop reducing the laser angle as soon as the fluorescence is just visible enough for quality acquisition (*see* **Note 10**). If the slide has been correctly mounted, the laser angle and the focus can be kept unchanged over the entire surface of the slide if the microscope is equipped with a device that automatically maintains the focus (e.g., Nikon "PFS," Zeiss "Definite Focus", etc.). Such autofocus systems are extremely useful when using this type of mounting since the thickness of the agarose pad will gradually change due to water evaporation, directly increasing the objective/sample distance.
3. First, use continuous acquisition while varying the exposure time to get a sense of the dynamics of the particles. Then, adapt the acquisition parameters: very fast dynamics requires a high sampling rate, while low-intensity signals require longer exposure times. High sampling rate and long exposures both decrease the "photon budget" of the fluorophores.
4. Change the field of view to find new cells, and image particle dynamics using the optimized frame rate and illumination. Typically, for *B. subtilis* fluorescently tagged MreB proteins, a rate of 0.5–2 frames/s is used when imaging exponentially growing cells at 37 °C (when fluorescent particles travel at 50–100 nm/s) [3, 6].

4 Notes

1. Starting a culture directly from a frozen stock (without pre-culture) is not appropriate because there is no control of the number of (living) cells inoculated, and thus of the number of generations that will occur during the experiment. It is possible to streak the strain on a fresh plate instead, but this is less convenient to control the cell density when starting the culture (*see* **Note 3**). Also, do not use old streaks (>24 h plate) because the cells will enter a variety of stationary-phase processes including sporulation and lysis, generating a broad spectrum of cellular states.
2. Keep in mind that turbidity (optical density) (a) is a value related to the dry mass of the population (not to the number of cells or the cfu (colony-forming units)), (b) can vary between spectrophotometers depending on their calibration (thus an "$OD_{600} = 0.5$" gives little information on the exact growth phase of the bacterial culture without the corresponding growth curve), and (c) is proportional to cell mass in the range of linearity of the material, which must be predetermined. A growth curve of your reference strain must be done first using the spectrophotometer that will be used for the experiments. This will reveal the OD_{600} where growth phase change(s) occur(s), which should be constant for a given growth condition (temperature, medium, aeration) [11].
3. Start cultures at a low cell density to ensure that a sufficient number of generations in exponential growth will occur.
4. Ideally, the solvent used to melt the agarose should be the same medium used for growing the cells (e.g., LB), if it supports microwaving. If not, an agarose/water pad can be prepared and the water later exchanged with the medium by soaking the pad for half an hour in a petri dish filled with the medium. In some cases, cells are also observed directly on a water/agarose pad, but keep in mind that the cells will face a rapid dilution of their growth medium when added to the water pad, which can cause stress to the cells and consequently affect localization of the proteins of interest.
5. Melted agarose in medium should be prepared fresh of the day but can be kept several days at 55 °C when prepared in water. Agarose pads should not be kept more than a few hours at room temperature to avoid desiccation (in all cases) and contamination when prepared with medium.
6. Commercial frames (e.g., "gene frame" from Thermo Fisher, Ref. AB0577) can be used instead of lab tape spacers. However, the custom-made tape spacers are infinitely cheaper and allow preparing pads than can be cut into smaller pieces to image

multiple samples on a single slide (*see* Fig. 2D), minimizing cost and manipulations. The main benefit of commercial frames is that by sticking to both the glass and the coverslip, they limit dehydration of the agarose pad.

7. The mini pads should be used right away to limit the evaporation process. The uncut agarose pad covered with the coverslip can be kept for a couple hours prior to use. Place it in a petri dish containing extra moisture, especially if it is to be kept in a warm room or in the oven so that it is at the same temperature than the bacterial culture to be imaged.
8. To position the objective on top of the four micro-chambers (dotted circles on the zoomed-in area in Fig. 3), which are too small to be seen with the naked eye, use as landmark the holes (black circles) that are surrounding them.
9. The micro-chambers contain six successive areas with a step-like, decreasing height of the microfabricated elastic silicone ceiling. The cells enter through the most spacious area and are pushed by the flow into the direction of the low-ceiling zones until they become immobilized, physically trapped in the area with height similar to their diameter. The B04A plate has heights of 4.0, 3.0, 2.0, 1.1, 0.9, and 0.7 μm. *B. subtilis* is usually found on areas 4 or 5.
10. Signal intensity will progressively increase as the angle is reduced, as some (and more and more) photons in addition to the evanescent waves will penetrate the sample. This intermediate state in which the cell is partially illuminated by photons with high incident angles is HILO, also referred to as "dirty TIRF" (Fig. 1). The transition from TIRF to HILO illumination is gradual when decreasing the laser angle and thus difficult to perceive. The goal is to optimize as much as possible the TIRF angle in order to minimize the penetration depth as well as bleaching and background noise.

Acknowledgment

This work was supported by a Consolidator grant from the European Research Council (ERC COG) under the European Union's Horizon 2020 research and innovation program (grant agreement No. 772178) to R.C.-L.

References

1. Axelrod D, Burghardt TP, Thompson NL (1984) Total internal reflection fluorescence. Annu Rev Biophys Bioeng 13:247–268. https://doi.org/10.1146/annurev.bb.13.060184.001335
2. Yao Z, Carballido-Lopez R (2014) Fluorescence imaging for bacterial cell biology: from localization to dynamics, from ensembles to single molecules. Annu Rev Microbiol

68:459–476. https://doi.org/10.1146/annurev-micro-091213-113034
3. Domínguez-Escobar J, Chastanet A, Crevenna AH, Fromion V, Wedlich-Soldner R, Carballido-López R (2011) Processive movement of MreB-associated cell wall biosynthetic complexes in bacteria. Science (New York, NY) 333(6039):225–228. https://doi.org/10.1126/science.1203466
4. Billaudeau C, Chastanet A, Yao Z, Cornilleau C, Mirouze N, Fromion V, Carballido-Lopez R (2017) Contrasting mechanisms of growth in two model rod-shaped bacteria. Nat Commun 8:15370. https://doi.org/10.1038/ncomms15370
5. Billaudeau C, Yao Z, Cornilleau C, Carballido-Lopez R, Chastanet A (2019) MreB forms subdiffraction nanofilaments during active growth in *Bacillus subtilis*. MBio 10(1):e01879-18. https://doi.org/10.1128/mBio.01879-18
6. Garner EC, Bernard R, Wang W, Zhuang X, Rudner DZ, Mitchison T (2011) Coupled, circumferential motions of the cell wall synthesis machinery and MreB filaments in *B. subtilis*. Science (New York, NY) 333(6039):222–225
7. Strahl H, Hamoen LW (2010) Membrane potential is important for bacterial cell division. Proc Natl Acad Sci U S A 107(27):12281–12286
8. Colville K, Tompkins N, Rutenberg AD, Jericho MH (2010) Effects of poly(L-lysine) substrates on attached *Escherichia coli* bacteria. Langmuir 26(4):2639–2644. https://doi.org/10.1021/la902826n
9. Katsu T, Tsuchiya T, Fujita Y (1984) Dissipation of membrane potential of *Escherichia coli* cells induced by macromolecular polylysine. Biochem Biophys Res Commun 122(1):401–406
10. Harwood CR, Cutting SM (1990) Molecular biological methods for *Bacillus*. John Wiley & Sons, New York
11. Stevenson K, McVey AF, Clark IBN, Swain PS, Pilizota T (2016) General calibration of microbial growth in microplate readers. Sci Rep 6:38828. https://doi.org/10.1038/srep38828

Chapter 9

Processing TIRF Microscopy Images to Characterize the Dynamics and Morphology of Bacterial Actin-Like Assemblies

Cyrille Billaudeau, Arnaud Chastanet, and Rut Carballido-López

Abstract

Total internal reflection fluorescence (TIRF) microscopy allows the visualization of the dynamic membrane-associated actin-like MreB filaments in live bacterial cells with high temporal resolution. This chapter describes computerized analysis methods to quantitatively characterize the dynamics and morphological properties of MreB assemblies. These include how to (1) segment bacterial cells, (2) perform single-particle tracking (SPT) of MreB filamentous structures, (3) classify their dynamic modes using mean squared displacement (MSD) analysis, and (4) measure their dimensions and orientation.

Key words Image analysis, MreB, Single-particle tracking (SPT), Automated dynamic classification, Mean squared displacement (MSD), Kymograph, 2D anisotropic Gaussian fitting

1 Introduction

MreB proteins are actin homologs essential for rod-shaped determination in bacteria [1, 2]. They polymerize into membrane-associated short filaments that are distributed along the sidewalls and exhibit several dynamic behaviors including processive circumferential motion around the short axis of the cell, random diffusion, and constrained diffusion [3–5]. Characterization of the localization and dynamics of MreB assemblies in the membrane requires fluorescence microscopy techniques providing high optical sectioning, high spatial resolution, fast acquisitions, and low photobleaching during acquisitions. Total internal reflection fluorescence microscopy (TIRFM), an elegant optical technique that selectively excites fluorophores in a thin region (<200 nm) at the coverslip/sample interface [6, 7], successfully allows quantitative analysis of MreB in live cells [3–5] (*see* Chapter 8 and **Note 1**).

The first quantifications of the velocity of MreB assemblies exhibiting directed motion around the cell periphery were

Helder Maiato (ed.), *Cytoskeleton Dynamics: Methods and Protocols*, Methods in Molecular Biology, vol. 2101,
https://doi.org/10.1007/978-1-0716-0219-5_9, © Springer Science+Business Media, LLC, part of Springer Nature 2020

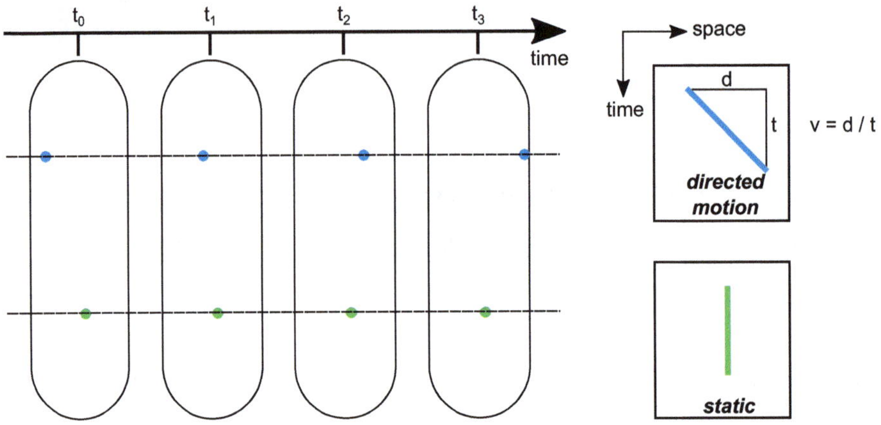

Fig. 1 Principle of kymograph analysis. Kymograph analysis allows to represent a dynamic process in a single image (kymogram). A line (dotted lines) is drawn on a time-lapse sequence of images, and the intensities are reported in a time-space plot. The resulting kymograms depend on the particle dynamics: tilted lines for directed motions (blue) or vertical lines (parallel to the time axis) for static objects (green). Speed (*v*) of directed particles is calculated from the line slope of the distance (*d*) over time (*t*)

performed using kymograph analysis [4, 5], a simple but efficient method to represent the dynamics of moving structures on a two-dimensional image (a kymogram). Kymograms allow a direct readout of speed, direction, and intensity of motile structures. Kymograph analysis of time-lapse TIRFM acquisitions is generated by drawing a line over a trajectory and plotting the intensity along this line for all the time frames of the stack. A dedicated software provides a space-time (x-t) plot where the vertical axis represents time and the horizontal axis corresponds to the intensity of the pixels along the length of the selected trajectory. Depending on the dynamic nature of the object, kymograms display vertical lines parallel to the time axis (static objects) or tilted lines (moving objects) (Fig. 1). The slope of moving objects gives a measurement of the velocity and direction of motion of the object (Fig. 1).

Although kymograph analysis is easy to implement, it only allows studying directed motions or static behaviors, whereas motility can be more diverse and include diffusive and mixed behaviors. Single-particle tracking (SPT) combined with mean squared displacement (MSD) analysis is an alternative approach that provides quantitative information of the dynamic behavior of individual objects and can help to determine whether a particle is freely diffusing, confined, or actively transported. In SPT, tracks (or particle trajectories) are obtained by detecting the individual particles in the acquired image sequences and by reconnecting them between consecutive time frames. Many open-source (e.g., u-track [8], TrackMate [9]) and customized tracking algorithms have been developed using different strategies to efficiently detect particles and take into account their dynamic properties. The mode

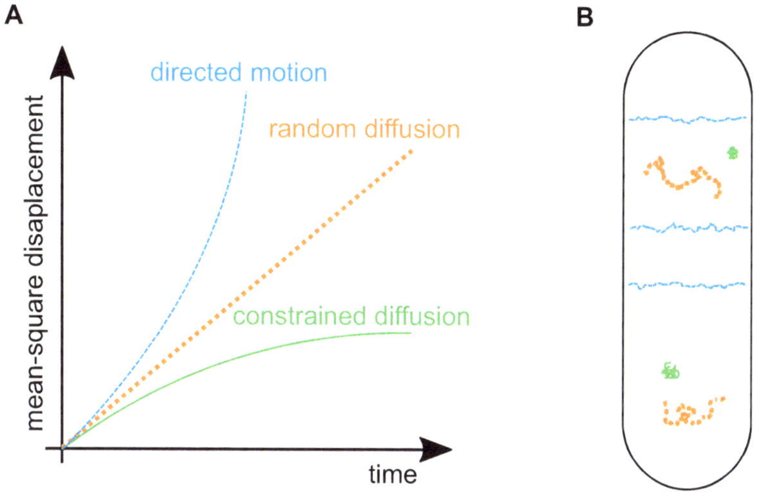

Fig. 2 Principle of mean square displacement analysis. (**A**) MSD curves allow the classification of the mode of movement: (1) Second-order polynomial fit indicates directed motion (blue): MSD(t) = $(\nu t)^2$, where ν is speed and t is time; (2) a linear plot indicates random diffusion (orange): MSD(t) = $4Dt$, where D is the diffusion constant; (3) when the MSD curve reaches a plateau, the system is undergoing constrained diffusion (green). (**B**) Representation of 2D-SPT trajectories (following the above-mentioned color code) of the three modes of MreB motion in the membrane of a rod-shaped cell

of displacement is determined from the resulting trajectories by calculating the so-called mean squared displacement. The MSD describes the average of the squared distances between the starting position of a particle and its end position over time for all time lags of certain length Δt within one trajectory. The properties of the resulting MSD versus t plots are indicative of the dynamic regime of the particle: linear MSD for random diffusion, quadratic for transported/directional movement, and asymptotic for bound/trapped/constrained motion, not freely diffusing (Fig. 2). The type of motion of a given particle is then automatically classified by a curve fitting.

In this chapter, we first describe how to segment cells automatically for the analysis of MreB assemblies. Next, we detail how to apply kymograph and MSD analysis to determine their dynamic behavior and to quantify their main features (velocity, direction, diffusion coefficient). Finally, we present two different approaches to measure the dimensions of MreB filamentous structures. The first method relies on 2D spatial fitting to extract the length and the width of objects of dimensions close to the spatial resolution. The second strategy is based on kymograph analysis and more adapted to determine the length of longer objects.

2 Materials

1. Time-lapse TIRFM acquisitions of MreB (or of your membrane protein or membrane-associated protein of interest) labeled with a fluorescent probe (i.e., fused to a fluorescent protein such as GFP or to an affinity tag binding a fluorescent ligand such as SNAP tag or Halo tag) in live bacterial cells (*see* Chapter 8). Frame rate, exposure time, and total duration of the time lapse should be adapted to the dynamics of the protein and to reduce photobleaching as much as possible during the image sequence acquisition (*see* **Note 2**). The sequence of images (stack) should be saved in TIFF format in 16-bit gray scale. Be sure that all acquisition settings, such as pixel size and time interval between consecutive frames in a movie, are accessible from the image metadata (*see* **Note 3**). If needed, correct the drift in the image sequence (*see* **Note 4**).
2. A computer with an installation of the Fiji open-source software (http://fiji.sc/#download [10]) and of Matlab (*see* **Note 5**).
3. Homemade script macros available on GitHub website: SegCells_TIRF.ijm and KymographAnalyzer.ijm (Fiji); patchDynamicClassification_MSD.m (Matlab).
4. u-track software (version 2.1.3, Matlab-based suite) [8, 11]: u-track offers different detection and reconnection approaches depending on the object shape (isotropic vs. anisotropic) and density (single vs. multiple 2D Gaussian fitting) and its dynamic properties (free diffusion, directed, disappearance, merge/spilt events, etc.).

3 Methods

3.1 Automated Cell Segmentation of TIRF Sections

Cell segmentation is an essential step in quantitative subcellular analysis. Many automated cell segmentation algorithms have been developed to determine the cell outline from phase contrast images or from membrane-stained cells (e.g., MicrobJ [12], Oufti [13], SuperSegger [14]). In TIRF microscopy, only the cell section located in the vicinity (~100–200 nm depth) of the coverslip is excited by the evanescent field (*see* Chapter 8). Because of the dimensions of bacterial cells (e.g., ~3 μm long × 1 μm wide for *Bacillus subtilis* or *Escherichia coli*), this optical section thickness is approximately the uppermost one-fifth of the cell diameter, and thus the TIRF visible cross section is smaller than the cell diameter. Therefore, the density of membrane particles would be systematically underestimated if cells are segmented from phase contrast images due to the estimation error on the cell area. Thus, it is recommended to segment cells directly from TIRF images, as follows:

1. Apply a Gaussian filter to the image, and generate a mask image resulting from the maximal intensity projection (MIP).
2. Identify and label single cells by thresholding the resulting MIP image, and if needed, (a) group over-segmented cells or (b) disconnect neighbored cells included in a single region of interest.

The resulting output should be a mask image where the TIRF sections of single cells visible in a field of view are specifically identified. Pixel values of the mask would be either null for background or a positive integer for the cell. The script macro SegCells_TIRF.ijm segments automatically the cell area under TIRF illumination. To run the macro in Fiji, go to "PluginsMacros/Run...," select the script file, and press the "Run" button in the script window.

3.2 Kymograph-Based Measurement of Particle Velocity

Kymograms can be generated with several tools available in the Fiji software (e.g., Reslice command, Multi Kymograph, KymographBuilder), but quantification of particle velocity has to be done manually. We developed for this the dedicated macro KymographAnalyser.ijm, which generates kymograms and automatically determines the speed of particles. Run the macro in Fiji (go to "PluginsMacros/Run...," select the file, and press the "Run" button in script windows). The macro will:

1. Ask the user to draw manually a line on each particle trajectory and add all lines in the ROI manager.
2. Automatically generate the corresponding kymograms (based on the "Reslice" command).
3. Detect automatically the intensity maxima in a user-defined rectangular selection of the kymogram and measure velocity as the line slope using a linear regression (Fig. 1).

Velocity should be measured on many cells from independent experiments in order to provide statistically significant datasets.

3.3 Automatic Classification of Particle Dynamics

1. Once the raw images have been correctly converted to a stack and drift-corrected, and the cells have been segmented, SPT is executed to obtain the trajectories of the objects, which will then be analyzed by MSD.
2. Run u-track on Matlab by typing movieSelectorGUI.m in the command window to launch the graphical user interface (GUI). A first window allows editing and saving the list of image stacks to be analyzed and to specify the acquisition settings (e.g., pixel size, objective numerical aperture, excitation/emission wavelength, exposure time).

Table 1
Main parameters used for single-particle tracking of MreB assemblies using the u-track Matlab-based package

Object detection (Comet detection)	Difference of Gaussian Watershed-based segmentation	$\sigma_1 = 1$ pixels, $\sigma_2 = 4$ pixels Minimum threshold = 4 standard deviation (s.d.) of image intensity, with a step size = 1 s.d.
Object linking	Particles exhibit random motion or directed movement No missing link allowed Splitting or merging of tracks not allowed	$0 <$ Brownian search radius < 3 pixels Maximal gap to close = 0

Pixel size = 64 nm

3. After validating and saving the images list, select "2D single-particle tracking" to configure the two main steps of SPT (all parameters are listed in Table 1). Object detection is performed using the comet detection method, which relies on structure enhancement (by a difference of Gaussian filter), followed by watershed segmentation. Object reconnection allows to link detections from frame to frame, assuming both diffusion and directional movement. The algorithm can also consider some more complex situations during reconnection as gap-closing events (spot disappearance for one frame because it moves out of focus or the detection fails) or splitting/merging events.
4. After automated particle tracking, it is important to carefully inspect the trajectories generated to verify that detections and reconnections are correct, and to modify analysis parameters when required.
5. Next, membrane particle density is estimated as the temporal average of detected objects on each image divided by the surface area under observation (based on the cell segmentation described in Subheading 3.1).
6. Dynamic classification is performed upon MSD analysis using the Matlab script "patchDynamicClassification_MSD.m" (to be typed and executed in the command window of Matlab). To characterize the dynamic properties of MreB assemblies, MSD curves are only computed on tracks of a sufficient minimal duration (i.e., minimal number of steps). Short trajectories are rejected from the analysis so that the main dynamic mode can be identified and the trajectory sampling is adapted to a robust fitting (*see* **Note 6**).
7. The following simple decision tree has been proposed to classify the dynamic mode: constrained, if MSD reaches a plateau below a threshold value to be determined experimentally using

fixed samples (e.g., fluorescent beads) imaged in the same imaging conditions than those used to generate TIRF movies, and mobile, if the maximal MSD value is higher than the threshold value.

8. The mode of movement is then determined based on the analysis of MSD curves using least-square fitting and appropriate models. Second-order polynomial fit ($\text{MSD}(t) = (vt)^2$ where v is speed and t is time) indicates directed motion. A linear plot ($\text{MSD}(t) = (Dt)/4$ (where D is the diffusion constant) indicates random diffusion. Decision between directed motion and free diffusion stands on the comparison of the goodness-of-fit R^2 obtained for each model. The behavior of a particle movement is classified as directed motion if $R^2_{\text{dir}} > 0.8$ and $R^2_{\text{dir}} > R^2_{\text{diff}}$, as random diffusion if $R^2_{\text{diff}} \geq 0.8$ and $R^2_{\text{diff}} > R^2_{\text{dir}}$, and unclassified if $R^2_{\text{dir}} < 0.8$ and $R^2_{\text{diff}} < 0.8$.
9. Upon determination of the type of motion, it is possible to characterize the population by quantifying the corresponding dynamic properties (e.g., velocities, diffusion coefficients, distribution of dynamic modes among the entire population or at the single-cell level).

3.4 Determination of Filament Dimensions and Orientation by 2D Gaussian Fit

This section describes the automatic quantification of the size and orientation of particles observed by TIRFM regardless of their apparent movement (*see* **Note 7**). It is only assumed that MreB assemblies display an anisotropic 2D Gaussian shape without preferential orientation relative to the cell or image axis. Dimensions (length and width) and orientation of filamentous structures (Fig. 3A) can be measured by fitting each particle in the image using the least-square method with a 2D Gaussian function as a model:

$$f(x,y) = B_0 + I_0 \exp\left[-\left(\frac{(x-x_0)\cos(\theta) + (y-y_0)\sin(\theta)}{\sigma_1}\right)^2 - \left(\frac{-(x-x_0)\sin(\theta) + (y-y_0)\cos(\theta)}{\sigma_2}\right)^2\right]$$

where B_0 and I_0 are the background intensity and amplitude of the Gaussian function centered at (x_0, y_0) with lateral widths (σ_1, σ_2) and θ is the orientation of the Gaussian in image coordinates. The length l and width w of filaments are defined as the maximum and minimum values, respectively, of the full width at half-maximum $\left(2\sigma_i\sqrt{(2\log 2)}, i = 1,2\right)$. To avoid quantification errors, it is recommended to perform the analysis only for isolated filaments and to discard values below the spatial resolution measured experimentally using fluorescent beads. It is important to keep in mind that under TIRF illumination, only ~1/5th of the bacterial cell is

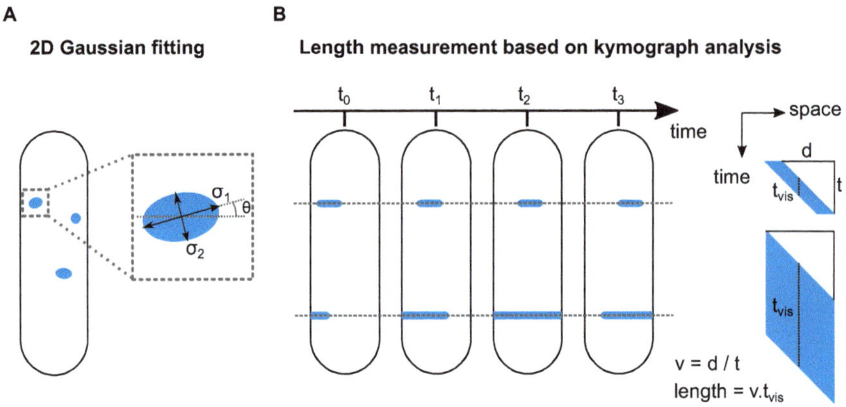

Fig. 3 Determination of the dimensions of filamentous structures. (**A**) Determination of filament dimensions and orientations by 2D Gaussian fit. The dimensions (length, σ_1, and width, σ_2) and orientation (θ) of the filamentous structure are measured by cropping the image area surrounding it and by fitting each object using a least-square method and a 2D anisotropic Gaussian function. (**B**) Quantification of the length of long filaments using kymograph analysis. Profile lines (1 pixel wide, dotted lines) are drawn along separated filament trajectories in order to generate kymographs. While spherical and ellipsoid structures appear as lines (*see* Fig. 1), elongated structures appear as parallelograms. Velocity (v) can be estimated as the angle of the parallelograms. t_{vis} is the time between the appearance of both extremities of a filament at a given position. The length of filaments is estimated as $v \times t_{vis}$

visible (*see* Subheading 3.1 above). Consequently, the length of structures spanning across the TIRF section can be underestimated. To overcome this problem, it is also possible to extract length information using kymographs analysis as described in the next section.

3.5 Determination of the Length of Dynamic Structures Using Kymograph Analysis

The length of structures can also be determined by kymograph analysis (Fig. 3B). This approach is essential for structures longer than the field of view, for which the 2D Gaussian fit method cannot be used. For MreB filaments exhibiting directed motion around the cell circumference, lines (1 pixel wide) are drawn on the trajectories of filaments spatially separated from each other. Intensity maxima are automatically detected along the kymograph and fitted using a linear function to quantify the filament velocity v. The length of each filament is estimated as $v \times t_{vis}$ where t_{vis} is the average time during which the filament is visible on each kymograph position. Only filaments entirely visible (i.e., appearing and disappearing) during the time of acquisition (and excluding the first and last five frames for certainty) are to be used for the analysis.

4 Notes

1. TIRF microscopy allowed single-particle tracking (SPT) of membrane-associated MreB assemblies, revealing their dynamic properties and heterogeneous distribution. TIRFM-

SPT studies based on fluorescent proteins are, however, limited by the density of the imaged proteins in the cell surface. Diffracted emission fluorescence prevents tracking of individual proteins that are separated by less than 0.5 μm. To bypass the limit of resolution of classical light microscopy, it is recommended to perform single-particle tracking with photoactivated localization microscopy (spt-PALM) instead, which is based on temporal emission decorrelation [15].

2. Reliable image quantification requires considering all aspects of the microscopy experiments, from the preparation of the sample to the acquisition settings. Among the parameters to pay particular attention to, exposure time and laser power have to be finely tuned to limit photobleaching but also to use the dynamic range of the camera in full (and avoid the loss of low- and high-signal intensities).
3. Most light microscopes cameras capture 12-bit images, while acquisition softwares mainly propose to save the acquired images in 16-bit or 8-bit formats. Although saving images in the 16-bit file format requires more storage space, it has the advantage of keeping all details of the images. In contrast, saving the acquired images in the 8-bit file format causes the loss of some fine details. This is unsuitable for quantifications requiring sensitivity such as imaging fluorescently labelled proteins in live bacteria (because of the low concentration of molecules in bacterial cells). Note that image files store both the image data (pixel values) and metadata (information on the data that give a sense to the pixel information, such as pixel dimensions, objective lens, excitation light source, emission and acquisition settings, etc.). Some specific file formats like OME-TIFF (hybrid format containing pixel data in TIFF format and metadata in XML format contained in the TIFF header) are well supported in current image processing software (e.g., Fiji using BioFormats) and avoid analysis mistakes by accessing directly to the acquisition settings.
4. During long experiments, temperature changes, dehydration of agarose pads, or mechanical relaxation effects can cause the drift of the sample or of the instrument. In modern microscopes, drift can be limited using an optical table, an environmental temperature-controlled chamber, and/or an automatic focus drift compensation system. Nevertheless, residual drift still occurs in most acquisitions and needs to be corrected. Many algorithms have been developed to correct drift and are readily available and easy to use (e.g., StackReg plugins in ImageJ). It is highly recommended to use one of them for time-lapse observations before processing the image sequence.

5. Requirements for computer: License for Matlab is required, with the Image Processing and Curve Fitting Matlab toolboxes. The software was tested with Matlab versions 2016B and 2018B.
6. A high number of rejected tracks might reflect that the acquisition settings (e.g., laser power, photobleaching, exposure time) and/or the detection and/or linking parameters in SPT have to be corrected. Settings should be optimized to minimize the number of rejected tracks.
7. Lateral resolution in TIRFM is diffraction-limited, and thus dimensions of structures below the limit of resolution (~250–300 nm) will be overestimated and appear as particles of 250–300 nm regardless of their length. Single-molecule localization microscopy allows pushing the spatial resolution down to the nanometer scale, but it requires fixed samples and the temporal information provided by TIRFM is lost. Structured illumination microscopy (SIM) combined with TIRFM is then a good compromise to achieve a spatial resolution of ~100–150 nm and fast imaging acquisition (more than one frame per second). SIM-TIRFM allowed the anisotropic morphology of the diffraction-limited MreB filamentous assemblies to be revealed in *B. subtilis* [16].

Acknowledgment

This work was supported by a consolidator grant from the European Research Council (ERC COG) under the European Union's Horizon 2020 research and innovation program (grant agreement No. 772178) to R.C.-L.

References

1. Jones LJ, Carballido-López R, Errington J (2001) Control of cell shape in bacteria: helical, actin-like filaments in *Bacillus subtilis*. Cell 104(6):913–922
2. van den Ent F, Amos LA, Löwe J (2001) Prokaryotic origin of the actin cytoskeleton. Nature 413(6851):39–44
3. Billaudeau C, Chastanet A, Yao Z, Cornilleau C, Mirouze N, Fromion V, Carballido-Lopez R (2017) Contrasting mechanisms of growth in two model rod-shaped bacteria. Nat Commun 8:15370. https://doi.org/10.1038/ncomms15370
4. Dominguez-Escobar J, Chastanet A, Crevenna AH, Fromion V, Wedlich-Soldner R, Carballido-Lopez R (2011) Processive movement of MreB-associated cell wall biosynthetic complexes in bacteria. Science 333 (6039):225–228. https://doi.org/10.1126/science.1203466
5. Garner EC, Bernard R, Wang W, Zhuang X, Rudner DZ, Mitchison T (2011) Coupled, circumferential motions of the cell wall synthesis machinery and MreB filaments in B. subtilis. Science 333(6039):222–225
6. Axelrod D, Burghardt TP, Thompson NL (1984) Total internal reflection fluorescence. Annu Rev Biophys Bioeng 13:247–268. https://doi.org/10.1146/annurev.bb.13.060184.001335
7. Fish KN (2009) Total internal reflection fluorescence (TIRF) microscopy. Curr Protoc Cytom Chapter 12:Unit12.18. https://doi.org/10.1002/0471142956.cy1218s50

8. Jaqaman K, Loerke D, Mettlen M, Kuwata H, Grinstein S, Schmid SL, Danuser G (2008) Robust single-particle tracking in live-cell time-lapse sequences. Nat Methods 5 (8):695–702. https://doi.org/10.1038/nmeth.1237
9. Tinevez JY, Perry N, Schindelin J, Hoopes GM, Reynolds GD, Laplantine E, Bednarek SY, Shorte SL, Eliceiri KW (2017) TrackMate: an open and extensible platform for single-particle tracking. Methods 115:80–90. https://doi.org/10.1016/j.ymeth.2016.09.016
10. Schindelin J, Arganda-Carreras I, Frise E, Kaynig V, Longair M, Pietzsch T, Preibisch S, Rueden C, Saalfeld S, Schmid B, Tinevez JY, White DJ, Hartenstein V, Eliceiri K, Tomancak P, Cardona A (2012) Fiji: an open-source platform for biological-image analysis. Nat Methods 9(7):676–682. https://doi.org/10.1038/nmeth.2019
11. Applegate KT, Besson S, Matov A, Bagonis MH, Jaqaman K, Danuser G (2011) plusTipTracker: quantitative image analysis software for the measurement of microtubule dynamics. J Struct Biol 176(2):168–184. https://doi.org/10.1016/j.jsb.2011.07.009
12. Ducret A, Quardokus EM, Brun YV (2016) MicrobeJ, a tool for high throughput bacterial cell detection and quantitative analysis. Nat Microbiol 1(7):16077. https://doi.org/10.1038/nmicrobiol.2016.77
13. Paintdakhi A, Parry B, Campos M, Irnov I, Elf J, Surovtsev I, Jacobs-Wagner C (2016) Oufti: an integrated software package for high-accuracy, high-throughput quantitative microscopy analysis. Mol Microbiol 99 (4):767–777. https://doi.org/10.1111/mmi.13264
14. Stylianidou S, Brennan C, Nissen SB, Kuwada NJ, Wiggins PA (2016) SuperSegger: robust image segmentation, analysis and lineage tracking of bacterial cells. Mol Microbiol 102 (4):690–700. https://doi.org/10.1111/mmi.13486
15. Manley S, Gillette JM, Patterson GH, Shroff H, Hess HF, Betzig E, Lippincott-Schwartz J (2008) High-density mapping of single-molecule trajectories with photoactivated localization microscopy. Nat Methods 5 (2):155–157. https://doi.org/10.1038/nmeth.1176
16. Billaudeau C, Yao Z, Cornilleau C, Carballido-Lopez R, Chastanet A (2019) MreB forms subdiffraction nanofilaments during active growth in Bacillus subtilis. MBio 10(1):e01879-18. https://doi.org/10.1128/mBio.01879-18

Chapter 10

Studying Centriole Duplication and Elongation in Human Cells

Catarina Peneda, Carla A. M. Lopes, and Mónica Bettencourt-Dias

Abstract

Centrioles assemble centrosomes and cilia/flagella, which are microtubule-based structures with key roles in cell division, polarity, motility, and signaling. Centriole biogenesis is a tightly regulated process, and deregulation of centriole numbers and structure can have dramatic consequences for cellular function and integrity. However, their small size poses a challenge to study them. Here, we describe protocols that allow the identification and assessment of true centrioles and that provide straightforward strategies to study the role of new candidate proteins in centriole duplication and elongation.

Key words Centrosome, Centriole, Centriole duplication, Centriole elongation, PLK4, CPAP

1 Introduction

The centrosome is the main microtubule-organizing center (MTOC) in most animal cells and has a role in a wide range of cellular processes such as cell migration, mitotic spindle formation, polarity, and cilia formation. In most animal cells, it is composed of two barrel-shaped centrioles surrounded by the pericentriolar material (PCM), which has a role in microtubule anchoring and nucleation (Fig. 1) [1, 2]. Each centriole is typically composed of nine microtubule triplets that have several posttranslational modifications (PTMs), such as acetylation, detyrosination, and polyglutamylation, which are thought to contribute to their high stability and resistance to drug-induced depolymerization [3–7]. The two centrioles within a centrosome are structurally different: the mother centriole has distal and subdistal appendages, while the daughter centriole does not [1]. These structural differences have important functional impact: the appendages have a role in cilia

Catarina Peneda and Carla A.M. Lopes contributed equally to this work.

Helder Maiato (ed.), *Cytoskeleton Dynamics: Methods and Protocols*, Methods in Molecular Biology, vol. 2101, https://doi.org/10.1007/978-1-0716-0219-5_10, © Springer Science+Business Media, LLC, part of Springer Nature 2020

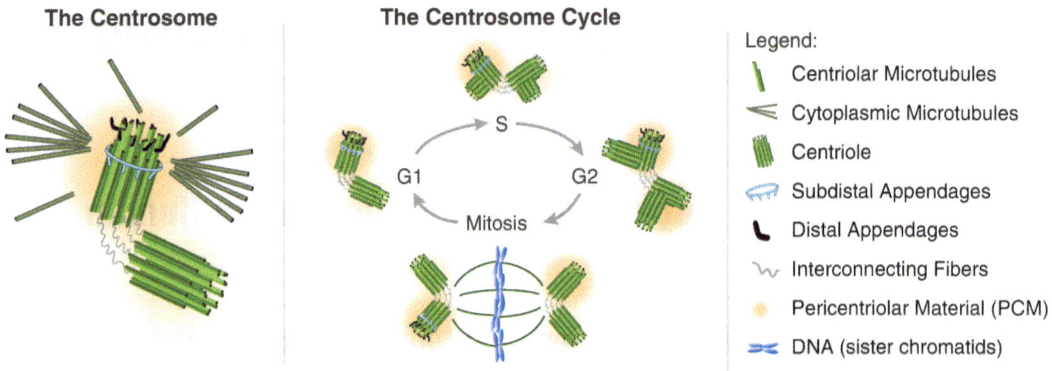

Fig. 1 The centrosome and the centrosome cycle. The centrosome is a small non-membranous organelle found in the cytoplasm and comprises a pair of centrioles surrounded by a matrix (PCM) that nucleates cytoplasmic microtubules. The two centrioles differ from one another as the older or mother centriole has two sets of nine appendages. In order to maintain the correct centrosome number over successive cell generations and to organize an effective bipolar mitotic spindle, a centrosome must duplicate only once in every cell cycle, and only one progeny centriole can be formed next to each parental centriole. At late G1/S phase of the cell cycle, a new centriole forms at the proximal end of each parental centriole. The two new centrioles then elongate throughout S and G2 phases. The two centrosomes then separate at the G2/M transition, each one migrating to one of the poles of the spindle

formation [8, 9] and confer the mother centriole with a higher ability to anchor microtubules than the daughter centriole [10].

Centriole biogenesis is a tightly regulated process that occurs only once per cell cycle in late-G1/S phase, in coordination with DNA replication (Fig. 1). Although the complete molecular mechanisms regulating this process are still not fully understood, several key players and events have already been elucidated [11]. Centriole duplication starts when PLK4, a known core protein in centriole biogenesis, is recruited to the proximal end of the existing centrioles, where there is a ring of CEP152 and CEP192 around the centriole [12–14]. PLK4 binds to and phosphorylates STIL, forming a complex that then triggers the recruitment of SAS6 to the centrosome [15, 16]. SAS6 loading is crucial for the formation of the cartwheel, the first observed structure during centriole assembly. Only one procentriole is formed adjacent to each of the preexisting centrioles [17–20]. Both procentrioles then elongate throughout S and G2, at which point they reach their full length. Generally, human centrioles in cultured cells range between 200 and 250 nm in diameter and 450 and 500 nm in length [11, 21]. After centrosome duplication, cells have two centrosomes, each one of them with two centrioles. However, only one of them has appendages, and it is only in this cell cycle phase, G2/M, that the second parent centriole acquires these structures which will promote PCM recruitment and centriole maturation [2, 22–24]. Dysregulation in centriole biogenesis can result in a change in centriole number, which has been associated with several consequences such as aneuploidy, cell cycle arrest, mitotic defects, cell death, and cancer [25–27].

2 Materials

2.1 Mammalian Cell Culture and Transfection

1. Plasticware: T75 flasks, 24-well plates.
2. Glass coverslips: 13 mm diameter, thickness no. 1.5.
3. Cell counter.
4. Cell lines (hTERT RPE-1, U2OS-myc-PLK4, or U2OS-CPAP-myc).
5. Trypsin or Tryple.
6. 1× Phosphate-buffered saline (PBS).
7. Opti-MEM Reduced Serum Medium (Gibco).
8. Lipofectamine RNAiMAX Reagent (Invitrogen).
9. RPE growth medium: 1:1 Mixture of DMEM and Ham's F-12 Nutrient Mix supplemented with 10% fetal bovine serum (FBS) heat inactivated, 0.348% sodium bicarbonate, 2 mM glutamine, 50 U/mL penicillin, and 50 μg/mL streptomycin.
10. U2OS growth medium: Advanced-DMEM supplemented with 2% Tet System Approved FBS (Takara), 2 mM glutamine, 50 μg/mL hygromycin, 1 mg/mL G418, 50 U/mL penicillin, and 50 μg/mL streptomycin.
11. Transfection medium: RPE growth medium or U2OS growth medium without penicillin/streptomycin.
12. HU medium: U2OS growth medium supplemented with 4 mM hydroxyurea (HU).
13. HU + TET medium: U2OS growth medium supplemented with 4 mM HU and 100 ng/mL tetracycline (TET).
14. Aph medium: U2OS growth medium supplemented with 1.6 μg/mL aphidicolin (Aph).
15. Aph + TET medium: U2OS growth medium supplemented with 1.6 μg/mL Aph and 1 μg/mL TET.
16. siRNA nucleotide sequences (stock concentration 100 μM): GL2 5′-CGT ACG CGG AAT ACT TCGA-3′, SAS6 5′-AAG CAC GTT AAT CAG CTA CAA-3′ [28], CEP120 5′-GAT GAG AAC GGG TGT GTAT-3′ [29].

2.2 Immunofluorescence

1. FBS.
2. Tweezers.
3. Glass slides for microscopy.
4. Nail polish.

Table 1
Primary antibodies used for IF staining

Antigen	Species	Dilution	Product manufacturer (Catalog #)
Centrin	Mouse	1:1000	Merck (04-1624)
Pericentrin	Rabbit	1:250	Abcam (ab4448)
CEP135	Rabbit	1:500	Abcam (ab75005)
γ-Tubulin (GTU88)	Mouse	1:500	Sigma (T5326)
Acetylated-α-tubulin	Mouse	1:500	Sigma (T7451)
Glutamylated tubulin (GT335)	Mouse	1:500	Adipogen (AG-20B-0020)

Table 2
Secondary antibodies used for IF staining

Antibody	Conjugate	Dilution	Product manufacturer (Catalog #)
Goat anti-mouse	Alexa Fluor 488	1:500	Molecular Probes (A11001)
Goat anti-mouse	Alexa Fluor 594	1:500	Molecular Probes (A11005)
Goat anti-rabbit	Alexa Fluor 488	1:500	Molecular Probes (A11008)
Goat anti-rabbit	Alexa Fluor 594	1:500	Molecular Probes (A11012)

5. 1× PBS.
6. Methanol.
7. VECTASHIELD Mounting Medium with DAPI.
8. Blocking solution: 10% FBS in 1× PBS.
9. Pre-extraction solution: 0.5% Triton X-100 in 1× PBS.
10. Antibodies and dilutions used for immunofluorescence (IF) (Tables 1 and 2).

3 Methods

All cell lines used in the following protocols, hTERT-RPE1, U2OS-myc-PLK4, and U2OS-CPAP-myc, should be cultured in the corresponding growth medium (*see* Subheading 2) at 37 °C in a 5% CO_2 atmosphere.

3.1 Centriole Duplication Assay

The aim of this assay is to quickly and easily test whether a candidate protein has a role in centriole duplication. Standard knockdown (or overexpression) of a candidate protein in normal cycling cells can be used to test its impact in canonical centriole duplication (*see* Subheading 3.1.1) (Fig. 2). In this protocol, we use hTERT-RPE1

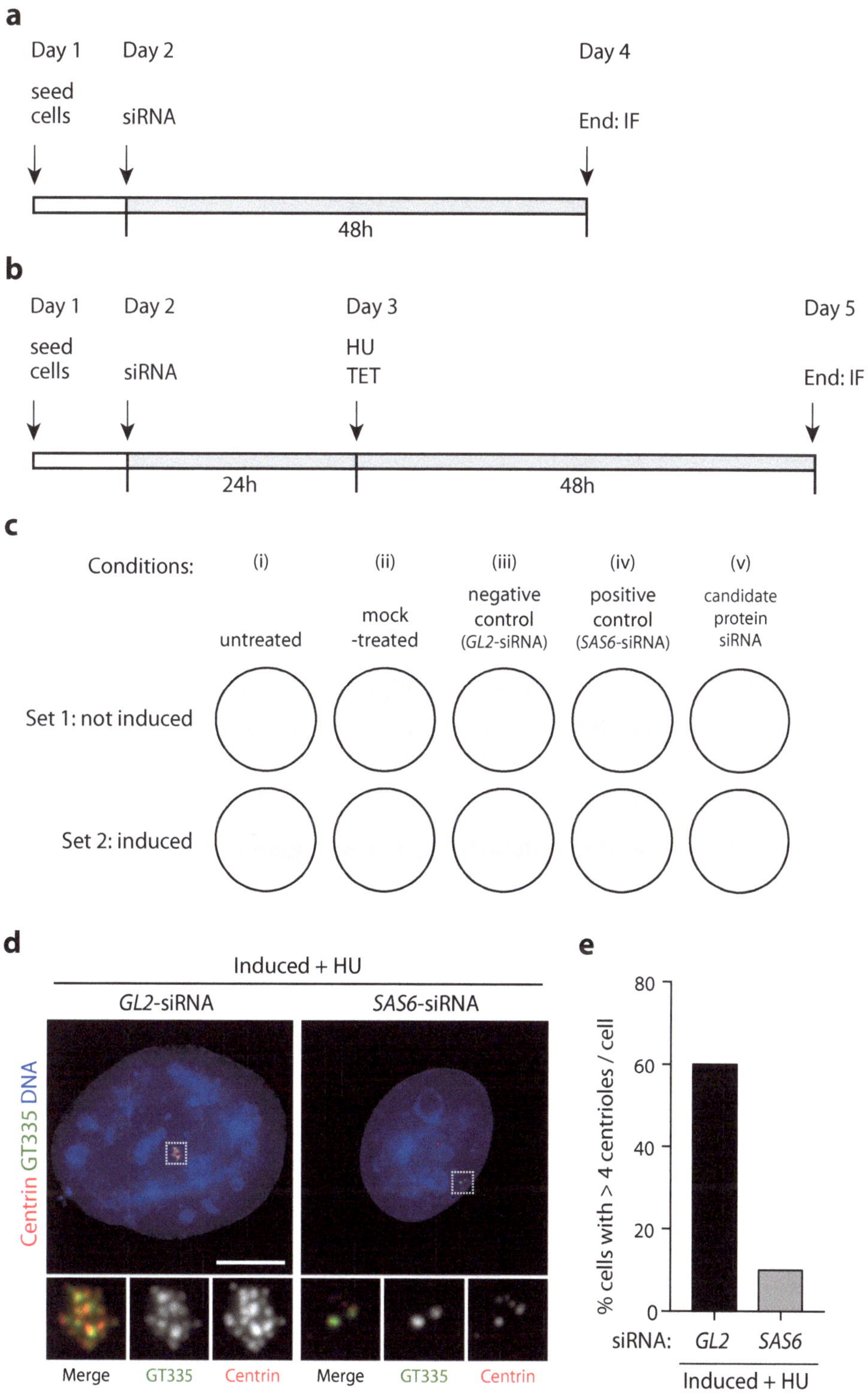

Fig. 2 Duplication assay. (**a**–**c**) Schematic representation of the "Asynchronous" (**a**) and "Arrested in S phase" (**b**) centriole duplication assay protocols and of the minimal experimental layout (**c**). (**a**, **b**) The protocols here described are designed for the analysis of centriole numbers by fluorescence microscopy. However, the same

(ATCC CRL-4000) cells, a common experimentally used non-transformed cell line. However, this phenotype may be difficult to detect. For this reason, we also describe a strategy using a system that can be pushed to form supernumerary centrioles and then test if the knockdown (or eventually the overexpression) of a candidate protein increases or impairs this, making it easier to detect and analyze the phenotype (*see* Subheading 3.1.2) (Fig. 2) [30, 31]. Moreover, many studies have shown that several proteins involved in centriole duplication also have roles in mitotic spindle assembly and function, either directly or indirectly through centrosome assembly [32, 33]. Therefore, in order to avoid mitotic arrest and decrease confounding effects, this assay is performed in cells arrested in S phase, which is a permissive phase for duplication. This protocol is optimized for a U2OS-myc-PLK4 cell line (such as the one used in [30]), in which myc-PLK4 overexpression can be induced upon addition of tetracycline (TET) to the culture medium (*see* **Note 1**). PLK4 overexpression leads to centriole overduplication, and depletion of molecules involved in centriole biogenesis affects this overduplication process, such as SAS6 (its depletion can be used as a positive control) [28, 34–36]. Also, by using hydroxyurea (HU)-arrested cells in S phase, the phenotype of supernumerary centrioles is enhanced as U2OS cells are permissive for overduplication upon HU treatment [37–39]. This protocol therefore provides a straightforward strategy to study the role of new candidate proteins in centriole duplication by rendering a more robust phenotype for quantification analysis.

3.1.1 Centriole Duplication Assay: Asynchronous

1. Place one coverslip per well and seed 5×10^4 hTERT RPE-1 cells/well in 500 μL of RPE growth medium in a 24-well plate. Per experiment, you should have one well for each of the following conditions: (i) untreated, (ii) mock treated, (iii) negative control (*luciferase(GL2)*-siRNA or equivalent), (iv) positive control (*PLK4*-siRNA or *SAS6*-siRNA), and

Fig. 2 (continued) experimental setup can be adapted to other cell culture formats (e.g., 6-well plates, 10 cm dishes) that are best suited to collect cells for flow cytometry analysis to check cell cycle arrest or for Western blotting to assess protein depletion efficiency, for example. Per experiment, one set of conditions will be subject to induction (with TET) and the other will not (without TET) (*see* **Note 3**). (**c**) The minimal conditions that should be included in both assays are (i)–(v). Note that the two sets (induced and not induced) only apply to the "Arrested in S phase" duplication assay protocol. (**d**, **e**) Representative results of a "Arrested in S phase" duplication assay. Cells were stained for centrioles (centrin and glutamylated tubulin, GT335) and DNA. (**d**) Representative IF images of U2OS-myc-PLK4 cells with overduplication (induced PLK4 overexpression, *GL2*-siRNA) and where overduplication was impaired by depletion of the key centriole biogenesis molecule SAS6 (induced PLK4 overexpression, *SAS6*-siRNA). Enlargements of centrioles (boxed areas) are shown. Scale bar, 10 μm. (**e**) Quantification of cells with centriole amplification (more than four centrioles per cell). To score centriole numbers, true centrioles were first identified by clear co-localization of both centriolar markers. Centrioles were then quantified according to centrin staining

(v) siRNA against your candidate protein. Incubate cells at 37 °C overnight.

2. Replace medium 30 min before transfection with 400 μL of transfection medium per well.
3. For each well, prepare the following mixes in 1.5 mL tubes:
   (a) Tube A: 50 μL of Opti-MEM + 0.25 μL of siRNA.
   (b) Tube B: 50 μL of Opti-MEM + 1 μL of Lipofectamine RNAiMAX.
4. Add Tube A to Tube B and mix by flicking the tube. Incubate at room temperature for 15 min.
5. Add 100 μL of the mix to each well, drop by drop.
6. Incubate cells at 37 °C for at least 48 h (extend this incubation time if 48 h is not enough for an efficient knockdown of your protein of interest).
7. Fix cells for IF (*see* Subheading 3.3) (*see* **Note 2**).

3.1.2 Centriole Duplication Assay: Arrested in S Phase

1. Place one coverslip per well and seed 5 × 10^4 U2OS-myc-PLK4 cells/well in 500 μL of U2OS growth medium in a 24-well plate. Per experiment, seed two sets of at least five wells, where one set will be subject to induction and the other will not (*see* **Note 3**). Each set should have one well for each of the following conditions: (i) untreated, (ii) mock treated, (iii) negative control (*GL2*-siRNA or equivalent), (iv) positive control (*PLK4*-siRNA or *SAS6*-siRNA), and (v) siRNA against your candidate protein. Incubate cells at 37 °C overnight.
2. Replace medium 30 min before transfection with 400 μL of transfection medium per well.
3. For each well, prepare the following mixes in 1.5 mL tubes:
   (a) Tube A: 50 μL of Opti-MEM + 0.25 μL of siRNA.
   (b) Tube B: 50 μL of Opti-MEM + 1 μL of Lipofectamine RNAiMAX.
4. Add Tube A to Tube B and mix by flicking the tube. Incubate at room temperature for 15 min.
5. Add 100 μL of the mix to each well, drop by drop.
6. Incubate cells at 37 °C for 24 h.
7. Replace the medium with 500 μL of either HU medium or HU + TET medium to the corresponding set of wells (*see* Fig. 2).
8. Incubate cells at 37 °C for 48 h.
9. Fix cells for IF (*see* Subheading 3.3).

3.2 Centriole Elongation Assay

The aim of this assay is to clearly and directly assess whether a candidate protein has a role in promoting or limiting centriole elongation. Similar to the centriole duplication assay (*see* Subheading 3.1), in which the system is pushed to create supernumerary centrioles, the system used in this protocol can be pushed to form elongated centrioles by inducing the overexpression of CPAP. This protocol is optimized for the U2OS-CPAP-myc cell line (established in [40]) in which CPAP overexpression can be induced upon addition of TET to the culture medium (*see* **Note 1**). It is then possible to test whether depletion (or overexpression) of a candidate protein synergizes with CPAP overexpression to create even longer centrioles (as does CP110 depletion [41]), or if it interferes with the formation of long centrioles promoted by CPAP (as does CEP120 knockdown, which can be used as a positive control [42]).

When studying centriole length, it is even more important to push the system to form longer centrioles. Since centrioles are very small and close to the resolution limit of most microscopy techniques, this protocol allows a more robust detection and analysis of the phenotypes, particularly that of shorter centriolar length. Allowing cells to undergo more than one cell cycle and, consequently, more than one centriole biogenesis and elongation cycle will allow centrioles to elongate significantly upon CPAP overexpression (*see* Subheading 3.2.1) (Fig. 3). However, given that depletion of your protein of interest may affect cell cycle progression and lead to mitotic defects, we also have a protocol that contemplates a synchronization process. This avoids the confounding effect by not allowing cells to progress through mitosis and yet allowing enough time to deplete the protein of interest before the elongation process takes place (*see* Subheading 3.2.2) (Fig. 3). In this protocol, the cells are first arrested in late G1/early S phase and then released into late G2, when centrioles are expected to reach their full length.

3.2.1 Centriole Elongation Assay: Asynchronous

1. Place one coverslip per well and seed 7×10^4 U2OS-CPAP-myc cells/well in 500 μL of U2OS growth medium in a 24-well plate. Per experiment, seed two sets of at least five wells, where one set will be subject to induction and the other will not (*see* **Note 3**). Each set should have one well for each of the following conditions: (i) untreated, (ii) mock treated, (iii) negative control (*GL2*-siRNA or equivalent), (iv) positive control (*CPAP*-siRNA or *CEP120*-siRNA), and (v) siRNA against your candidate protein. Incubate cells at 37 °C overnight.
2. Replace medium 30 min before transfection with 400 μL of transfection medium per well.
3. For each well, prepare the following mixes in 1.5 mL tubes:
   (a) Tube A: 50 μL of Opti-MEM + 0.25 μL of siRNA.
   (b) Tube B: 50 μL of Opti-MEM + 1 μL of Lipofectamine RNAiMAX.

4. Add Tube A to Tube B and mix by flicking the tube. Incubate at room temperature for 15 min.
5. Add 100 μL of the mix to each well, drop by drop.
6. Incubate cells at 37 °C for 24 h.
7. Replace the medium with 500 μL of either U2OS growth medium alone or U2OS growth medium supplemented with 1 μg/mL TET to the corresponding set of wells (*see* Fig. 3).
8. Incubate cells at 37 °C for 24 h.
9. Seal plate with parafilm. Incubate on ice for 2 h to depolymerize microtubules (*see* **Note 4**).
10. Fix cells for IF (*see* Subheading 3.3).

3.2.2 Centriole Elongation Assay: Synchronized in G2 Phase

1. Place one coverslip per well and seed 14 × 10^4 U2OS-CPAP-myc cells/well in 500 μL of U2OS growth medium in a 24-well plate. Per experiment, seed two sets of at least five wells, where one set will be subject to induction and the other will not (*see* **Note 3**). Each set should have one well for each of the following conditions: (i) untreated, (ii) mock treated, (iii) negative control (*GL2*-siRNA or equivalent), (iv) positive control (*CPAP*-siRNA or *CEP120*-siRNA), and (v) siRNA against the candidate protein. Incubate cells at 37 °C overnight.
2. Replace medium 30 min before transfection with 400 μL of transfection medium per well.
3. For each well, prepare the following mixes in 1.5 mL tubes:
   (a) Tube A: 50 μL of Opti-MEM + 0.25 μL of siRNA.
   (b) Tube B: 50 μL of Opti-MEM + 1 μL of Lipofectamine RNAiMAX.
4. Add Tube A to Tube B and mix by flicking the tube. Incubate at room temperature for 15 min.
5. Add 100 μL of the mix to each well, drop by drop. Incubate cells at 37 °C for 9 h.
6. Add 0.8 μL of a 1 mg/mL solution of aphidicolin (Aph) (*see* **Note 5**) to each well. Incubate cells at 37 °C for 24 h.
7. Replace medium with 500 μL of either Aph medium or Aph + TET medium to the corresponding set of wells (*see* Fig. 3). Incubate cells at 37 °C for 24 h.
8. Replace medium with 500 μL of U2OS growth medium (*see* Fig. 3). Incubate cells at 37 °C for 13 h.
9. Seal plate with parafilm. Incubate on ice for 2 h to depolymerize microtubules (*see* **Note 4**).
10. Fix cells for IF (*see* Subheading 3.3).

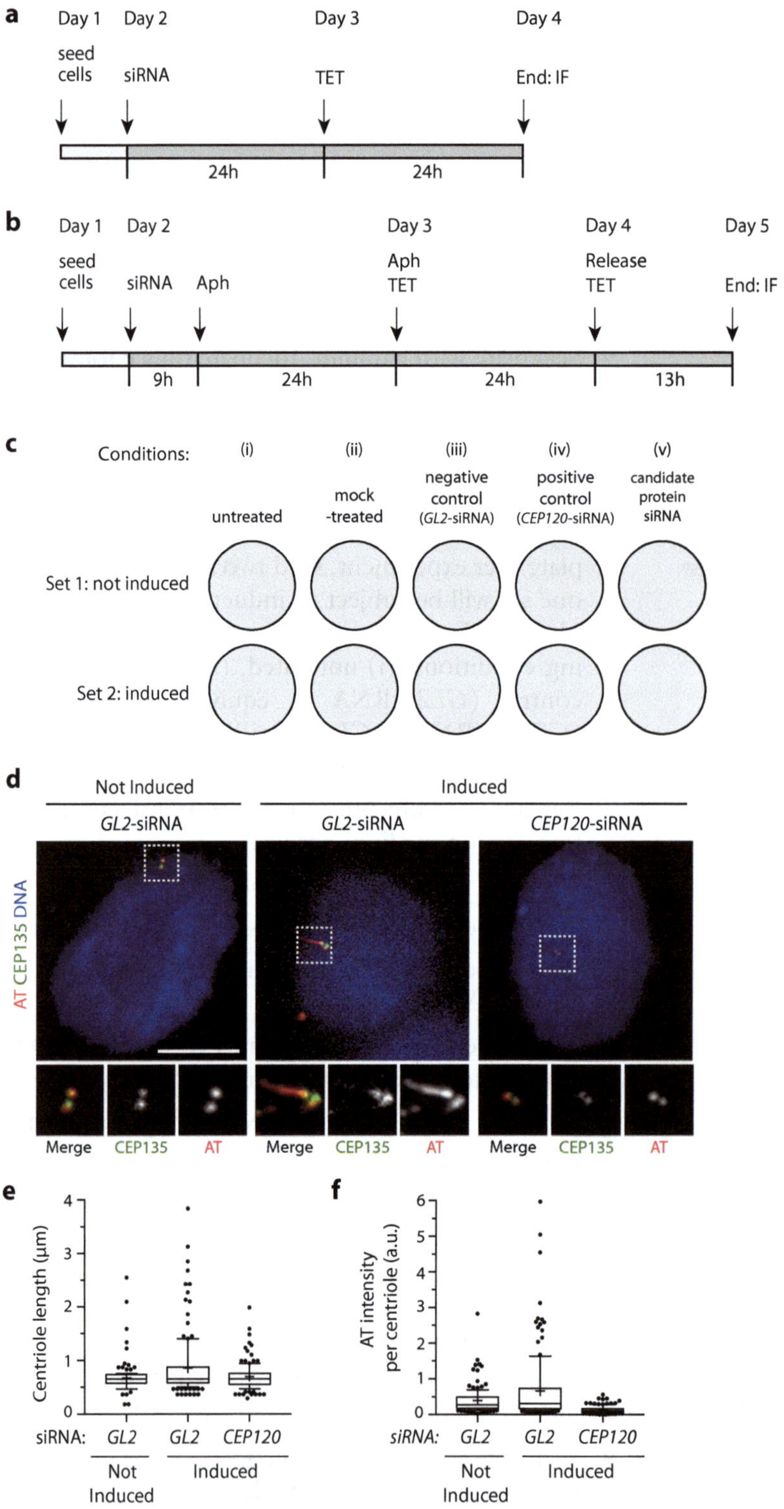

Fig. 3 Elongation assay. (**a–c**) Schematic representation of the "Asynchronous" (**a**) and "Synchronized in G2 phase" (**b**) centriole elongation assay protocols and of the minimal experimental layout (**c**). (**a**, **b**) The protocols

3.3 Cell Fixation and Indirect Immunofluorescence

Cells are then fixed and processed for IF using the specific centriole markers, for example, CEP135 and centrin (*see* Table 1). The analysis of two centriolar markers is essential for the identification of true centrioles by co-localization (*see* Subheading 3.4).

3.3.1 Cell Fixation

1. Replace medium with 500 μL of ice-cold methanol. Incubate plate at −20 °C for 10 min.
2. Wash cells three times with 1× PBS for 5 min in an orbital shaker (*see* **Note 6**).

3.3.2 Indirect Immunofluorescence

1. Block cells with 500 μL of blocking solution (*see* **Note** 7) per well. Incubate at room temperature for 1 h.
2. Remove all the blocking solution from each well.
3. Add 40 μL of primary antibodies (*see* Table 1) diluted in blocking solution to the middle of each coverslip (*see* **Note 8**). Incubate at room temperature for 1 h (*see* **Note 9**).
4. Wash three times with 1× PBS for 5 min in an orbital shaker.
5. Add 40 μL of secondary antibodies diluted in blocking solution to the middle of each coverslip (*see* **Note 8**). Incubate at room temperature for 1 h.
6. Wash three times with 1× PBS for 5 min in an orbital shaker.
7. Mount coverslips on slides using VECTASHIELD DAPI (*see* **Note 10**). Seal coverslips with nail polish. Store at 4 °C until image acquisition (*see* Subheading 3.4).

3.4 Imaging

Given that centrioles are very small structures with a size close to the diffraction-limited optical resolution, analysis of centriole numbers and size is challenging. Using double markers to correctly identify bona fide centrioles and defining the best image acquisition conditions are therefore critical steps in both duplication and elongation assays.

Fig. 3 (continued) here described are designed for the analysis of centriole size by fluorescence microscopy. However, the same experimental setup can be adapted to other cell culture formats (e.g., 6-well plates, 10 cm dishes) that are best suited to collect cells for flow cytometry analysis to check cell cycle synchronization or for Western blotting to assess protein depletion efficiency, for example. Per experiment, one set of conditions will be subject to induction (with TET) and the other will not (without TET) (*see* **Note 3**). (**d–f**) Representative results of an "Asynchronous" elongation assay. Cells were stained for centrioles (CEP135 and acetylated-α-tubulin, AT) and DNA. (**d**) Representative IF images of U2OS-CPAP-myc cells with regular-sized centrioles (not induced CPAP overexpression, *GL2*-siRNA) and cells with over-elongated centrioles (induced CPAP overexpression, *GL2*-siRNA) or in which over-elongation was impaired by depletion of CEP120 (induced CPAP overexpression, *CEP120*-siRNA). Enlargements of centrioles (boxed areas) are shown. Scale bar, 10 μm. (**e**, **f**) To assess centriolar size, true centrioles were first identified by clear co-localization of both centriolar markers. Centriolar size was then assessed by acetylated-α-tubulin staining by measuring either its length (in maximum projections of this channel) (**e**) or its intensity (in total sum projections of this channel) (**f**)

For both counting and measuring centrioles, at least two markers should be used to be able to identify true centrioles by co-localization. Also, for the elongation assay, it is critical to choose at least one marker that localizes along the centriolar wall, such as acetylated-α-tubulin [40]. As an alternative, it is also possible to use a proximal (e.g., CEP135) and a distal centriolar marker (e.g., CP110 or CEP97). Depending on the phenotypes observed upon perturbation of your protein of interest in any of these protocols, you may consider assessing other centriolar proteins to further dissect the mechanism through which your protein is affecting centriole biogenesis and maturation [2, 11].

When using confocal, spinning-disk, or even wide-field microscopy, it is important to acquire images with an objective that allows you to get the higher resolution possible (at least close to 200 nm) so that identification of single centrioles is possible. Image acquisition should always be performed, making sure that a complete series of optical cross sections (z-stacks) is acquired and that it includes the whole cell, to make sure that all the centrioles will be detected. Also, be sure that you acquire all the z-stacks necessary to include in your final capture all information from the microscope into the image. Whenever possible, using super-resolution techniques is a good alternative, especially for the elongation assay. In this case, this is the best way to have a readout of centriole length. This will allow you to better detect differences (longer or shorter centrioles), especially if they are smaller than 200 nm, that you wouldn't be able to distinguish by conventional microscopy. Using electron microscopy is also a good way to assess not only centriole number and length but also structural defects [41, 43].

For image acquisition, acquire z-stacks for each channel, making sure you acquire the whole cell. To determine cell limits, it is helpful to use both DNA channel as well as the background staining of centriole markers. Centriole markers usually also have a slight background in the cytoplasm that can be used to assess cell limits. We advise you to acquire images of at least 100 cells per condition (per experiment) for quantification. Deconvolution of images before image processing and assessment/quantification is also recommended.

Regarding data analysis, we recommend a manual analysis. To assess centriole number and length, you should first analyze the merged images with all the channels used to identify true centrioles within each cell by assessing the co-localization of both centriolar makers. As maximum projection images can be misleading in assessing co-localization, it is helpful to go through the z-stack to ensure true co-localization. Also, as centrioles can be clustered together (especially in conditions where there is significant overduplication), this will also be helpful to distinguish individual centrioles. After a first manual analysis, we recommend automation of the image analysis process, for example, by developing macros using Fiji/ ImageJ. This will make the process much quicker and more reproducible.

4 Notes

1. Both U2OS-myc-PLK4 and U2OS-CPAP-myc were established using the T- Rex system, as described in [32, 39]. Briefly, these cells express a TET repressor that acts upon the two TET operator 2 (TetO2) sites, inhibiting the expression of myc-PLK4 or CPAP-myc under the CMV promoter. When TET is added, it will bind the repressor and induce a conformational change that will impair its binding to the TetO2 sites. Therefore, TET addition will allow the induction of myc-PLK4 or CPAP-myc expression under the CMV promoter. The system responds to different concentrations of TET in the medium.
2. The number of centrioles in a cell is tightly regulated throughout the cell cycle: two centrioles (i.e., one centrosome) in G1 and four centrioles (i.e., two centrosomes) in S, G2, and mitosis. Therefore, to correctly and accurately detect centriole number defects (either more or less than normal), you need to be able to identify the cell cycle stage a given cell is in within the asynchronous population. For example, you can co-stain with an antibody against cyclin D to identify cells in G1 or with an antibody against cyclin A to identify cells in S and G2 phases [44]. Alternatively, you can examine centriole numbers only in mitotic cells, where the expected correct number of centrioles is four. Mitotic cells can be easily distinguished from the remaining interphase cells by looking at DNA staining alone [45, 46].
3. In the "Arrested in S phase" centriole duplication and centriole elongation assays, we suggest having a set of conditions where PLK4 or CPAP overexpression, respectively, is not induced. With these sets of conditions, some valuable conclusions can also be reached. If depletion of a candidate protein results in less (duplication assay) or shorter (elongation assay) centrioles, these phenotypes will be more easily detected and assessed in an overexpression scenario (i.e., induced). On the other hand, if depletion of a candidate protein results in more (duplication assay) or longer (elongation assay) centrioles, this may be more evident in a non-overexpression scenario (i.e., not induced).
4. As an optional extra step, after microtubule depolymerization you can replace the medium with pre-extraction solution for 40 s (while keeping the plate on ice!) and then immediately fix the cells for IF staining (*see* Subheading 3.3). This will clear

your preparation, especially from microtubule remnants, making it easier to correctly identify centrioles.

5. In the "Synchronized in G2 phase" centriole elongation assay protocol, we used Aph to synchronize the cells. A similar result can also be obtained with thymidine. However, we do not advise the use of HU since it causes overduplication in these cell lines, which can cause confounding effects in the elongation assay.
6. After fixing the cells, the plates with the coverslips can be kept in 1× PBS with 0.05% sodium azide at 4 °C and sealed with parafilm for several weeks until needed to perform the IF staining.
7. This protocol describes the use of 10% FBS in 1× PBS as a blocking solution for IF. However, the use of 3% BSA in 1× PBS works equally well with a wide range of antibodies. The use of different antibodies might require protocol optimization.
8. As an optional step, centrifuge the diluted antibodies at 20,000 × *g*, at 4 °C for 10 min to get rid of aggregates. Use only the supernatant.
9. Alternatively, primary antibodies incubation can be done overnight at 4 °C in a humidified chamber to prevent the coverslips from drying. Good results can be obtained for a wide range of antibodies, but you should always optimize the staining protocol for your system and specific antibodies.
10. In this protocol, DNA is stained by DAPI, which is present in the mounting medium. You can also use a different mounting medium without any dye that stains DNA. In this case, DNA staining should be performed, for example, with DAPI or Hoechst, during or after secondary antibodies incubation, before mounting the coverslip on the slide, following standard protocols as described by the manufacturer. Also, mounting medium choice should be made taking into account the type of microscopy that will be used. For example, some hardening-mounting medium might not be a good option for some super-resolution techniques (e.g., Dako Mounting Medium is not a good option for 3D-SIM).

Acknowledgments

We are grateful to E. Nigg for cell lines. C.A.M. Lopes and C. Peneda are funded by Fundação para a Ciência e a Tecnologia (SFRH/BPD/124127/2016 and PD/BD/128004/2016, respectively). The M. Bettencourt-Dias laboratory is funded by Fundação Calouste Gulbenkian and the European Research Council (ERC-2015-CoG-683258).

References

1. Bornens M (2002) Centrosome composition and microtubule anchoring mechanisms. Curr Opin Cell Biol 14(1):25–34
2. Fu J, Lipinszki Z, Rangone H, Min M, Mykura C, Chao-Chu J, Schneider S, Dzhindzhev NS, Gottardo M, Riparbelli MG, Callaini G, Glover DM (2016) Conserved molecular interactions in centriole-to-centrosome conversion. Nat Cell Biol 18(1):87–99
3. Piperno G, LeDizet M, Chang XJ (1987) Microtubules containing acetylated alpha-tubulin in mammalian cells in culture. J Cell Biol 104(2):289–302
4. Gundersen GG, Bulinski JC (1986) Distribution of tyrosinated and nontyrosinated alpha-tubulin during mitosis. J Cell Biol 102 (3):1118–1126
5. Eddé B, Rossier J, Le Caer JP, Desbruyères E, Gros F, Denoulet P (1990) Posttranslational glutamylation of alpha-tubulin. Science 247 (4938):83–85
6. Bobinnec Y, Khodjakov A, Mir LM, Rieder CL, Eddé B, Bornens M (1998) Centriole disassembly in vivo and its effect on centrosome structure and function in vertebrate cells. J Cell Biol 143(6):1575–1589
7. Bobinnec Y, Moudjou M, Fouquet JP, Desbruyères E, Eddé B, Bornens M (1998) Glutamylation of centriole and cytoplasmic tubulin in proliferating non-neuronal cells. Cell Motil Cytoskeleton 39(3):223–232
8. Bettencourt-Dias M, Hildebrandt F, Pellman D, Woods G, Godinho SA (2011) Centrosomes and cilia in human disease. Trends Genet 27(8):307–315
9. Tanos BE, Yang HJ, Soni R, Wang WJ, Macaluso FP, Asara JM, Tsou MF (2013) Centriole distal appendages promote membrane docking, leading to cilia initiation. Genes Dev 27 (2):163–168
10. Piel M, Meyer P, Khodjakov A, Rieder CL, Bornens M (2000) The respective contributions of the mother and daughter centrioles to centrosome activity and behavior in vertebrate cells. J Cell Biol 149(2):317–330
11. Nigg EA, Holland AJ (2018) Once and only once: mechanisms of centriole duplication and their deregulation in disease. Nat Rev Mol Cell Biol 19(5):297–312
12. Sonnen KF, Gabryjonczyk AM, Anselm E, Stierhof YD, Nigg EA (2013) Human Cep192 and Cep152 cooperate in Plk4 recruitment and centriole duplication. J Cell Sci 126 (Pt 14):3223–3233
13. Cizmecioglu O, Arnold M, Bahtz R, Settele F, Ehret L, Haselmann-Weiss U, Antony C, Hoffmann I (2010) Cep152 acts as a scaffold for recruitment of Plk4 and CPAP to the centrosome. J Cell Biol 191(4):731–739
14. Hatch EM, Kulukian A, Holland AJ, Cleveland DW, Stearns T (2010) Cep152 interacts with Plk4 and is required for centriole duplication. J Cell Biol 191(4):721–729
15. Dzhindzhev NS, Tzolovsky G, Lipinszki Z, Schneider S, Lattao R, Fu J, Debski J, Dadlez M, Glover DM (2014) Plk4 phosphorylates Ana2 to trigger Sas6 recruitment and procentriole formation. Curr Biol 24 (21):2526–2532
16. Arquint C, Gabryjonczyk AM, Imseng S, Böhm R, Sauer E, Hiller S, Nigg EA, Maier T (2015) STIL binding to Polo-box 3 of PLK4 regulates centriole duplication. Elife 4. https://doi.org/10.7554/eLife.07888
17. Ohta M, Ashikawa T, Nozaki Y, Kozuka-Hata H, Goto H, Inagaki M, Oyama M, Kitagawa D (2014) Direct interaction of Plk4 with STIL ensures formation of a single procentriole per parental centriole. Nat Commun 5:5267
18. Moyer TC, Clutario KM, Lambrus BG, Daggubati V, Holland AJ (2015) Binding of STIL to Plk4 activates kinase activity to promote centriole assembly. J Cell Biol 209 (6):863–878
19. Ohta M, Watanabe K, Ashikawa T, Nozaki Y, Yoshiba S, Kimura A, Kitagawa D (2018) Bimodal binding of STIL to Plk4 controls proper centriole copy number. Cell Rep 23 (11):3160–3169.e4
20. Arquint C, Nigg EA (2016) The PLK4-STIL-SAS-6 module at the core of centriole duplication. Biochem Soc Trans 44(5):1253–1263
21. Azimzadeh J, Marshall WF (2010) Building the centriole. Curr Biol 20(18):R816–R825
22. Vorobjev IA, Chentsov YS (1982) Centrioles in the cell cycle. I. Epithelial cells. J Cell Biol 93 (3):938–949
23. Kong D, Farmer V, Shukla A, James J, Gruskin R, Kiriyama S, Loncarek J (2014) Centriole maturation requires regulated Plk1 activity during two consecutive cell cycles. J Cell Biol 206(7):855–865
24. Uzbekov R, Alieva I (2018) Who are you, subdistal appendages of centriole? Open Biol 8 (7):180062
25. Marthiens V, Rujano MA, Pennetier C, Tessier S, Paul-Gilloteaux P, Basto R (2013) Centrosome amplification causes microcephaly. Nat Cell Biol 15(7):731–740

26. Levine MS, Bakker B, Boeckx B, Moyett J, Lu J, Vitre B, Spierings DC, Lansdorp PM, Cleveland DW, Lambrechts D, Foijer F, Holland AJ (2017) Centrosome amplification is sufficient to promote spontaneous tumorigenesis in mammals. Dev Cell 40(3):313–322.e5
27. Godinho SA, Pellman D (2014) Causes and consequences of centrosome abnormalities in cancer. Philos Trans R Soc Lond Ser B Biol Sci 369(1650):20130467
28. Leidel S, Delattre M, Cerutti L, Baumer K, Gönczy P (2005) SAS-6 defines a protein family required for centrosome duplication in C. elegans and in human cells. Nat Cell Biol 7 (2):115–125
29. Comartin D, Gupta GD, Fussner E, Coyaud É, Hasegan M, Archinti M, Cheung SW, Pinchev D, Lawo S, Raught B, Bazett-Jones DP, Lüders J, Pelletier L (2013) CEP120 and SPICE1 cooperate with CPAP in centriole elongation. Curr Biol 23(14):1360–1366
30. Kleylein-Sohn J, Westendorf J, Le Clech M, Habedanck R, Stierhof YD, Nigg EA (2007) Plk4-induced centriole biogenesis in human cells. Dev Cell 13(2):190–120
31. Habedanck R, Stierhof YD, Wilkinson CJ, Nigg EA (2005) The Polo kinase Plk4 functions in centriole duplication. Nat Cell Biol 7 (11):1140–1146
32. Kimura M, Yoshioka T, Saio M, Banno Y, Nagaoka H, Okano Y (2013) Mitotic catastrophe and cell death induced by depletion of centrosomal proteins. Cell Death Dis 4(4): e603
33. Wang G, Jiang Q, Zhang C (2014) The role of mitotic kinases in coupling the centrosome cycle with the assembly of the mitotic spindle. J Cell Sci 127(Pt 19):4111–4122
34. Bettencourt-Dias M, Rodrigues-Martins A, Carpenter L, Riparbelli M, Lehmann L, Gatt MK, Carmo N, Balloux F, Callaini G, Glover DM (2005) SAK/PLK4 is required for centriole duplication and flagella development. Curr Biol 15(24):2199–2207
35. Holland AJ, Fachinetti D, Zhu Q, Bauer M, Verma IM, Nigg EA, Cleveland DW (2012) The autoregulated instability of Polo-like kinase 4 limits centrosome duplication to once per cell cycle. Genes Dev 26 (24):2684–2689
36. Coelho PA, Bury L, Shahbazi MN, Liakath-Ali K, Tate PH, Wormald S, Hindley CJ, Huch M, Archer J, Skarnes WC, Zernicka-Goetz M, Glover DM (2015) Over-expression of Plk4 induces centrosome amplification, loss of primary cilia and associated tissue hyperplasia in the mouse. Open Biol 5(12):150209
37. Balczon R, Bao L, Zimmer WE, Brown K, Zinkowski RP, Brinkley BR (1995) Dissociation of centrosome replication events from cycles of DNA synthesis and mitotic division in hydroxyurea-arrested Chinese hamster ovary cells. J Cell Biol 130(1):105–115
38. Kuriyama R, Terada Y, Lee KS, Wang CL (2007) Centrosome replication in hydroxyurea-arrested CHO cells expressing GFP-tagged centrin2. J Cell Sci 120 (Pt 14):2444–2453
39. Loncarek J, Hergert P, Magidson V, Khodjakov A (2008) Control of daughter centriole formation by the pericentriolar material. Nat Cell Biol 10(3):322–328
40. Tang CJ, Fu RH, Wu KS, Hsu WB, Tang TK (2009) CPAP is a cell-cycle regulated protein that controls centriole length. Nat Cell Biol 11 (7):825–831
41. Schmidt TI, Kleylein-Sohn J, Westendorf J, Le Clech M, Lavoie SB, Stierhof YD, Nigg EA (2009) Control of centriole length by CPAP and CP110. Curr Biol 19(12):1005–1011
42. Lin YN, Wu CT, Lin YC, Hsu WB, Tang CJ, Chang CW, Tang TK (2013) CEP120 interacts with CPAP and positively regulates centriole elongation. J Cell Biol 202(2):211–219
43. Kirkham M, Müller-Reichert T, Oegema K, Grill S, Hyman AA (2003) SAS-4 is a C. elegans centriolar protein that controls centrosome size. Cell 112(4):575–587
44. Pihan GA (2013) Centrosome dysfunction contributes to chromosome instability, chromoanagenesis, and genome reprograming in cancer. Front Oncol 3:277
45. Lopes CAM, Mesquita M, Cunha AI, Cardoso J, Carapeta S, Laranjeira C, Pinto AE, Pereira-Leal JB, Dias-Pereira A, Bettencourt-Dias M, Chaves P (2018) Centrosome amplification arises before neoplasia and increases upon p53 loss in tumorigenesis. J Cell Biol 217(7):2353–2363
46. Marteil G, Guerrero A, Vieira AF, de Almeida BP, Machado P, Mendonça S, Mesquita M, Villarreal B, Fonseca I, Francia ME, Dores K, Martins NP, Jana SC, Tranfield EM, Barbosa-Morais NL, Paredes J, Pellman D, Godinho SA, Bettencourt-Dias M (2018) Over-elongation of centrioles in cancer promotes centriole amplification and chromosome missegregation. Nat Commun 9(1):1258

Chapter 11

Assaying Microtubule Nucleation

Artur Ezquerra, Ricardo Viais, and Jens Lüders

Abstract

Assaying microtubule nucleation is essential to understand the organization of microtubule networks in any cell type. In this chapter we describe methods for measuring nucleation activity at centrosomes and at mitotic chromatin in cell lines, to study interphase and mitotic microtubule organization, and for measuring non-centrosomal nucleation in cultured primary neurons, to study microtubule organization in the absence of a microtubule organizing center. While a number of different approaches and variations thereof have been reported in the literature, here we aim to keep the methodology as simple as possible and thus accessible to most research laboratories.

Key words Microtubules, Nucleation, MTOC, Centrosome, γ-Tubulin, Augmin

1 Introduction

Microtubule nucleation, the de novo formation of microtubules, is fundamental for the organization of microtubule arrays in both proliferating and postmitotic, differentiated cells. The main microtubule nucleator is the γ-tubulin ring complex (γTuRC), but additional γTuRC-independent mechanisms may exist. By controlling nucleation in space and time cells regulate where, when, and in what orientation new microtubules are formed. Typically, nucleation occurs at microtubule organizing centers (MTOCs) such as the centrosome, the Golgi, or the mitotic chromatin [1–4]. However, it is now well established that microtubules can also be nucleated in a manner that is independent of an MTOC. The best characterized mechanism involves the multi-subunit augmin complex, which recruits γTuRC to nucleate microtubules from the lattice of preexisting microtubules [5–7]. In mammalian cells augmin-dependent nucleation has been described to increase microtubule density within the spindle during mitosis [5, 6, 8] and throughout all cellular compartments in differentiated

Artur Ezquerra and Ricardo Viais contributed equally to this work.

Helder Maiato (ed.), *Cytoskeleton Dynamics: Methods and Protocols*, Methods in Molecular Biology, vol. 2101, https://doi.org/10.1007/978-1-0716-0219-5_11, © Springer Science+Business Media, LLC, part of Springer Nature 2020

postmitotic neurons [9, 10]. In the latter, nucleation mediated by augmin is additionally important for controlling correct microtubule polarity. However, it is likely that augmin-dependent nucleation is also more widely employed by other cell types [11].

In most model systems we currently lack the tools to directly detect nucleation events within complex microtubule arrays. Exceptions are live cell imaging-based methods in yeast and in plant cells. In yeast cells this is possible due to the relatively low number of microtubules [12, 13] and in plant cells the specific configuration of the cortical microtubule array and its amenability to imaging by total internal reflection fluorescence (TIRF) microscopy has allowed direct visualization of nucleation events [11, 14–17]. However, even in cases where direct visualization by live imaging is possible, it is not nucleation per se that is detected, but rather the growth of new microtubules from presumed nucleation sites.

In this chapter we describe methods for assaying microtubule nucleation in cultured mammalian cells and provide advice on both technical and cell biological challenges. We present two main and complementary types of assays: an end-point assay, based on microtubule depolymerization/repolymerization followed by chemical fixation, and a steady-state assay, based on imaging of growing microtubule plus-ends in live cells. The first method is suitable for assaying nucleation at MTOCs and, as examples, we will describe how to measure nucleation from centrosomes and, during mitosis, from chromatin. The second approach is only an approximation to assessing nucleation activity, but is particularly useful in cases where there is no evidence for the involvement of a specific structure that functions as MTOC or where depolymerization/repolymerization of microtubules is not an option (e.g., for technical or cell biological reasons). Differentiated neurons, for example, display a highly stable microtubule network that is resistant to treatments commonly used to depolymerize microtubules (exposure to drugs such as nocodazole, incubation on ice, combination of the former two treatments). As an example, we describe how to measure augmin-dependent nucleation in cultured mouse neurons [9, 10]. Importantly, while we can describe only a few specific applications in detail, with appropriate modifications these assays may also be adapted to other MTOCs/cell types.

2 Materials

2.1 Microtubule Regrowth Assay in Cell Lines

1. RPE-1 or U2OS cells for studying interphase centrosomal nucleation; HeLa cells for studying chromatin-mediated nucleation during mitosis (*see* **Note 1**).

2. Media suitable for the culture of the chosen cell line. We use DMEM +10% fetal calf serum for U2OS and HeLa cells and DMEM/F12 + 10% fetal calf serum for RPE-1 cells.
3. Cell culture incubator at 37 °C, 5% CO_2.
4. 6- and 12-well cell culture plates.
5. 12 mm #1.5 round coverslips, poly-L-lysine coated.
6. Temperature-controlled water bath, preheated to 37 °C.
7. 50 mL Falcon tube.
8. Timer.
9. Forceps suitable for handling 12 mm coverslips.
10. Ice-cold methanol (stored at −20 °C) or, alternatively, 37% formaldehyde stock solution.
11. 1× Phosphate-buffered saline (PBS).
12. Primary antibodies against microtubules (e.g., DM1A, Sigma-Aldrich) and centrosomes.
13. Fluorescence microscope equipped with 63× or 100× oil immersion objective of highest available numerical aperture (e.g., 1.4) and camera.

2.2 Microtubule Plus-End Growth Assay in Neurons

1. Primary mouse hippocampal neuronal culture.
2. Cell culture incubator at 37 °C, 5% CO_2.
3. Temperature-controlled water bath.
4. 35 mm Glass-bottom cell culture dishes (14 mm Microwell, #1.5 cover glass; MatTek).
5. Poly-D-lysine.
6. 0.1 M Borate buffer pH 8.5: 0.04 M Boric acid, 0.019 M sodium tetraborate in H_2O.
7. Neuron-plating medium: Dulbecco's modified Eagle medium (DMEM) with high glucose and L-glutamine, 10% fetal bovine serum (1% concentration in the medium) and penicillin/streptomycin 10,000 U/mL (100 U/mL concentration in the medium).
8. Neuron-maintaining medium: Neurobasal medium, (+) D-glucose solution at 200 g/L (final concentration in the medium 2 g/L), Glutamax Supplement 100× (1× concentration in the medium), B27 Supplement 50× (1× concentration in the medium), and penicillin/streptomycin 10,000 U/mL (concentration of 100 U/mL in the medium).
9. Transfection medium: Neurobasal medium, (+) D-glucose solution at 200 g/L (concentration of 2 g/L in the medium), Glutamax Supplement 100× (1× concentration in the medium), B27 Supplement 50× (1× concentration in the medium).

10. Transfection reagent: Lipofectamine 2000 (Invitrogen).
11. Plasmid expressing EB3-tdtomato [18] or alternative reporters [10].
12. Spinning disk microscope equipped with 100× objective and temperature- and CO_2-controlled incubation chamber, automated z-drive, and enabled for time-lapse imaging.

2.3 Quantification

1. Computer with ImageJ/Fiji software installed.
2. For analysis of growing plus-ends in neurons: ImageJ/Fiji with plug-in tsp050706.txt installed. This plug-in can be downloaded from the following website link: https://www.embl.de/eamnet/html/kymograph.html.

3 Methods

3.1 Microtubule Regrowth Assay in Cell Lines

3.1.1 Microtubule Depolymerization and Regrowth

1. Grow cells on poly-L-lysine-coated coverslips. Place 1 coverslip per well in a 12-well plate or multiple coverslips per well of a 6-well plate (*see* **Note 2**). Plan your experiment (e.g., recombinant protein expression, RNAi, or drug treatment) so that on the day of the assay cells are at 50–80% confluence.
2. 30 min before starting the assay warm up the water bath to 37 °C and place a 50 mL Falcon tube with culture medium in it.
3. Reserve bench space next to the water bath for placement of an ice-water bath and a timer.
4. Remove multi-well plates with cells from the incubator and float the dishes on ice with added water for 30 min to depolymerize microtubules (*see* **Note 3**).
5. During the incubation prepare a multi-well plate with fixative. For each coverslip and/or condition fill a well with ice-cold methanol (about half of the volume of the well) and float it on the ice-water bath. Methanol may be substituted with formaldehyde as fixative. In this case prepare 3.7% formaldehyde solution in PBS, add it to the wells, and keep plates at RT on the bench (*see* **Note 4**).
6. For each sample/condition use forceps (*see* **Note 5**) to remove one coverslip from the multi-well plate in the ice-water bath and quickly place it directly in a well with fixative. This coverslip (0 s of regrowth) will be used to confirm that the microtubules were completely depolymerized (*see* **Note 6**).
7. Set the timer to 0 and use forceps to remove a coverslip from the multi-well plate floating in the ice-water bath. To induce microtubule regrowth, quickly submerge the coverslip in the warm medium in the 50 mL Falcon tube and start the timer.

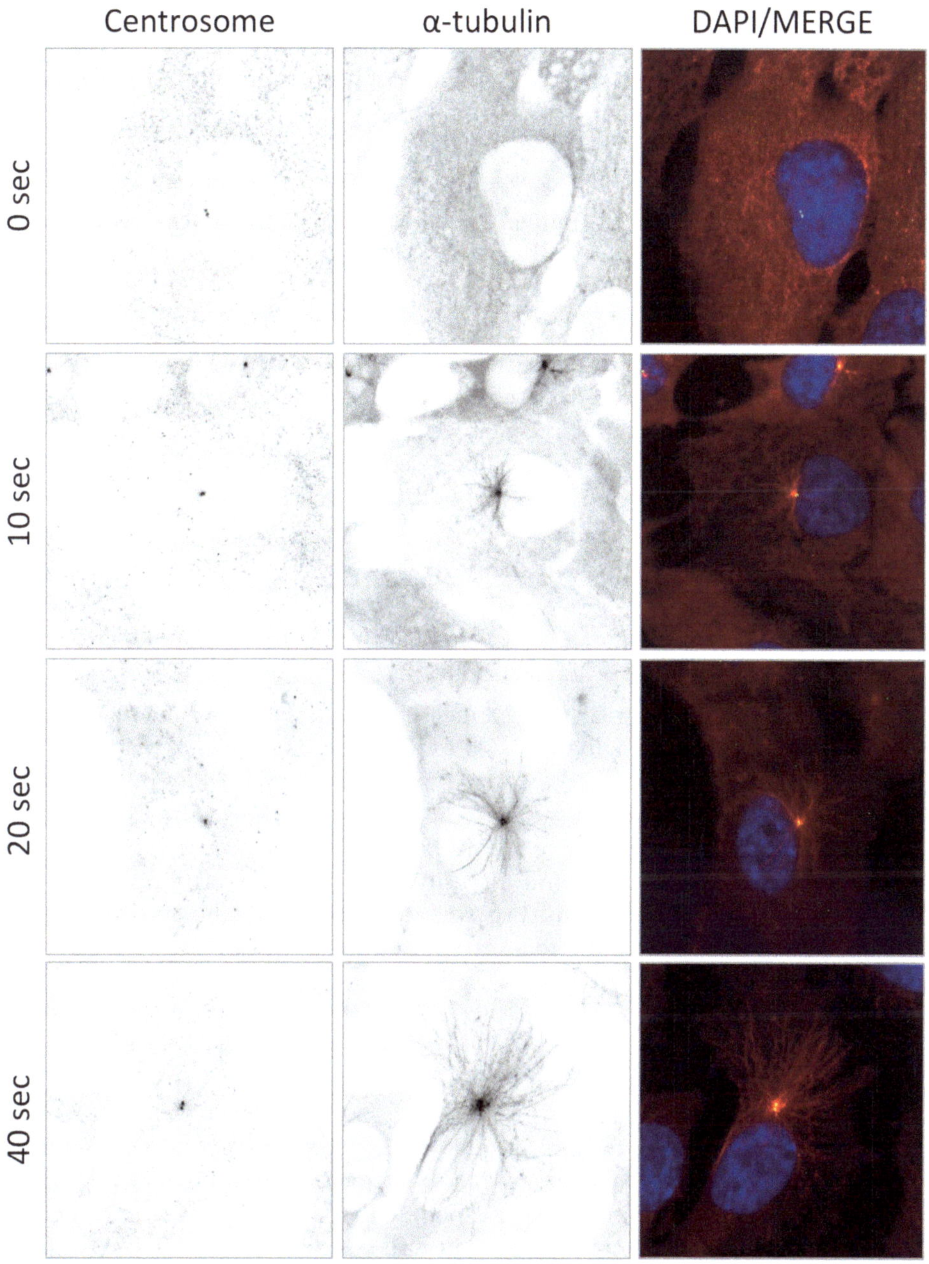

Fig. 1 Centrosome nucleation during interphase. RPE-1 cells were incubated on ice for 30 min and microtubule regrowth was allowed to occur at 37 °C for the indicated time periods. Cells were fixed in methanol, and stained with antibodies labeling centrosome and microtubules. DNA was stained with DAPI

Keep the coverslip submerged, measure the desired incubation time as accurate as possible, and then rapidly transfer the coverslip into a well with fixative. We suggest to start with incubation times between 10 and 20 s to obtain robust but well-isolated centrosomal microtubule asters (*see* Fig. 1 and **Note 7**). Repeat this procedure with the remaining coverslips, one at a time.

8. Fixation in ice-cold methanol should be done for 10 min on ice before processing for immunofluorescence microscopy. Fixation in formaldehyde solution should proceed for 10 min at room temperature.
9. Stain coverslips with any standard immunofluorescence protocol to label centrosomes and microtubules. For microtubules we recommend the mouse monoclonal DM1A antibody, and for centrosomes a PCM marker such as pericentrin may be used (*see* **Note 8**).
10. Mount coverslips on slides and observe with a fluorescence microscope using a 1.4 NA 63× or 100× oil immersion objective. Select cells and focus on the centrosomal areas using the centrosome marker. Image both channels (centrosomes and microtubules). For each condition take at least 20 images of centrosomes from different cells.

3.1.2 Quantification of Centrosomal Microtubule Asters

1. Open the raw image files with the ImageJ software (*see* **Note 9**).
2. Generate a ROI by drawing a circle of 4–6 μm diameter using the oval selection tool.
3. Display the channel showing the centrosome marker and place the circular ROI around the centrosome of individual interphase cells. Switch to the channel showing microtubule staining and measure the intensity of the selected region (*see* **Note 10**). Move the ROI to a suitable nearby region with similar background intensity but without microtubules and measure the intensity as well (*see* **Note 11**). This second value will be used for background subtraction (*see* Fig. 2). All measured values will be displayed in a table generated by ImageJ. Repeat this procedure for all cells/centrosomes to be quantified. Copy and paste the values in an Excel spreadsheet or equivalent, and then for each measured aster correct the intensity by subtracting the corresponding background intensity. Since the absolute staining intensities are variable, in particular between experiments, we suggest to normalize values measured for all centrosome of an experimental condition by the mean value determined for asters in control cells of the same experiment. Normalized intensities should be averaged from at least three different experiments.

3.1.3 Assaying Microtubule Regrowth in Mitotic Cells

Since mitotic cells in a non-synchronized culture are at various different stages of mitosis and thus display variable levels of nucleation activity we recommend synchronization of cells in prometaphase. The remaining procedure is similar to the protocol for interphase cells.

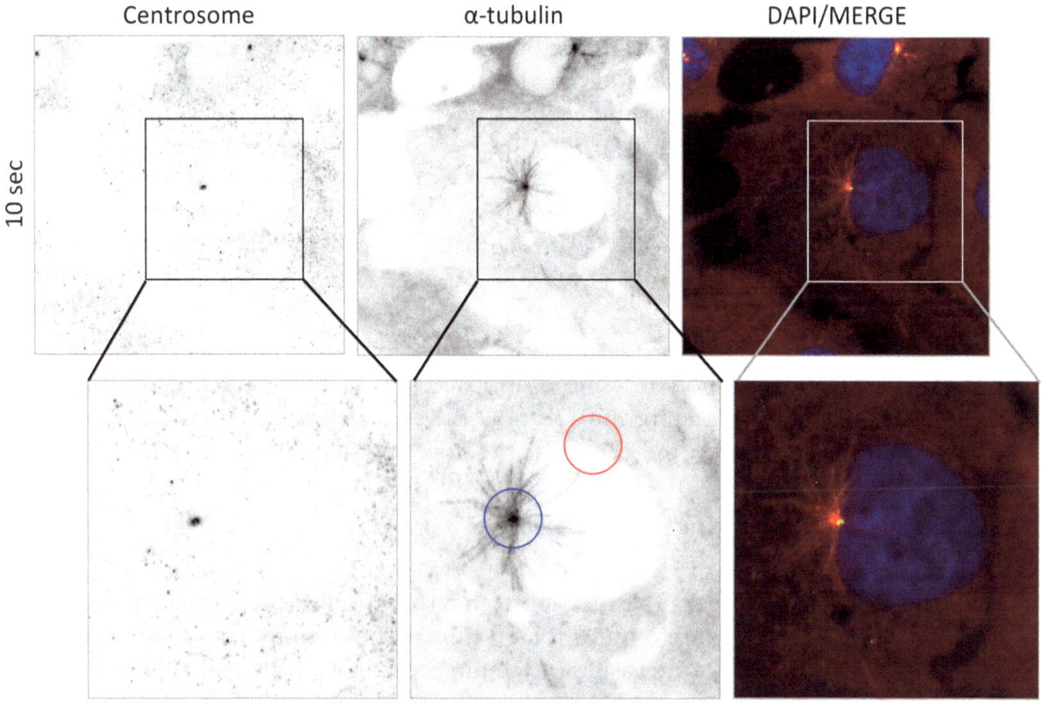

Fig. 2 Quantification of centrosomal nucleation. For the cell of the 10-s time point in Fig. 1, a circular region of interest (ROI) is used to measure the intensity of the centrosomal microtubule aster (blue) and is then moved to a nearby region that does not contain microtubules but displays similar background intensity to measure the background signal (red circle)

1. Split cells on poly-L-lysine-coated coverslips to get a 50% confluence density (*see* **Note 12**).
2. Block cells in S-phase by the addition of 2 mM thymidine and incubation for 20 h.
3. Wash cells 3× in PBS and add fresh medium. Return to incubator.
4. After 5 h add 0.1 μg/mL nocodazole and return to incubator. After 4–5 h, most cells are arrested in prometaphase of mitosis.
5. Wash out the nocodazole with three washes of ice-cold medium, add fresh ice-cold medium, and incubate cells on an ice-water bath for 30 min.
6. Induce microtubule regrowth by submerging the coverslips in warm medium followed by fixation exactly as described in the previous section for interphase cells but extending the regrowth time (*see* **Note 13**).
7. Stain coverslips and image microtubule regrowth in mitotic cells. Note that in the same experiment chromatin-associated nucleation can be analyzed as well. For this stain chromatin with DAPI.

8. For the quantification follow the same procedure as described for interphase cells. For quantifying chromatin-associated nucleation use larger ROIs encircling the multiple foci of microtubules that have formed within the chromatin mass (*see* **Note 14**).

3.2 Microtubule Plus-End Growth Assay in Neurons

During all the procedure avoid maintaining neuron primary cultures outside of the incubator for prolonged periods of time.

3.2.1 Preparing Neurons on Glass-Bottom Dishes

1. Before starting the neuron primary culture, perform an extra poly-D-lysine coating of the 35 mm glass-bottom dishes by adding 350 μL poly-D-lysine at 0.1 mg/mL in borate buffer to the center of the coverslip at the bottom of the dish (*see* **Note 15**). Place the lids on the plates.
2. Incubate for at least 3 h or overnight in the cell culture incubator at 37 °C.
3. Remove glass-bottom dishes from the incubator and wash with sterile water for 10 min at room temperature. Perform three extra washes of 10 min each.
4. Sacrifice the pregnant female mice, remove and sacrifice mouse embryos, and dissect embryonic hippocampus as previously described [9, 19] to obtain single-cell suspensions of neuronal cells.
5. Plate neurons in previously prepared 35 mm glass-bottom dishes at a density of 90–120 cells/mm^2 (*see* **Note 16**). We plate neurons in neuron-plating medium and 2 h later (when cells are attached) medium is changed to neuron-maintaining medium.
6. On the day you wish to transfect cells, prepare transfection medium (*see* **Note 17**). For 10 mL of transfection medium mix 100 μL of (+) D-glucose solution at 200 g/L, 100 μL of Glutamax Supplement 100×, 200 μL of B27 Supplement 50×, and 9.4 mL of neurobasal medium in a 15 mL Falcon tube. Transfecting one 35 mm glass-bottom dish requires 1 mL of transfection medium.
7. Warm up the transfection medium in a water bath at 37 °C for at least 15 min before proceeding to the next step.
8. Remove and save the medium from the 35 mm glass-bottom dishes containing the cultured neurons that you wish to image by time-lapse microscopy. Keep this medium (conditioned medium) in a Falcon tube at 37 °C. This conditioned medium contains factors essential for neuron growth and survival and will be required at the end of the transfection procedure to be added back to the cultured neurons.

9. To each 35 mm glass-bottom dish with cells add 1 mL of transfection medium and return the dishes to the incubator.
10. Prepare two separate reagent mixes. For transfecting one 35 mm glass-bottom dish prepare 250 μL Lipofectamine mix by diluting 2 μL of Lipofectamine 2000 in 250 μL of neurobasal medium and 250 μL DNA mix by mixing 2 μg of EB3-tdtomato plasmid with 250 μL of neurobasal medium. Scale up the amount of each mix depending on the number of dishes that need to be transfected.
11. Incubate both Lipofectamine mix and DNA mix for 5 min at room temperature.
12. Combine DNA and Lipofectamine mixes and mix well. Incubate Lipofectamine:DNA mix at room temperature for 15 min.
13. To each 35 mm glass-bottom dish with transfection medium, add 500 μL of Lipofectamine:DNA mix.
14. Place dishes in the incubator at 37 °C, 5% CO_2, for 90 min. When transfecting neurons that are more mature than 7DIV reduce this incubation period to 30 min (*see* **Note 18**).
15. Replace the transfection medium with 2 mL of the conditioned medium collected in **step 3**.
16. Incubate cells at 37 °C, 5% CO_2, until imaging by time-lapse microscopy (between 24 and 48 h after transfection) (*see* **Note 19**).

3.2.2 Imaging of Microtubule Plus-End Growth in Neurons

In this protocol we focus on imaging in dendrites and axons, which is simpler and more straightforward in terms of imaging and analysis when compared to the soma.

1. Before starting the experiment adjust the environmental control chamber of the microscope to 37 °C and 5% CO_2. Once the temperature and CO_2 levels are correct, proceed to imaging.
2. Start by imaging neuron cultures using transmitted light to find the correct focal plane. Switch to fluorescence imaging and move along the dish until you find a transfected, tdtomato-positive cell (*see* **Note 20**).
3. Once a transfected cell is found, focus on the desired axonal or dendritic region (*see* **Notes 21** and **22**).
4. Set the parameters for imaging z-stacks covering the entire thickness of the region that you want to analyze. For imaging EB3 comets in axons/dendrites acquire image stacks with a step size of 0.2 μm every second for a total of 2.5 min.
5. Once imaging is finished for one region, image another cell compartment (if desired) or look for another transfected cell and image it by repeating **steps 3** and **4**. We recommend to

image and analyze between 10 and 15 cells per condition and to repeat each experiment at least three times.

6. The obtained time-lapse data should be saved in .TIF format for proceeding to data analysis.

3.2.3 Quantification of EB Comet Number

1. In case the acquired images have a bit depth different than 8-bit, convert them to 8-bit and save as a new file (*see* **Note 23**). Open 8-bit time-lapse .TIF files with ImageJ software.
2. Perform a maximum intensity projection of the z-stacks imaged for each time point. For this select in ImageJ: Image>Stacks>Tools>Grouped Z-project. This will open a z-project window. As "Projection Method" select "Max Projection" and in the field "Group size": introduce the number of planes taken per stack in each time point. For example, if you imaged a region with a stack containing three planes (separated by 0.2 μm), insert the value 3. Save the projected time lapse in . TIF format.
3. To obtain the kymographs for each time lapse use the kymograph plug-in tsp050706.txt (*see* **Note 24**).
4. Open the z-projection files generated in **step 2** and draw a straight line on top of the region that you want to analyze (*see* Fig. 3b, g) (you may do so in any of the frames of the time lapse). The line's length should measure at least 20 μm and its thickness should cover the entire axon/dendrite width. Always start drawing this line from the region closest to the soma, to the most distal region. It is essential if you want to also analyze microtubule polarity. Add it to the ROI manager list of ImageJ (Edit>Selection>Add to Manager).
5. Select the line that you have just drawn and run the kymograph plug-in (command: Plugins>Macros>kymograph). A window will open where you should introduce the thickness (in pixels) of the line you drew. This will generate a kymograph like the ones presented in Fig. 3c (axon) and Fig. 3h (proximal dendrite). Each line in the kymograph represents one EB comet and consequently a growing microtubule plus-end (*see* **Note 25**).
6. As an approximate readout for nucleation activity determine the total comet number and calculate the number of comets/s/μm (*see* **Note 26**). For the analysis with ImageJ draw a straight line on top of each EB comet trace present in the kymograph and add it to the ROI manager list. Once you have finished, your ROI manager box will contain information on all the EB comet tracings observed in the selected region during the entire imaging period (Fig. 3e).

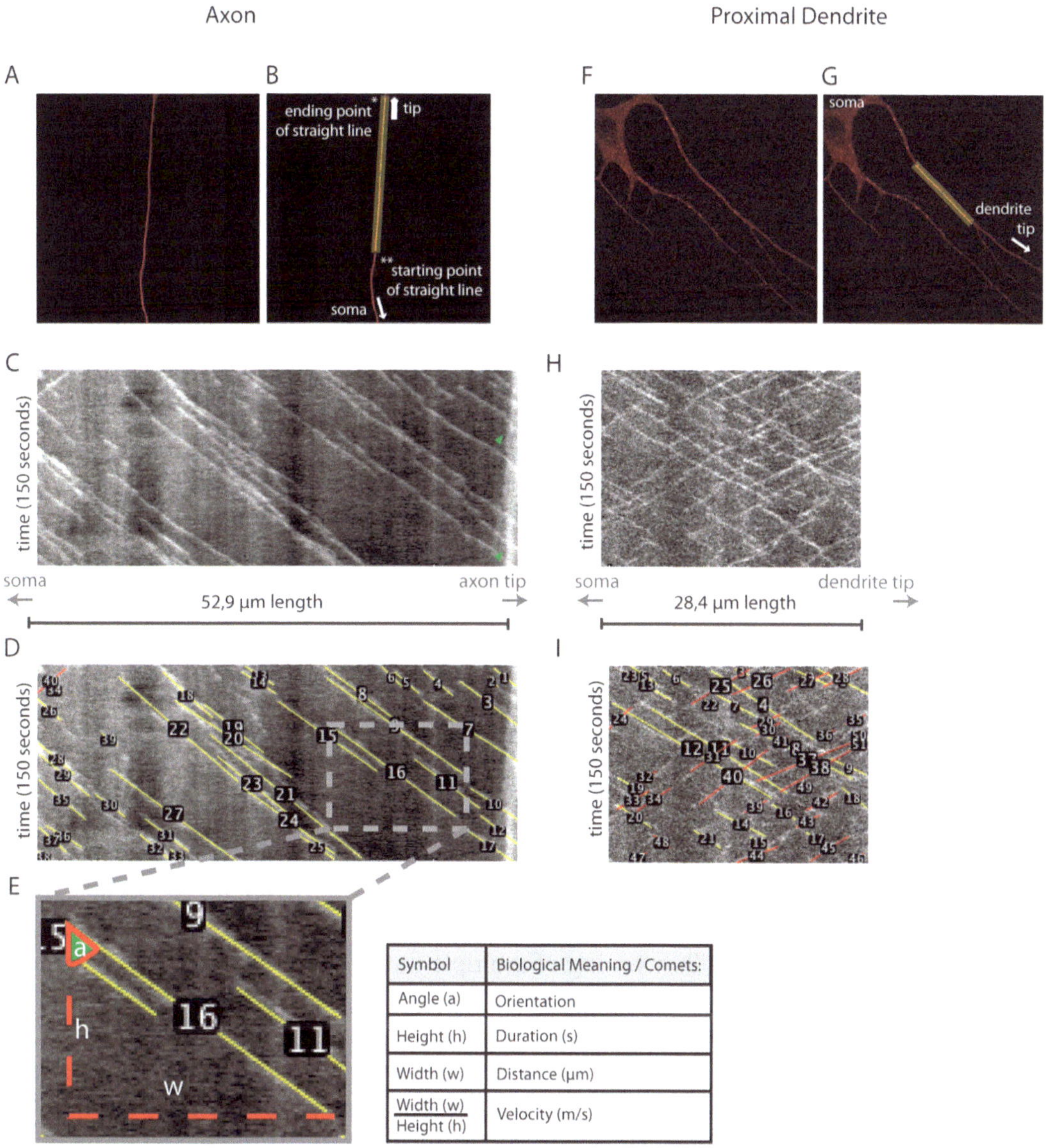

Symbol	Biological Meaning / Comets:
Angle (a)	Orientation
Height (h)	Duration (s)
Width (w)	Distance (µm)
Width (w) / Height (h)	Velocity (m/s)

Fig. 3 Time-lapse z-projections of a series of image stacks showing EB3-tdtomato signals in an axonal region (**a**, **b**) and a proximal dendrite region (**f**, **g**) of hippocampal neurons cultured for 4 days in vitro. In **b** and **g** the yellow straight line represents the region selected to generate the kymographs of the EB3 comets in the axon and proximal dendrite, respectively (**c**, **h**). In **b** it is indicated that the yellow straight line was drawn from the soma-proximal point labeled with two asterisks to the soma-distal point labeled with one asterisk. All kymographs were generated from time-lapse image series with duration of 150 s. (**d**, **i**) Tracings of the comets in axons and proximal dendrites, respectively. One can observe the uniform anterograde microtubule polarity (yellow tracings) that characterizes axons and the mixed polarity of dendrites with anterograde (yellow tracings) and retrograde polarity (red tracings) that characterizes the proximal dendrite. Green arrowheads in *c* point to lines that do not correspond to EB comets. (**e**) Representation of the values obtained for the EB comets tracing measurements, which can be used to obtain further information on microtubule dynamics and polarity

7. To determine the comet density in an imaged region count the total number of tracings in the kymograph (comet number) and divide it by the imaging period (e.g., 150 s as described above) and the length of the region used to generate the kymograph (kymograph width in pixels multiplied by the pixel size). This will generate the comet density in comet #/s/μm.
8. In order to obtain additional information about comet dynamics start by setting up the measurements using the following commands on ImageJ/Fiji: Analyze>Set Measurements and select Area and Bounding Rectangle. Then click on Measure in the ROI Management Box and a new window will open containing additional parameters for the traced EB comets that will allow to derive information about comet run length, duration, and orientation.

4 Notes

1. In our hands U2OS and RPE-1 cells with their flat morphology give good results for interphase nucleation assays; HeLa cells are well suited for mitotic assays. However, other cell types and combinations may be used. Occasionally we observed reduced nucleation activity upon extended passaging. Thus it is recommended to use fresh, low-passage cultures.
2. We recommend to use multiple coverslips per experimental condition, since even with careful handling occasionally one may accidentally lose a single coverslip. Moreover this will allow testing of multiple time points or staining with different combinations of antibodies.
3. The ice-water bath ensures good contact with the bottom of the dishes and thus efficient cooling. Make sure not to trap any air bubbles under the dish. We prefer to use cold-induced depolymerization rather than microtubule-depolymerizing drugs such as nocodazole. We found that even after careful washout of nocodazole, regrowth is significantly slower than after cold-induced depolymerization, indicating the presence of residual drug inside the cells.
4. For antibody staining of centrosomes methanol fixation is recommended. However, depending on the requirements of the antibodies used for the analysis formaldehyde fixation may also be used.
5. Suitable forceps should be identified (forceps with flat or pointed tips both may be used, depending on the experimenter preference) and the secure handling of the small coverslips should be practiced before performing the first assay.

6. The 0 s of regrowth time point is only strictly needed for method establishment. For most cell lines centrosomal microtubules are completely depolymerized after a 30-min incubation on ice. Once this is confirmed there is no need to include this time point in every assay.
7. The optimal timeframe for regrowth needs to be empirically determined. In some cell lines such as RPE-1 microtubules will also be nucleated from Golgi membranes under these conditions, although typically with somewhat slower kinetics. In this case, since the proximity of the Golgi to the centrosome may interfere with the analysis of centrosomal nucleation, we recommend to stop the assay at earlier rather than later time points.
8. Staining centrosomes is important for identifying the region to be used for quantifying nucleation activity, in particular if in a given experimental condition centrosomal nucleation is compromised and no or only weak microtubule asters form.
9. For quantification of fluorescence intensity it is crucial to use raw images without any contrast or level adjustment applied. Export files from your imaging software as .TIF format or use the importer of ImageJ, which allows direct import of files generated by imaging software of major brand microscopes.
10. The ROI should encircle only the most intense region of the microtubule aster, close to the centrosome. Some long microtubules may extend beyond the boundaries of the ROI.
11. Selecting a suitable position for background measurement is important for accurate quantification. For example, the background of the nuclear area may be darker than the background in the cytoplasmic area. Most reliable results are obtained when the same ROI (dimension and shape) is used for both microtubule aster and background measurements.
12. This protocol works well for HeLa cells, but may be adapted to other cell lines.
13. Regrowth may take slightly longer than in the interphase protocol due to the presence of residual nocodazole. Allowing regrowth for 1 min is a good starting point. If regrowth occurs for too long the centrosomal and chromatin-associated microtubules will interconnect, making separate quantification impossible.
14. It is recommended to choose cells in which the two centrosomes are well separated from the mass of the mitotic chromatin.
15. This is a crucial step as, unlike for other cell types, the original poly-D-lysine coating of MatTek 35 mm glass-bottom dishes is not sufficient for neurons to properly adhere to the glass surface.

16. Since the transfection efficiency of cultured primary neurons is considerably low, it is crucial to culture neurons at high cell densities (90–120 cells/mm^2) in order to obtain enough cells for imaging and analysis.
17. Consider that, for the specific purposes of your experiment, you can combine different types of cellular treatments such as [1] addition of drugs to the culture medium, [2] prior infection with virus expressing shRNA, or [3] co-transfection of EB3-tdtomato with other plasmids that you may require.
18. Neurons are very sensitive to transfection reagents and strongly depend on the presence of survival factors (some of which are not present during transfection). Therefore, if you observe high neuronal cell death after transfection, try to reduce the incubation period with the transfection mix.
19. Overexpression of EB3-tdtomato for periods longer than 48 h may cause a high background of EB3-tdtomato fluorescence and identifying EB3 comets may become difficult or impossible. Therefore we recommend imaging neurons from 24 to 48 h after transfection. Alternatively, other plus-end tracking reporters that overcome this limitation may be used [10].
20. EB comet number and polarity may strongly vary depending on the axonal/dendritic region that you are imaging. Try to be consistent when selecting the region for imaging. For example, for axons we recommend to analyze a region of the axon that is close to the soma but at least at 60–70 μm away from it in order to avoid imaging the axon initial segment. For imaging the proximal dendrites select a region ~50 μm away from the soma; for distal dendrites select a region ~50 μm away from the tip. In general one should select regions where the axons/dendrites are as straight as possible. This will make it easier to obtain and interpret the kymographs during the data analysis procedures.
21. Randomly select cells that express intermediate levels of EB3-tdtomato. Very low expression may not generate sufficiently strong signal for the entire period of the time lapse due to photo-bleaching and very high expression will result in too much background signal, making it hard to distinguish EB comets. A good selection criterion is whether EB comets can be clearly observed in the soma.
22. If you image a region where the relative position of the soma is unclear, make a lower magnification snapshot image of the desired region and take notes on the relative position of the soma. This is essential so that, during the analysis procedures, you are able to identify the proximal and distal region of the axon/dendrite.
23. For converting Images with different bit depths to 8-bit use the ImageJ commands: Image>Type>8-bit.

24. After downloading the plug-in (https://www.embl.de/eamnet/html/kymograph.html), install it (Plugins > Install) and it will appear in your Plug-in list.
25. Thick static lines or lines in the kymograph that strongly change direction and polarity are not comets. Do not consider them (Fig. 3c, green arrowheads).
26. It is important to note that this assay simply measures growing microtubule plus-ends and does not distinguish whether these are derived from nucleation events or from rescue/repolymerization events. Therefore it is crucial to complement this assay with other types of analyses before concluding that an experimental condition affects nucleation rather than microtubule dynamics. Similarly, while microtubule polarity may be altered due to changes in nucleation activity, such effects may also be the result of selective stabilization/destabilization of microtubules [20].

Acknowledgments

J.L. acknowledges support by grants BFU2012-33960, BFU2015-69275-P, and PGC2018-099562-B-I00 (Ministerio de Ciencia, Innovación y Universidades, Spain; FEDER). A.E. and R.V. were supported by PhD fellowships BES-2013-066571 and SVP-2014-068770, respectively (Ministerio de Ciencia, Innovación y Universidades, Spain; Fondo Social Europeo).

References

1. Paz J, Lüders J (2018) Microtubule-organizing centers: towards a minimal parts list. Trends Cell Biol 28:176–187
2. Lüders J, Stearns T (2007) Microtubule-organizing centres: a re-evaluation. Nat Rev Mol Cell Biol 8:161–167
3. Prosser SL, Pelletier L (2017) Mitotic spindle assembly in animal cells: a fine balancing act. Nat Rev Mol Cell Biol 18:187–201
4. Sanchez AD, Feldman JL (2017) Microtubule-organizing centers: from the centrosome to non-centrosomal sites. Curr Opin Cell Biol 44:93–101
5. Goshima G, Mayer M, Zhang N et al (2008) Augmin: a protein complex required for centrosome-independent microtubule generation within the spindle. J Cell Biol 181:421–429
6. Lawo S, Bashkurov M, Mullin M et al (2009) HAUS, the 8-subunit human Augmin complex, regulates centrosome and spindle integrity. Curr Biol 19:816–826
7. Petry S, Groen AC, Ishihara K et al (2013) Branching microtubule nucleation in Xenopus egg extracts mediated by augmin and TPX2. Cell 152:768–777
8. Goshima G, Kimura A (2010) New look inside the spindle: microtubule-dependent microtubule generation within the spindle. Curr Opin Cell Biol 22:44–49
9. Sánchez-Huertas C, Freixo F, Viais R et al (2016) Non-centrosomal nucleation mediated by augmin organizes microtubules in post-mitotic neurons and controls axonal microtubule polarity. Nat Commun 7:12187
10. Cunha-Ferreira I, Chazeau A, Buijs RR et al (2018) The HAUS complex is a key regulator of non-centrosomal microtubule organization during neuronal development. Cell Rep 24:791–800

11. Sánchez-Huertas C, Lüders J (2015) The augmin connection in the geometry of microtubule networks. Curr Biol 25:R294–R299
12. Flor-Parra I, Iglesias-Romero AB, Chang F (2018) The XMAP215 ortholog Alp14 promotes microtubule nucleation in fission yeast. Curr Biol 28:1681–1691.e4
13. Borek WE, Groocock LM, Samejima I et al (2015) Mto2 multisite phosphorylation inactivates non-spindle microtubule nucleation complexes during mitosis. Nat Commun 6:7929
14. Ehrhardt DW, Shaw SL (2006) Microtubule dynamics and organization in the plant cortical array. Annu Rev Plant Biol 57:859–875
15. Liu T, Tian J, Wang G et al (2014) Augmin triggers microtubule-dependent microtubule nucleation in interphase plant cells. Curr Biol 24:2708–2713
16. Chan J, Sambade A, Calder G et al (2009) Arabidopsis cortical microtubules are initiated along, as well as branching from, existing microtubules. Plant Cell 21:2298–2306
17. Kirik A, Ehrhardt DW, Kirik V (2012) TONNEAU2/FASS regulates the geometry of microtubule nucleation and cortical array organization in interphase Arabidopsis cells. Plant Cell Online 24:1158–1170
18. Straube A, Merdes A (2007) EB3 regulates microtubule dynamics at the cell cortex and is required for myoblast elongation and fusion. Curr Biol 17:1318–1325
19. Sanchez-Huertas C, Rico B (2011) CREB-dependent regulation of GAD65 transcription by BDNF/TrkB in cortical interneurons. Cereb Cortex 21:777–788
20. Freixo F, Martinez Delgado P, Manso Y et al (2018) NEK7 regulates dendrite morphogenesis in neurons via Eg5-dependent microtubule stabilization. Nat Commun 9:2330

Chapter 12

Detection of Microtubule Nucleation Hotspots at the Golgi

Roslin J. Thoppil, Anna A. W. M. Sanders, and Irina Kaverina

Abstract

Cell polarization is important for multiple physiological processes. In motile cells, microtubules (MTs) are organized as a polarized array, which is to a large extent comprised of Golgi-derived MTs (GDMTs), which asymmetrically extend toward the cell front. We have recently found that GDMT asymmetry is based on a nonrandom positioning of spatially restricted nucleation hotspots, where MTs form in a cooperative manner. Here, we summarize methods used for GDMT identification including microtubule regrowth after complete drug-induced depolymerization and tracking of growing microtubules using fluorescent MT plus-end-tracking proteins (+TIPs) in living cells, and subsequent detection of those GDMTs that originate from the nucleation hotspots. These approaches can be used for quantification of the spatial distribution of MT nucleation events associated with the Golgi or another large structure.

Key words Microtubules, MTOC, Golgi, Nucleation, Hotspots

1 Introduction

The polarity of the MT network is important for directional migration and establishing polarity in a motile cell; both MT distribution and dynamics require asymmetry. To a large extent, MT asymmetry is achieved by MTs that nucleate from the Golgi apparatus [1, 2]. As opposed to radial centrosomal MT arrays, Golgi-derived MTs (GDMTs) comprise a major non-centrosomal asymmetric MT array. Our lab has shown that GDMTs enable directional persistence of cell migration serving as tracks for polarized trafficking and supporting organization of the Golgi itself [1–3]. These features likely underlie essential GDMT functions in neurons and motile and secretory cells [4–8].

While GDMTs bear essential cellular functions, it is still unclear what defines the distribution of MT nucleation sites at the Golgi, and what underlies GDMT asymmetry. We have recently shown that GDMTs are distributed at the Golgi surface in a nonrandom

Roslin J. Thoppil and Anna A. W. M. Sanders contributed equally with all other contributors.

Helder Maiato (ed.), *Cytoskeleton Dynamics: Methods and Protocols*, Methods in Molecular Biology, vol. 2101,
https://doi.org/10.1007/978-1-0716-0219-5_12,

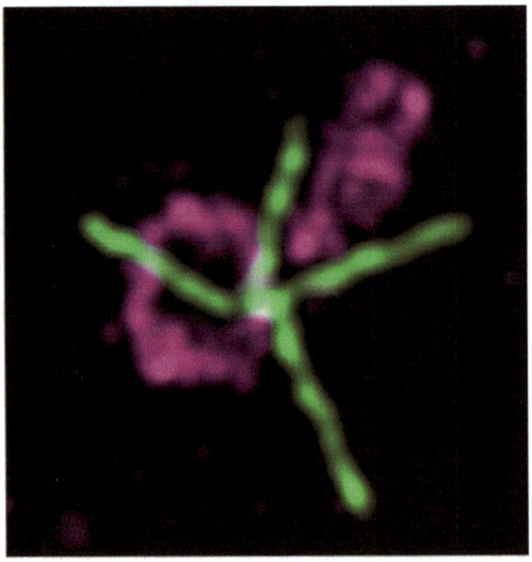

Fig. 1 Example of hotspots with GDMTs in MRC-5 cells 40 s after nocodazole washout. SIM image maximum-intensity projections. Tubulin, green. Giantin, magenta (Reproduced from [9] with permission from [MBoC])

manner. Rather, they emerge from a small number of nucleation hotspots [9] (Figs. 1 and 2a). Interestingly, GDMT array polarity reflects the distribution of the hotspots, highlighting the significance of this specific nucleation location. Hotspots likely encompass clusters of highly concentrated active molecules ensuring efficient MT nucleation at the Golgi. Our results indicate that MTs within the hotspot are nucleated in a cooperative manner, suggesting that recruitment and/or activation of molecular factors governing MT nucleation at the Golgi is locally facilitated by a preexisting GDMT.

MT nucleation from the centrosome has been studied by tracking fluorescently labeled +TIPs such as CLIP170 and EB1 [10, 11]. As +TIPs associate only with newly nucleated MTs and move away from the MT organizing center (MTOC), as a MT polymerizes, the emergence of one +TIP spot from the MTOC is considered as one nucleation event. Using this principle, we analyze MT nucleation at the Golgi by tracking Emerald-EB3-positive comets in interphase RPE1 or MRC-5 cells at steady state. A complementary approach is the complete disassembly of the existing MTs by nocodazole and observing the initiation of MT nucleation from the resulting fragments of the Golgi after nocodazole washout. In undifferentiated RPE1 cells, these approaches have previously allowed us to identify the Golgi apparatus as an alternative MTOC [4, 9]. Here, we use similar approaches to identify and further describe how to quantify microtubule nucleation hotspots at the Golgi.

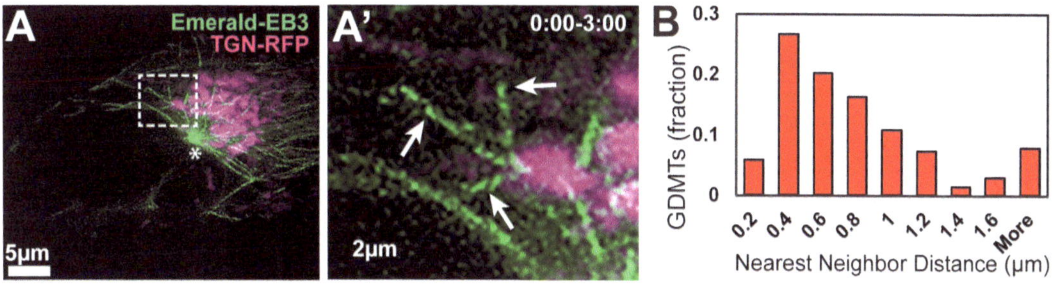

Fig. 2 Microtubules are nucleated at specific sites within the Golgi ribbon. (**a**) An RPE1 cell expressing Emerald-EB3 (green, MT +TIP marker) and TGN-RFP (magenta, Golgi marker). A maximum-intensity projection of a confocal spinning disk microscopy sequence over a 3-min period and *Z* thickness 3 μm is shown (**a**, **a′**). Inset in **a** is enlarged in **a′**, showing newly formed GDMTs formed at the same site (arrows); asterisk indicates centrosome position. (**b**) Quantification of nearest-neighbor distance between GDMT nucleation sites, based on 3D live-time imaging (Reproduced from [9] with permission from [MBoC])

2 Materials

2.1 Cell Culture and Transfection

1. RPE1 (Clontech) and MRC-5 (American Type Culture Collection) cells were maintained in DMEM/F12 with 10% FBS and MEM with 10% FBS (Lonza) media, respectively, and grown at 37 °C with 5% CO_2.
2. 0.25% Trypsin solution with EDTA, without sodium bicarbonate, calcium, and magnesium.
3. Ectopic expression of fluorescent markers achieved via transfection using Fugene 6 (Roche), or nucleofection (Amaxa).
4. Plasmids: As MT +TIP markers, Emerald-EB3 (kind gift from M. Davidson; Addgene plasmid # 54076) or another fluorescently tagged +TIP marker can be used. GFP-centrin (kind gift of Dr. M. Bornens, Paris; Addgene plasmid# 72641) can be used as a centrosome marker. To mark the Golgi apparatus, pmCherry-N1-GalT (kind gift from Lei Lu, Addgene plasmid# 83727), TGN-RFP [12], or GM130-GFP (kind gift from C. Sütterlin, University of California, Irvine [13]) can be used.

2.2 Live-Cell Imaging and Immunofluorescence

1. Cell culture dishes for live-cell imaging: 35 mm Glass-bottom dish (e.g., MatTek Corporation) with No. 1.5 coverslips.
2. Human fibronectin is dissolved in 1 M urea up to 2.27 μM (1 mg/mL) stock concentration.
3. Phosphate-buffered saline (PBS): Prepare and autoclave 10× stock solution: 1.37 M NaCl, 27 mM KCl, 80 mM Na_2HPO_4, and 20 mM KH_2PO_4, store at room temperature. The stock solution is diluted into 1× working solution before use.
4. Fixative: Methanol stored at −20 °C in the presence of 3 Å molecular sieves (Sigma) to remove water.

5. Nocodazole was prepared at 16.6 mM in DMSO (cell culture grade) and stored at −20 °C.
6. Extraction buffer: 60 mM PIPES, 25 mM HEPES, 10 mM EGTA, 2 mM $MgCl_2$, 0.1% saponin, adjust pH to 6.9. Store at 4 °C [14].
7. Blocking and antibody dilution buffer: 1% Bovine serum albumin (BSA)/PBS supplemented with 5% donor horse serum (DHS).
8. Primary antibodies: Mouse anti-α-tubulin (1:1000, DM1A; Sigma-Aldrich), mouse anti-GM130 (1:300; BD Transduction Laboratories), rabbit anti-Giantin (1:1000; Abcam), rabbit anti-β-tubulin (1:1000, Abcam), sheep anti-TGN46 (1:5000; AbD Serotec; now BioRad).
9. Secondary antibodies: Alexa488-, Alexa568-, and Alexa647-conjugated highly cross-absorbed secondary antibodies (Invitrogen), CF405M-conjugated highly cross-absorbed secondary antibodies (Biotium).
10. Coverslips are mounted in Vectashield Mounting Medium (Vector Labs) and can be stored at −20 °C.

2.3 Confocal Microscopy

1. Laser scanning confocal microscope: Nikon A1r with advanced photokinetic capacity, based on a TiE Motorized Inverted Microscope with 405, 488, 568, and 647 nm excitation lasers; ×20, ×40, ×60, and ×100 lenses; and DIC module, equipped with a high-speed resonant scanner (30 fps), run by NIS Elements C software.
2. A 4-color spinning disk Yokogawa CSU-X1 confocal based on a Nikon Eclipse Ti-E inverted microscope with 405, 488, 568, and 647 nm excitation lasers, Photometrics 95B Prime Back-illuminated sCMOS camera for high-resolution fast multicolor 3D live-cell imaging, run by NIS Elements C software. Apochromat lens ×100 NA 1.49 was used.
3. Tokai Hit Incubation Package 1 STXG incubation system for small dishes; 35 mm, 50/60 mm, chamber slide, chambered cover glass and slide only for Ti/Ti2 Piezo Z stage. This system includes all dish holder and dish fixing as standard. It includes temperature controller with built-in digital gas mixer for 100% CO_2 use.
4. Software packages such as Nikon Elements Software, Imaris (Bitplane), and Fiji (NIH) can be used for image analysis.

2.4 Super-Resolution Microscopy

For quantitative analysis of hotspots, structural illumination microscopy should be used. For example, images can be acquired using either DeltaVision OMX (GE Technology) or N-SIM (Nikon) systems.

1. DeltaVision OMX with 405, 488, 568, and 642 nm lasers, a ×60 Plan-Apo NA1.42 lens, and a CMOS camera.
2. Nikon N-SIM with 405, 488, 561, and 647 nm lasers, an SR Apo TIRF ×100 NA1.49 lens, and an EMCCD camera (DU-987, Andor Technology).

2.5 Calculation of Distance Between GDMT Minus Ends

GDMT minus-end distances can be calculated using Excel, Matlab, or any other analysis software. Distance D between GDMT minus ends is calculated in 3D by using the distance formula:

$$D_{(a,b)} = \sqrt{\left((X_a - X_b)^2 + (Y_a - Y_b)^2 + (Z_a - Z_b)^2\right)}$$

3 Methods

GDMT nucleation hotspots are defined as spheres with a diameter of 0.4 μm containing both Golgi membranes and multiple ($n > 1$) MT minus ends. For hotspot analysis, it is necessary to determine positions of the minus ends of emerging MTs and the Golgi surface.

Three methods can be used to identify GDMT nucleation sites; all methods are briefly described below; for detailed methods refer to [15].

The first method involves transfection of cells with a +TIP marker and a Golgi marker followed by live-cell imaging to record MT tip movement in steady state (Fig. 2a). This method allows identification of nucleation sites in cells with unchanged MT organization. Imaging those cells in the appropriate cell cycle stage where the Golgi and centrosome are physically separated will help to clearly distinguish between centrosomal MTs and GDMTs and further accurately measure GDMT hotspot nucleation.

An alternative and more definitive method involves monitoring MT re-polymerization after nocodazole washout. When MTs are completely depolymerized upon nocodazole treatment, the Golgi subsequently fragments into ministacks which are distributed all over the cell. Imaging cells shortly after nocodazole washout can distinctly identify those MTs which are nucleated from the Golgi. The second method thus involves live imaging of cells expressing a +TIP marker and a Golgi marker in the course of a nocodazole washout (Fig. 3a, b).

For the purpose of statistical evaluation of MT number organized at the Golgi, a suggested third method involves nocodazole washout experiments in non-transfected cells with fixation at specific time points and immunostaining to visualize MTs and the Golgi.

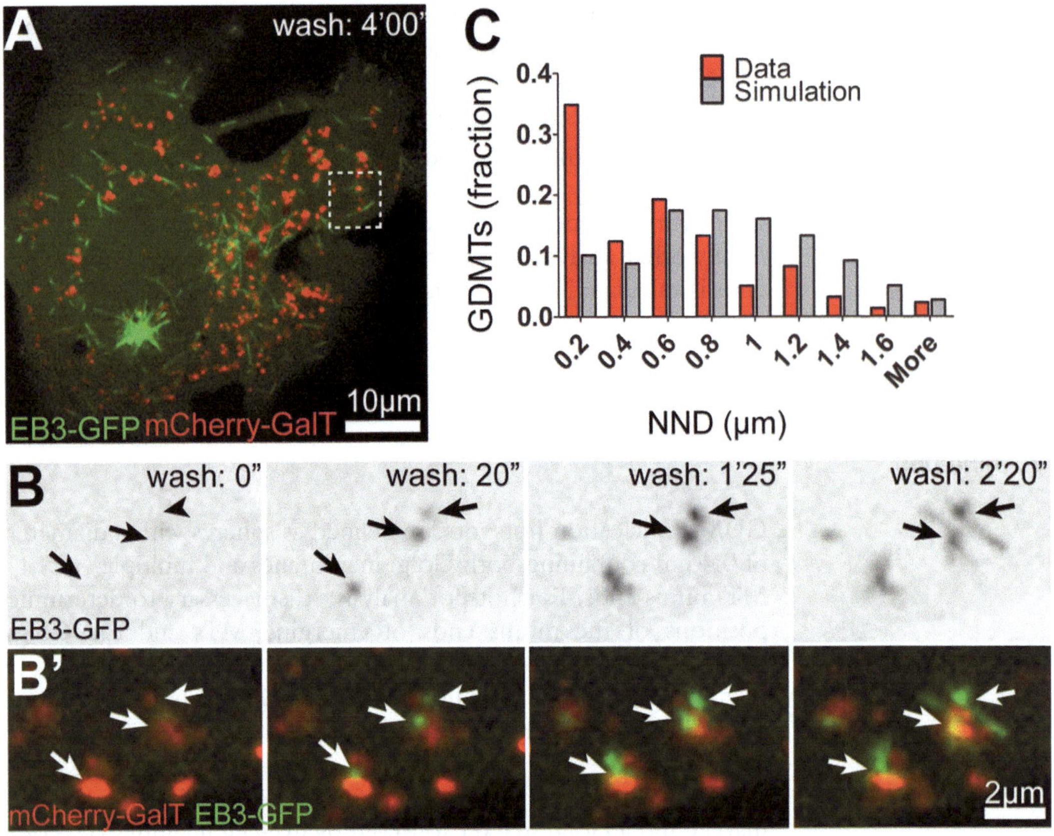

Fig. 3 GDMT nucleation is spatially restricted to distinct hotspots. (**a**) RPE1 cell expressing EB3-GFP (green) and mCherry-GalT (red, Golgi marker) 4 min after nocodazole washout. Single-plane confocal spinning disk microscopy. (**b**, **b′**) Example of simultaneous multiple GDMT nucleation events (arrows) at Golgi fragments following nocodazole washout. Frames from a time-lapse image sequence. (**b**) EB3-GFP, inverted grayscale image. (**b′**) EB3-GFP (green) and mCherry-GalT (red, Golgi marker). Time from the start of the movie, minutes: seconds. (**c**) Newly formed GDMTs are distributed nonrandomly on Golgi fragments following nocodazole washout. Nearest-neighbor distances of GDMT minus ends were calculated for each fragment associated with multiple GDMTs (Reproduced from [9] with permission from [MBoC])

3.1 Live-Cell Imaging of MT Nucleation in Steady State

1. Cell maintenance and preparation for transfection: RPE1 and MRC-5 cells are transfected on 35 mm tissue culture dishes at 80% confluency.
2. Transfection: Cells are transfected overnight using Fugene 6 or nucleofection according to the manufacturer's instructions. For live-cell imaging of steady-state MT dynamics, cells can be transfected with Emerald-EB3 and TGN-RFP (RPE1) or EB3-mCherry and GTN-GFP (MRC-5) (*see* **Note 1**). Post-transfection, cells are plated and grown on 35 mm glass-bottom dishes that have been coated with fibronectin (1:100 dilutions in PBS, 30-min incubation at room temperature) and used for imaging.

3. Image stacks with 0.3 μm between slices can be recorded to cover the entire Golgi (over 3–4 μm) for 5 min with maximal frame rate (~4–5 s per stack) using the spinning disk confocal microscope (Fig. 2a).

3.2 Live-Cell Imaging of MT Nucleation After Nocodazole Washout

1. Cells are maintained and transfected as described in the above procedure (*see* **steps 1** and **2** in Subheading 3.1).
2. Nocodazole treatment: To disassemble MTs, the nocodazole stock is diluted with complete medium at 1:2000 dilution, added to the cells, and incubated at 37 °C for 2 h.
3. Nocodazole washout: Cells incubated with nocodazole are directly placed on the microscope stage for imaging. Cells are briefly imaged prior to nocodazole washout to obtain accurate pre-washout images (evidence of whole MT network disruption).
4. Cells are washed at least ten times with ice-cold medium at the stage, to remove all traces of nocodazole (*see* **Note 2**).
5. For live-cell imaging of nocodazole washouts, image stacks with 5 s intervals are taken using the spinning disk confocal microscope (Fig. 3a, b) for at least 10 min.

3.3 Immunofluorescence Analysis of MT Nucleation After Nocodazole Washout

1. Cells are maintained as mentioned previously (*see* **step 1** in Subheading 3.1).
2. Cells are plated and grown on fibronectin-coated coverslips (*see* above) (*see* **Note 3**).
3. Nocodazole treatment: Cells are treated with nocodazole as previously described (*see* **step 2**, Subheading 3.2).
4. Nocodazole washout for immunofluorescence: Two 60 mm tissue culture dishes containing 5 mL complete medium and extraction buffer are placed on a floating raft in a 37 °C water bath. To remove all traces of nocodazole while preventing MT repolymerization during washing, nocodazole-treated cells are placed directly on ice and the cells are washed with ice-cold plain medium for minimum ten times. Coverslips are quickly immersed in the 60 mm dish containing complete medium with gentle swirling of the dish for either 30, 45, or 60 s to allow MT regrowth. The coverslips are then immersed into extraction buffer for 15 s with gentle swirling to extract free tubulin from the cytoplasm. Cells are then subjected to fixation.
5. Cells are fixed with ice-cold 100% methanol at −20 °C for 10 min; coverslips are triple washed with PBS and allowed to rehydrate prior to blocking for at least 1 h at room temperature or overnight at 4 °C.
6. Cells are blocked with DHS/BSA for 1 h at room temperature.

7. Post-blocking, cells are incubated with primary antibody diluted in blocking buffer solution overnight at 4 °C.
8. Post-incubation with primary antibody, cells are triple washed with 1× PBS for 5 min, and incubated in secondary antibody diluted in blocking buffer solution for 1 h at room temperature.
9. Post-washing with PBS, coverslips are mounted on slides using Vectashield, sealed with nail polish, and dried and are then ready to be imaged. To retain fluorescence signal long-term, slides can be stored at −20 °C.
10. Fixed preparations are imaged by SIM microscopy for the best resolution outcome.

3.4 Detection of MT Nucleation-Site Positions

3.4.1 Steady-State Live Imaging

1. ImageJ (NIH), Imaris (Bitplane), or any other appropriate software package can be used for processing. Image stack sequences covering the entire Golgi during 3 min are used. To subtract background, time-lapse images are processed by rolling ball background correction. Brightness, contrast, and gamma settings are then adjusted to make both the major and minor structures visible.
2. All MT plus tip tracks are determined in the Z- and T-dimensions. Newly formed Emerald-EB3 comets (growing MT plus ends) in 3D can be tracked using the ImageJ plug-in MtrackJ [16] or Imaris particle tracking. To quantify only newly nucleated MTs throughout the movie, tracks starting in the bottom or top slices of a stack or in the first time frame are excluded. Similarly, all the plus tip signals are tracked through Z-planes to determine the true origin. The plus tip signals existing in the upper or lowest plane of the stack are not considered as the sites of origin because those MTs might be crossing the stack in Z-direction. All the other tracks initiated during the imaging have to be tracked back to the sites of origin.
3. The starting points of MT tracks are colocalized with the Golgi marker and roughly assigned as GDMTs or non-GDMTs.
4. A short (25 s–3 min) video starting at the first detection time point of each emerging GDMT is used to produce a maximum-intensity projection of GDMT.

3.4.2 Live Imaging of Nocodazole Washouts

1. In nocodazole washout video sequences, background subtraction and manual 3D tracking of emerging MTs can be utilized as in steady-state experiment above, and the putative GDMT subset identified.
2. A short (2–3 min) video starting at MT regrowth onset is used to produce maximum-intensity projections of GDMTs (Fig. 3a, b).

3.4.3 Fixed Immunostained Preparations After Short-Term Nocodazole Washout

3D stacks acquired by SIM imaging are used for fixed-cell nocodazole washout analysis. For approximate detection of GDMT subset, MT ends are manually colocalized with the Golgi.

3.4.4 Precise Detection of GDMT Minus-End Position

1. Z-stacks produced via time projections of live imaging (*see* above) or fixed cells are used.
2. In each Z-plane where GDMT minus end is detectable, a straight-line scan plot is made using ImageJ. The line scans are used to find the intensity maxima at the MT minus end (the PSF center) and compared between Z-planes to find the outmost minus-end position in Z. Resulting X, Y, and Z coordinates of each GDMT minus end are recorded for further analysis.

3.5 Analysis of the Golgi Surface and Golgi Fragment Size

The Golgi is detected by a fluorescently labeled or immunostained Golgi marker (*see* above) (*see* **Note 4**).

1. The Golgi surface can be detected using the surface mapping function in Imaris (Bitplane).
2. In nocodazole washout experiments, the Golgi fragment size can be determined using ImageJ (Fiji by Ferret's diameter (*XY*) and minimum and maximum range (*Z*), resulting in center of mass coordinates (*XYZ*) and the radius of the fragment, which can then be used to determine the Golgi surface coordinates) (*see* **Note 5**).

3.6 Detection of the Analyzed Golgi Surface and Golgi Fragment Size

GDMT hotspots are quantified to contain the Golgi surface and at least two MT minus ends within 0.4 μm to each other.

1. The distance between each GDMT minus end and the Golgi surface is calculated using nearest neighbor algorithm in Matlab. Only MTs that are with 0.4 μm of the Golgi marker are considered true GDMTs.
2. The distances between all true GDMT minus ends are calculated. This can be performed in Excel or Matlab or using any other suitable software. A histogram is generated to evaluate the distribution of distances, and those within 0.4 μm to each other are demarcated as hotspot GDMTs (Figs. 2b and 3c).
3. Histograms are compared to a histogram of randomly distributed points within 0.4 μm of the Golgi fragment surface (Fig. 3c).

4 Notes

1. It is critical to select cells with appropriate expression levels, e.g., when overexpressed, +TIPs decorate the entire MT network which can affect MT dynamics. Conversely, weak fluorescence signal requires high laser intensity and/or long exposure time, which can induce photodamage to the cells.
2. Use a pump for efficient aspiration (avoid touching the chamber with the pipet tip!). Alternatively, a large-volume flow chamber can be used. Image recording is resumed immediately after washing.
3. For each experimental condition, consider using a minimum of three coverslips.
4. To detect the Golgi surface in fixed preparations, it is recommended to use Giantin as a Golgi marker. Giantin is a transmembrane protein (recognizes aa 1–469, cytoplasmic side) which presents a "defined" Golgi signal lending itself well to surface mapping.
5. For analysis, the entire image file can be used; however cropping images is helpful to keep track of Golgi fragments throughout analysis process.

Acknowledgments

This work was supported by a National Institutes of Health (NIH) grant 5T32 CA 119925-10 and American Heart Association (AHA) POST33990271 (to R.J.T.), AHA 17POST32650000 (to A.A.W.M.S.), NIH R01GM078373, NIH R35-GM127098, and NIH R01-DK106228 (to I.K.).

References

1. Vinogradova T, Miller PM, Kaverina I (2009) Microtubule network asymmetry in motile cells: role of Golgi-derived array. Cell Cycle 8 (14):2168–2174. https://doi.org/10.4161/cc.8.14.9074
2. Miller PM, Folkmann AW, Maia AR, Efimova N, Efimov A, Kaverina I (2009) Golgi-derived CLASP-dependent microtubules control Golgi organization and polarized trafficking in motile cells. Nat Cell Biol 11 (9):1069–1080. https://doi.org/10.1038/ncb1920
3. Vinogradova T, Paul R, Grimaldi AD, Loncarek J, Miller PM, Yampolsky D, Magidson V, Khodjakov A, Mogilner A, Kaverina I (2012) Concerted effort of centrosomal and Golgi-derived microtubules is required for proper Golgi complex assembly but not for maintenance. Mol Biol Cell 23(5):820–833. https://doi.org/10.1091/mbc.E11-06-0550
4. Efimov A, Kharitonov A, Efimova N, Loncarek J, Miller PM, Andreyeva N, Gleeson P, Galjart N, Maia AR, McLeod IX, Yates JR 3rd, Maiato H, Khodjakov A, Akhmanova A, Kaverina I (2007) Asymmetric CLASP-dependent nucleation of noncentrosomal microtubules at the trans-Golgi network. Dev Cell 12(6):917–930. https://doi.org/10.1016/j.devcel.2007.04.002
5. Yalgin C, Ebrahimi S, Delandre C, Yoong LF, Akimoto S, Tran H, Amikura R, Spokony R, Torben-Nielsen B, White KP, Moore AW

(2015) Centrosomin represses dendrite branching by orienting microtubule nucleation. Nat Neurosci 18(10):1437–1445. https://doi.org/10.1038/nn.4099

6. Zhu X, Hu R, Brissova M, Stein RW, Powers AC, Gu G, Kaverina I (2015) Microtubules negatively regulate insulin secretion in pancreatic beta cells. Dev Cell 34(6):656–668. https://doi.org/10.1016/j.devcel.2015.08.020
7. Ori-McKenney KM, Jan LY, Jan YN (2012) Golgi outposts shape dendrite morphology by functioning as sites of acentrosomal microtubule nucleation in neurons. Neuron 76(5):921–930. https://doi.org/10.1016/j.neuron.2012.10.008
8. Rivero S, Cardenas J, Bornens M, Rios RM (2009) Microtubule nucleation at the cis-side of the Golgi apparatus requires AKAP450 and GM130. EMBO J 28(8):1016–1028. https://doi.org/10.1038/emboj.2009.47
9. Sanders A, Chang K, Zhu X, Thoppil RJ, Holmes WR, Kaverina I (2017) Nonrandom gamma-TuNA-dependent spatial pattern of microtubule nucleation at the Golgi. Mol Biol Cell 28(23):3181–3192. https://doi.org/10.1091/mbc.E17-06-0425
10. Piehl M, Tulu US, Wadsworth P, Cassimeris L (2004) Centrosome maturation: measurement of microtubule nucleation throughout the cell cycle by using GFP-tagged EB1. Proc Natl Acad Sci U S A 101(6):1584–1588. https://doi.org/10.1073/pnas.0308205100
11. Komarova YA, Vorobjev IA, Borisy GG (2002) Life cycle of MTs: persistent growth in the cell interior, asymmetric transition frequencies and effects of the cell boundary. J Cell Sci 115(Pt 17):3527–3539
12. Deora AA, Diaz F, Schreiner R, Rodriguez-Boulan E (2007) Efficient electroporation of DNA and protein into confluent and differentiated epithelial cells in culture. Traffic 8(10):1304–1312. https://doi.org/10.1111/j.1600-0854.2007.00617.x
13. Kodani A, Sutterlin C (2008) The Golgi protein GM130 regulates centrosome morphology and function. Mol Biol Cell 19(2):745–753. https://doi.org/10.1091/mbc.e07-08-0847
14. Zhai Y, Borisy GG (1994) Quantitative determination of the proportion of microtubule polymer present during the mitosis-interphase transition. J Cell Sci 107(Pt 4):881–890
15. Zhu X, Kaverina I (2011) Quantification of asymmetric microtubule nucleation at subcellular structures. Methods Mol Biol 777:235–244. https://doi.org/10.1007/978-1-61779-252-6_17
16. Meijering E, Dzyubachyk O, Smal I (2012) Methods for cell and particle tracking. Methods Enzymol 504:183–200. https://doi.org/10.1016/B978-0-12-391857-4.00009-4

Chapter 13

Ultrastructural Analysis of Microtubule Ends

J. Richard McIntosh, Eileen O'Toole, Cynthia Page, and Garry Morgan

Abstract

Microtubules can be detected in light microscopes, but the limited resolution of these instruments means that the polymers appear as lines whose width is defined by the diffraction of light. Much important work on microtubule dynamics has been accomplished by light microscopy, but the details of microtubule end structure are not accessible in such studies. Slight variations in fluorescence intensity, etc. have been used to comment on the structure of dynamic ends, and the combination of light microscopy with laser tweezers has provided insight into aspects of microtubule elongation. However, for views that reveal structural details of the pathways for microtubule growth and shortening, electron microscopy has been of great value. Here, we describe methods for using electron microscopes to look at the ends of microtubules as they grow and shrink, both in vivo and in vitro. The key problems to be overcome for ultrastructural study of microtubule dynamics are those of reliable sample preparation. Dynamic microtubules are labile and can therefore be modified by preparative methods. Our chapter follows the premise that rapid freezing, which converts sample water into vitreous ice, is the best approach for sample preparation. Therefore, all of the methods described involve finding optimal conditions for sample vitrification, and then getting the frozen sample into a form suitable for electron microscopy. We also posit that the end of a microtubule must be considered in three dimensions, so we employ electron tomography as a way to get the necessary information. The methods described for the study of microtubules in cells employ rapid freezing, freeze-substitution fixation, plastic embedding, serial sectioning, and tomography of stained samples. The methods for following microtubule growth in vitro employ sample preparation on holy grids, blotting, and plunge-freezing, followed by electron cryo-tomography. Quantification of structure from both approaches is accomplished by segmentation and analysis of graphic models.

Key words Microtubule, Protofilament, Polymerization, Depolymerization, Electron microscopy, Electron tomography, Electron cryo-tomography, Three-dimensional, Frozen-hydrated, Freeze-substitution

1 Introduction

Electron microscopy (EM) requires that hydrated samples be prepared to withstand the vacuum of the microscope column, ~10^{-7} torr ($\approx 10^{-7}$ mmHg, 1.33×10^{-5} Pa, or 1.33×10^{-7} mbar). When water is cooled to the temperature of liquid nitrogen (-196 °C or ~77 K), its vapor pressure is so low that its rate of sublimation is negligible. Therefore, the most direct method for

Helder Maiato (ed.), *Cytoskeleton Dynamics: Methods and Protocols*, Methods in Molecular Biology, vol. 2101, https://doi.org/10.1007/978-1-0716-0219-5_13, © Springer Science+Business Media, LLC, part of Springer Nature 2020

EM sample preparation is rapid freezing to approximately this temperature. However, when water is cooled, it tends to crystallize, and the formation of crystals damages biological structure, so freezing must be accomplished without crystallization. If a sample is very thin, cooling without crystallization can be accomplished by plunging the sample into a very cold liquid that is far from its temperature of vaporization, e.g., liquid ethane cooled to its freezing point (−183 °C) by liquid nitrogen. However, samples thicker than a few micrometers will not freeze fast enough with this method to avoid crystallization, so one must do something to inhibit ice crystal formation. The most effective way to do this for biological samples is by applying a very high hydrostatic pressure immediately before freezing [1]. Thus, samples of isolated, dynamic microtubules (MTs) in vitro can be made thin enough for good freezing by blotting away excess liquid before plunging the sample into liquid ethane [2, 3], but most cells must be frozen at high pressure to avoid ice crystal damage [4]. Subheading 3 treats first the preparation of MTs assembled in vitro, and then the preparation of cells for optimal fixation of dynamic MTs in vivo. In each case, the issue of possible sample alteration by the preparative method used is addressed.

Most eukaryotic cells contain dynamic MTs, but one cannot usually learn from a static picture whether a given MT was growing or shrinking at the time it was frozen. If one wishes MTs of only one dynamic state, it is common to apply an experimental perturbation, so the state of all MTs in the preparation can be known [5]. Alternatively, some biological processes are sufficiently well described and stereotyped that one can know the state of the MTs under view by inspection, e.g., kinetochore MTs during anaphase A or interpolar MTs during anaphase B. When the MT's state is known, the MT ends can then be examined in preparations that preserve the desired biology.

A single electron micrograph of a thin section of dynamic MTs is usually insufficient to characterize a MT end because sections are commonly >40 nm thick, and the diameter of a MT is ~25 nm. Thus, even an image of a perfectly positioned section will only show the MT end in a single projected view; the individual ~13 protofilaments (PFs) of the MT cannot then be distinguished with clarity. Electron tomography (ET) improves on single projection imaging by providing three-dimensional (3D) information with roughly isotropic resolution [6–8]. The details of this method are a big subject that will not be treated in this chapter, but can be found in several references [9, 10].

1.1 Studying MT Ends In Vitro

Samples of MTs polymerizing in vitro are relatively easy to prepare, given that one can either make [11, 12] or buy (e.g., Cytoskeleton, Denver, CO) mammalian brain tubulin for biochemical study. Note, however, that brain tubulin is a complex mixture of the products of multiple tubulin genes, and each of these tubulins is subject to several kinds of posttranslational modifications. These variations may have an impact on the details of tubulin polymerization. Whatever tubulin one uses, however, the study of its polymerization starts with the making of a solution in which tubulin polymerization will occur, placing a droplet of the solution on an appropriate EM sample grid, blotting it to remove excess liquid, and freezing it by plunging into a very cold liquid. There are, however, subtleties of tubulin polymerization that suggest setting up samples in better defined ways. The initiation of MT growth is slower than MT elongation, and it requires a higher concentration of soluble tubulin [13, 14]. If one uses a sufficient concentration of protein, polymerization will initiate and MTs will form, but the processes of dynamic instability [15] mean that the population of polymer will soon include shrinking as well as growing MTs, a situation that can lead to confusion. Moreover, the two ends of a MT polymerize and depolymerize at different rates, faster at the "plus" end [13]. With high-quality imaging of frozen-hydrated MTs, one can identify the polarity of each MT [16], so this ambiguity can be resolved, but an alternative that simplifies subsequent image study is to use a structure that will nucleate tubulin polymerization in a defined orientation. Preexisting MTs are excellent nucleators of tubulin polymerization, though one must be mindful of the fact that the structure of the "seed" influences the structure of the polymer that grows. Many people use short segments of MTs made by polymerizing tubulin in the nucleotide analogue, GMPCPP, which renders the polymers quite stable. These polymers are then fragmented by shear and used as seeds (*see* several protocols at https://hymanlab.mpi-cbg.de/hyman_lab/wp-content/uploads/2012/08/Tubulin-Protocols-Mitchison.pdf). This approach is well established, but if the number of PFs in the resulting MTs is of concern, it may not be the best approach. Most of the seeds will contain 12, 14, or 15 PFs, which may affect the behavior and shape of the growing MT end; the 13 PF polymers that seem to predominate in vivo have a seam, whereas other PF numbers generally do not [17]. In this chapter we describe a method that uses flagellar axonemes to provide seeds for MT nucleation. This results in the polymerization of mostly 13 PF polymers and provides a useful way to identify MT plus ends.

1.2 Studying MT Ends In Vivo

Since all growing eukaryotic cells contain MTs, there is a wide range of samples from which to select an object for study. Usually, specimen choice is defined by the scientific interests of the investigator, so the issues to solve are those of good preservation of the chosen cell type and designing an experimental setup that will let one know which MTs are growing and which are shortening. The latter issue can be solved by using a drug that binds tubulin and blocks its polymerization, and then removing this perturbation so all visible

MTs are elongating [5]. The method of sample preparation is equally important, because molecular processes, such as tubulin polymerization, can change in milliseconds, whereas fixation of cells by chemical cross-linkers is much slower, thanks to the problem of reagent diffusion. For very thin samples, such as the periphery of a tissue culture cell, plunge freezing of cells grown on EM grids can provide excellent immobilization of biological processes and is thus the method of choice. Whole-mount microscopy and ET may then give excellent images. Most cells are, however, too thick to freeze well or to be imaged directly in an EM; thin sections must be cut. This means an alternative method of freezing, such as the use of very high hydrostatic pressure to retard ice crystal formation and fixation by freeze-substitution to make samples that can be embedded and sectioned to the desired sample thickness [1].

1.3 Analyzing PF Structure

Once tomograms of MT ends have been prepared, there remains the challenge of visualizing them with sufficient clarity to see and record the shapes of individual PFs. Since these structures are chains of tubulin dimers, they are only 4–5 nm thick and can be many nanometers long. As such, they are slender, of only modest contrast, and somewhat flexible; they are therefore hard to see in the context of background material. Moreover, individual PFs are not sufficiently similar in shape to be easy to align and average, which would improve the image signal-to-noise ratio. With a tomographic reconstruction of a MT end, one can sample the volume in various orientations to maximize the clarity of the PFs, as described below, but even under these conditions, obtaining an accurate representation of each PF is a challenge. Our group has relied upon the human eye as a reliable interpreter of complex images; others have used software to track these sinuous objects [18]. Both methods will be mentioned and their strengths/limitations described.

2 Materials

2.1 Studying MT Ends In Vitro

2.1.1 Tubulin

Tubulin can be purified from mammalian brain by well-established methods (https://hymanlab.mpi-cbg.de/hyman_lab/wp-content/uploads/2012/08/Tubulin-Protocols-Mitchison.pdf) or purchased from commercial sources (e.g., Cytoskeleton, Denver, CO). We have used Cytoskeleton catalogue no. t238p. This material contains many isoforms of both alpha and beta tubulin, as well as diverse posttranslational modifications, so it is not ideal. Simpler samples of tubulin can be prepared from specific organisms [12, 19] or purchased (https://www.abcam.com/recombinant-human-beta-tubulin-protein-ab70187.html). Making the right chromatography columns and obtaining material that is polymerization active in one's own lab is a significant investment of time and money, so purchase can be a good approach.

2.1.2 Seeds for Polymerization

Seeds that will nucleate tubulin polymerization can be prepared from GMPCPP tubulin fragments as mentioned above. Methods for the preparation of axonemes from sea urchin sperm tails [20], from *Tetrahymena* [21], or from *Chlamydomonas* [22, 23] are also available. We have found that the latter work best, in part because they disperse at their distal ends, leaving nine individual doublet MTs to serve as growth initiators at different places on the EM sample support.

2.1.3 Polymerization Buffers

Tubulin will polymerize in a variety of buffers. We have used 0.8 mM PIPES (piperazine-N,N′-bis(2-ethanesulfonate)) at pH 6.9, supplemented with 1 mM GTP, $MgCl_2$, and EGTA (ethylene glycol-bis(β-aminoethyl ether)-*N,N,N′,N′*-tetraacetic acid), often known as BRB80. In this buffer, tubulin purchased from Cytoskeleton, Inc. polymerizes well at concentrations >15 μM. With *Chlamydomonas* axonemes as seeds, it polymerizes at concentrations >7 μM. In comparison with tubulin made in local laboratories, these numbers are quite high, suggesting that approximately one-half the protein in the commercial samples was not capable of efficient polymerization. However, the material forms beautiful MTs.

2.1.4 Electron Tomography

The preparation of cellular samples for study by electron tomography (ET) is a big subject, nicely described in many reviews, so it will not be treated further here. Many relevant references are available at https://mcdbiology.colorado.edu/facilities/ems/docs/methods.shtml. The preparation of in vitro samples for electron cryotomography (cryoET) requires EM grids that are coated with a stable film, such as carbon that is perforated by many holes, one or a few micrometers in diameter. Sections of in vivo samples can conveniently be picked up as ribbons of serial sections on a slot grid coated with a film of Formvar. These grids can all be made in the lab, but it is easier to buy them from a supplier, such as Electron Microscopy Sciences (https://www.emsdiasum.com/microscopy/products/grids/cflat.aspx).

For electron microscopy of sections cut from plastic-embedded cellular samples, an instrument with higher than common voltage (>120 keV) is desirable, because samples can then be thicker than those used for conventional EM sections without a loss of resolution, due to plural and inelastic electron scattering. We have used a 300 keV instrument from FEI (Hillsboro, OR). For imaging frozen-hydrated samples, like those prepared from MTs polymerizing in vitro, a field-emission gun is highly desirable, because it produces an electron beam with sufficient coherence to give good phase contrast. We have commonly used a 200 keV microscope, the F-20 from FEI. Image recording for tomography should be done on an electronic camera of good quality. We have used CCD cameras from Gatan Inc. (Pleasanton, CA) and more recently a direct electron detector (K3, Gatan Inc.); many other good cameras are available. For image processing, we have used the IMOD software package [6], which is freely available from https://bio3d.colorado.edu/imod/.

2.1.5 A Plunge Freezer to Prepare Samples from MTs Polymerized In Vitro

Tubulin polymerization is sensitive to cold temperatures and high hydrostatic pressures. It is therefore necessary to maintain each sample at atmospheric pressure, elevated temperatures, and high humidity during the time the MTs are growing, right up until the moment they are frozen. Obviously, freezing must be as rapid as possible. Two commercial devices accomplish these goals nicely, the Vitrobot from FEI Corp. and the EM GP2 (Leica Microsystems, Buffalo Grove, IL). Alternatively, one can place any plunge freezer in a warm room (37 °C) whose atmosphere has been brought to ~100% relative humidity, e.g., by boiling water in an electric tea kettle. Some workers employing this approach have found a bathing suit to be appropriate attire.

2.1.6 Colloidal Gold to Use as Fiducial Marks for Alignment of Tilted Views

Colloidal gold can be purchased (e.g., Electron Microscope Services, Hatfield, PA). Particles of 10 or 15 nm diameter are suitable for easy viewing and rapid alignment of tilted views, following imaging for ET.

2.2 Studying MT Ends In Vivo

2.2.1 Sample Preparation

For most biological materials, a high-pressure freezer will provide EM specimens of the highest quality. Appropriate instruments are made by Leica Microsystems and Wohlwend (Sennwald, Switzerland). We have found the latter instrument to give the best preservation of biological structure. For freeze-substitution fixation, the Leica EMAFS2 works very well, but alternatively one can carry out this process with homemade equipment. Embedding requires plastic resins, such as Epon or Araldite. The preparation of sections requires a microtome; we have used the Reichert-Leica instrument with success. Thick sections can then be picked up on slot grids coated with a thin film of strong plastic, such as Formvar, stained with heavy metals and decorated with particles of colloidal gold to serve as fiducial markers in the alignment of tilted views, as is necessary for EM tomography.

2.2.2 Imaging Samples

For finding MT ends in cells, the thicker the plastic section, the more likely it is to contain an end. On the other hand, the thicker the section the higher the electron voltage required for images of good resolution. A practical compromise is a 300 kilovolt (keV) instrument, such as those made by the Japanese Electron Optical Laboratories (JEOL Ltd. Tokyo, Japan) or FEI Corporation. If ribbons of serial sections have been prepared and picked up on a Formvar-coated slot grid, one can find corresponding places on successive sections and collect tilt series at each place, accumulating the information to construct a cellular tomogram of good resolution. For image acquisition, one can use software sold by the EM venders, but we have used the SerialEM software package [24], which may be downloaded at no cost. For more information, *see* http://bio3d.colorado.edu.

2.3 Analyzing PF Structure

Once images that contain MT ends have been obtained, they must be used to construct tomograms that provide the necessary 3D information. Any fast computer can do this job. Software for tomogram construction can be purchased from any EM manufacturer; alternatively, the IMOD software suite can be downloaded free from http://bio3d.colorado.edu. Manuals that come with each of these software packages will provide instructions for aligning the multiple tilted views obtained with the EM and generating the needed 3D image. We have used the weighted back projection from the IMOD package, followed by filtering to maximize image signal-to-noise ratio.

Tracing PFs at MT ends can be done by hand, using a tablet digitizer or the mouse/cursor found on most computers. Computer-facilitated segmentation and analysis can be achieved with the same instrument used for computing the tomogram, but segmentation software that can trace objects as faint and sinuous as PFs is hard to find and use.

3 Methods

3.1 Studying MT Ends In Vitro

3.1.1 Preparing Tubulin for Polymerization

1. Tubulin purchased from Cytoskeleton is in a BRB 80 buffer at a concentration of 100 μM. Aliquot this into samples of 10 μL and immediately flash-freeze in liquid nitrogen. When stored at −80 °C, these samples will be good for months.
2. Shortly before beginning an experiment, dilute a tube of the stock tubulin to the desired concentration with BRB80 at 0 °C, adding fresh GTP to 1 mM and planning for the addition of 1 μL of colloidal gold particles, which should be added cold to the 0 °C tubulin sample, immediately before initiating polymerization.

3.1.2 Preparing the Plunge Freezer

The incubation chamber of the freezer must be brought to the appropriate temperature and humidity before beginning a freezing run. It takes ~20 min for the chamber to reach 37 °C and to become fully humidified, following the manufacturer's instructions.

3.1.3 Preparing Grids

EM grids, such as C-flats with 4 μm diameter holes, spaced by 2 μm (Electron Microscopy Sciences), are good for viewing the tips of MTs growing from axonemes. These should be glow-discharged shortly (within an hour) before use, so the liquid applied to them will spread evenly.

3.1.4 Preparing Axonemes for Use as Seeds

To prepare axonemes from *Chlamydomonas*, cells in mid-log phase are subjected to pH shock as described in ref. 22. The sample is then chilled to 0 °C and maintained on ice or in a 4 °C cold room thereafter. Cell bodies are pelleted by centrifugation, leaving the flagella in the supernatant liquid. The axonemes are then pelleted and resuspended in 10 mM HEPES buffer at pH 7.4, supplemented with 5 mM MgSO4, 0.1 mM EDTA, 1 mM EGTA, 25 mM NaCl, and 0.1 mg/mL each of pepstatin, aprotinin, and leupeptin with 0.1% NP40 added to dissolve the membranes. After 1 h in the detergent-containing buffer, the axonemes are pelleted and resuspended in the same buffer without detergent and kept on ice overnight. They may be concentrated into a soft pellet by spinning in a microfuge at top speed (~14,000 rpm ~16kg) for 10–15 min, then diluted by the addition of BRB80 to a useful concentration, aliquoted, and flash-frozen in liquid nitrogen for storage at −80 °C until use.

Dilute a sample of axonemes to the point that the axonemes adhering to the glow-discharged grid will be spaced by ~20 μm. Apply 2–3 μL of this solution to the grid (Fig. 1a) and allow ~30 s for the axonemes to adhere to the film on the grid (*see* **Note 1**).

3.1.5 Making a Sample

1. Use a small piece of no. 1 filter paper to blot off excess liquid from the drop that contained axonemes (Fig. 1b), and immediately add 5 μL of tubulin at the desired concentration in polymerization buffer, supplemented with 10 nm colloidal particles (*see* **Note 2**).
2. Draw the sample up into the pre-warmed incubation chamber of the plunge-freezing device and allow an appropriate time for the tubulin to polymerize (Fig. 1c). The polymerizing MTs should be long enough at the time of blotting and freezing to be easily distinguished from axonemal MTs (either doublets or singlets from the central pair), so one can be confident that the MT end being imaged is growing, not the static end of an axonemal MT (Fig. 2) (*see* **Note 3**).
3. As soon as the MTs are likely to be long enough, initiate plunge-freezing. The instruments mentioned will automatically blot the sample to remove excess liquid and plunge it into liquid ethane that has been cooled to its freezing point by liquid nitrogen (*see* **Note 4**).
4. The frozen sample may then be removed from liquid ethane and stored under liquid nitrogen until it is time for viewing.

3.1.6 Acquiring Images for CryoET

1. Make a low-magnification map of a grid to find places where the ice is of appropriate thickness, where there is no contamination, and where axonemes can be seen.

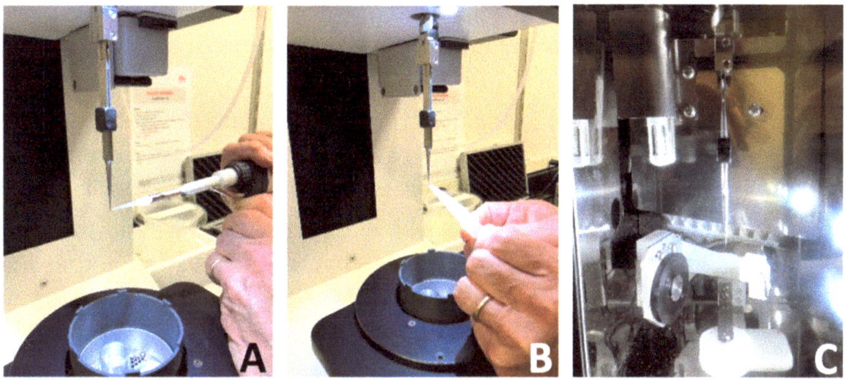

Fig. 1 Loading a sample for plunge-freezing. (**a**) Axonemes can be added to the surface of a glow-discharged EM grid by pipetting on 2–3 μL of a sample with appropriate titer. (**b**) After ~30 s for the axonemes to adsorb to the carbon film, excess liquid is removed with filter paper. About 5 μL of the tubulin polymerization solution is added quickly to prevent drying of the axonemes. (**c**) The sample is then drawn up into the incubation chamber of the freezing device, where the warm environment will support polymerization and the hydrated air will prevent evaporation

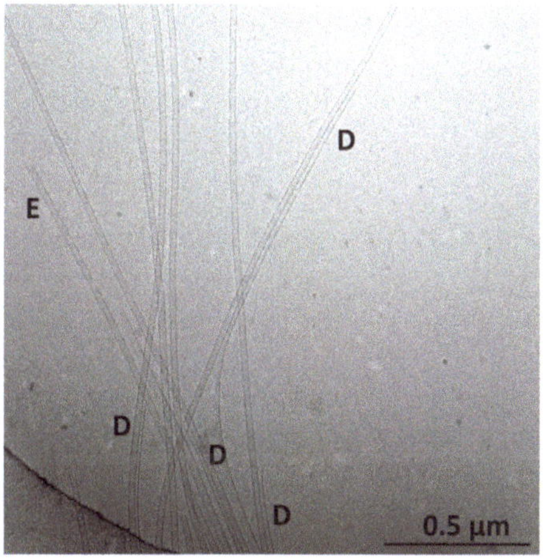

Fig. 2 A frozen-hydrated axoneme with MTs growing from the tips of some of its MTs. The image shows about one-sixth of a hole in the film of a C-flat grid to which axonemes were adsorbed and on which tubulin has polymerized, elongating the axonemal MTs. An edge of the hole is visible in the lower left corner of the picture. *D*s mark axonemal doublet MTs; *E* marks the end of a MT that elongated from a doublet

2. Mark these areas for future examination at higher magnification. These operations can be performed with software provided by the microscope manufacturer or with the Navigator software from http://bio3d.colorado.edu/SerialEM/.
3. At moderate magnification (×2200–×20,000) and using low-dose conditions, examine each promising area for the following conditions:

(a) The plus end of an axoneme shows clear evidence of elongation by polymerization of tubulin on at least some of the doublet MTs.

(b) These elongations are long enough to assure that the MT tips you see are assembled from the added tubulin dimers, not simply pieces of a broken axoneme (Fig. 2).

(c) There are MT ends lying over a hole in the grid, surrounded by ice that is thick enough to preserve MT structure but thin enough to permit imaging with good quality (30–100 nm).

(d) Ideally, there are sufficient numbers of gold particles nearby (5–10) for use as alignment fiducials.

(e) The area of interest is far enough from a grid bar that both a tilt series and an appropriate area for focusing away from the imaging region will be available.

4. If all these criteria are met, then collect a tilt series for a cryo-tomogram (*see* **Note 5**). First, set up low dose so that the focus area is on the tilt axis and at least 2 μm from the Record area. A low-dose Trial area is typically the same as the focus area and is used for tracking during tilt series acquisition. A Preview area is the same area as a Record image with a much lower dose. This is useful for imaging what is in the Record and determining if MT plus ends are in the Record area.

5. Adjust Record exposure so that each record image has a dose of ~1 e^-/A^2. SerialEM has a tilt series setup window to adjust these acquisition parameters. We have used bidirectional, single-axis tilt series, starting at zero tilt, and going up to about 60° in increments of 2°, then returning to zero tilt, and collecting a similar series up to −60°. Exposure can also be varied at different tilts to give a higher dose at the higher tilts, where the electron beam passes through a thicker sample. The total dose for the entire tilt series should be kept <80 e^-/A^2 (*see* **Notes 6** and 7).

3.2 Studying MT Ends In Vivo

3.2.1 Preparing Cells for Electron Microscopy of MT Ends

Cells or samples of tissue should be fixed by the best procedures available, with special attention to the temperature before freezing, since MT dynamics are affected by changes in this parameter. A good manual for methods of chemical fixation can be found at https://electron-microscopy.hms.harvard.edu/methods. These methods, properly applied, will preserve the flared ends of MTs, but the details of sample structure are generally less good than when materials are frozen and then fixed by freeze-substitution [1]. If one's cellular samples are thick enough to require high-pressure freezing (>4 μm), one can worry that the high hydrostatic pressure, applied briefly before freezing, will induce changes in the structure of something as labile as a MT end. We have previously compared the PFs at growing MT ends in fission yeasts, which are thin enough to be plunge frozen without inducing ice crystals, and we found no difference between the lengths and curvatures of PFs in the plunge-frozen sample, compared with a comparable high-pressure frozen sample, suggesting that this issue not a problem [26].

After embedding in resin, sections can be cut at thickness ranging from 200 to 300 nm thickness, if an electron microscope of intermediate voltage (300 keV) is available. If not, limit sections to about 100 nm. The thicker the section, the greater the likelihood that an entire MT end will lie within the section. Formvar-coated slot grids are typically used for tomography of serial plastic sections because when the ribbon of slices lies on a plastic film, there are no grid bars to obscure regions of interest during tilt series acquisition. Post-staining of the sections can be carried out by any good method, but for ET be sure to affix colloidal gold particles (10–15 nm diameter) to the section surface. These are useful for aligning multiple tilted views.

3.2.2 Imaging Sections of Cells to Reveal MT End Structure

Hunting out MT ends by EM of interphase cells can be time consuming. In a well-formed MT array, like the mitotic spindle, the task is easier, but in either case ends can be found. In all cases, serial sections can be useful for providing information about whether an apparent end is a true end or simply a place where the MT passes out from the section being viewed. If the sample is in a tilt and rotate holder, then it is useful to rotate the sample so the MT axis is parallel with the microscope's tilt axis. With embedded and stained samples, low-dose conditions are not necessary, but given the number of images that must be recorded for a tilt series that will make a good tomogram, minimizing beam exposure is good practice.

With fixed and stained material, the sample is sufficiently robust that dual-axis tomography is definitely preferred for obtaining tomograms with the best isotropy of resolution [27]. Once the tilted views from the first axis have been obtained, the grid is rotated 90° and a second tilt series is acquired. Tomograms can be calculated with several useful programs, some available from microscope companies. We have used the IMOD program suite and reconstruction by weighted back projection, which is particularly good for dual-axis tilt series.

Once these tomograms have been obtained, extracting MT ends and analyzing their structure can be carried out by the same methods that are useful for cryoET of isolated MTs as described below.

3.3 Analyzing PF Structure

3.3.1 Extracting MT Ends for Structural Analysis

1. To characterize the structure of an MT end, particularly to describe the arrangements of its PFs, each MT end must be examined in 3D. This is facilitated by cutting out volumes that contain each MT end and orienting the MT axis conveniently. If you are using the IMOD software package, the program mtrotlong will greatly facilitate this process.
2. To use this program, first use the modeling tools in the IMOD viewing program, 3Dmod, to draw a line that is collinear with the MT axis, and that ends at the MT tip. Mtrotlong and the companion program, rotallends, will then use this line to extract a volume of a size you specify and align the MT axis along the *Y*-axis of the display while putting the MT end in the center of the volume extracted (*see* **Note 8**).
3. In one good tomogram, one can often find several MT ends. Each of these should be extracted as a stack of planes in *Z*. If one keeps those stacks in a single directory, IMOD can work with them very efficiently.
4. One can then use the Slicer module of 3dmod to refine that alignment and carry out detailed study of each MT end, as described below.

3.3.2 Tracing Protofilaments

The clarity of individual protofilaments in slices cut from a tomogram is highly dependent on image quality, the orientation of the MT axis relative to the axis of tilt during data collection, the thickness of the ice or section, the absence of contamination, and the quality of the alignment of the tilt series images prior to reconstruction. Even with an optimal specimen, tracing PFs is not easy; the true path of a PF can be obscured by a nearby MT, by other structures in the cytoplasm, by material that leached from an axoneme in vitro, and by tubulin oligomers. One might imagine that the best way to look closely at the MT would be to step through the thickness of the polymer, using planes parallel to its axis. Our experience has taught us that this is harder than it sounds, because the PFs flare out from the MT axis in many directions. Following them with serial slices is very difficult. A more effective approach is to sample the end of the MT rotationally, making the axis of rotation collinear with the MT axis so one can see the PFs in planes as they flare out from the MT wall. The following protocol is one we have followed to maximize the chances of doing this job well.

3.3.3 Aligning the MT with the Coordinate System in Slicer

1. Align the MT axis with the *Y*-axis of slicer's display window by a two-step process. First, use the edge of the slicer window as a guide for orientation in the *XY* plane (Fig. 3a); then step up and down in *Z* to the two surfaces of the MT and rotate about the *X*-axis to get the PFs of the MT wall persisting in the field

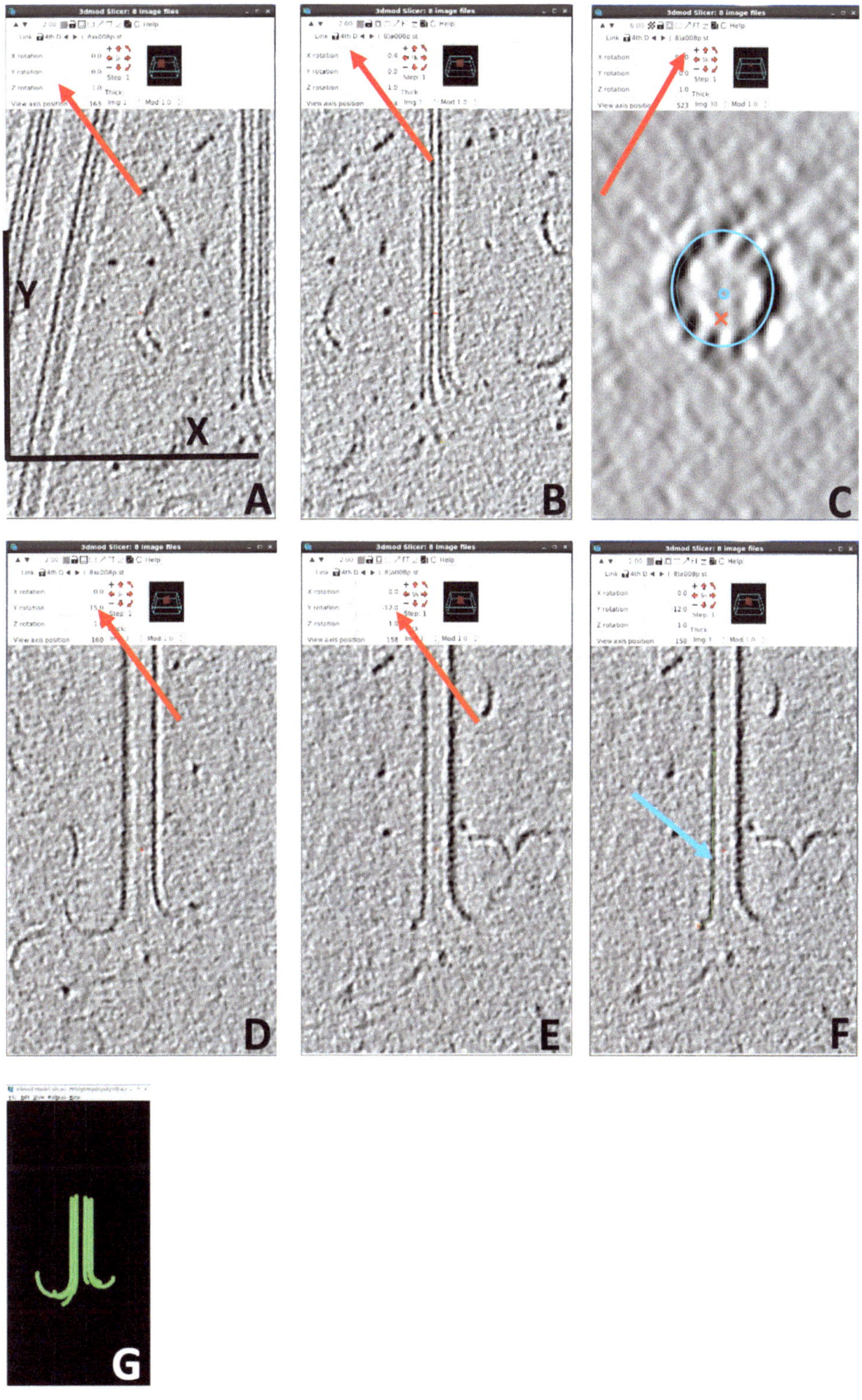

Fig. 3 Using slicer to orient a MT for analysis of its end. (**a**) Several MTs are seen at a level where the sampling plane grazes their surfaces. One MT has been moved to the right edge of the field so its orientation can be changed with the "*Z*-axis slider" (red arrow indicates the control for rotation about the *Z*-axis, which is

of view throughout their length. The MT axis is then correct in the *YZ* plane (Fig. 3b).

2. Position the center of the coordinate system at the center of the MT. This is best done by rotating the already aligned MT by 90° about the *X*-axis, so one is looking along the MT axis.
3. Increase the magnification of the display until the MT is large, perhaps 4 cm in outside diameter.
4. Increase the number of voxels being projected into the slicer window using the display of slices in the Image Thickness box in the slicer window. We have used a thickness of 30 voxels, each 0.9 nm on an edge, which averages the positions of individual PFs in the MT wall and makes them easy to see.
5. Place a model point in the center of the MT, using the "scattered object" type of model objects and a diameter for the point that is approximately the same as the diameter of the MT (Fig. 3c) (*see* **Note 9**).
6. Position of the MT center must be estimated as best as one can.
7. Once the model is centered on the MT, move the center of the model (blue circle in Fig. 3c) to lie over the origin of the slicer window (red cross).
8. Rotate the MT around the *X*-axis, back to the orientation of the plane of the slicer window. The MT axis is now collinear with the *Y*-axis of the slicer window so one can view the shape of each PF simply by rotating around the *Y*-axis.

Fig. 3 (continued) perpendicular to the plane shown). The MT now lies parallel to the vertical edge, which is the *Y* direction. (**b**) The same MT is now near the center of the field, and the *X*-axis slider (red arrow) has been altered to make the PFs in the MT wall run within the plane of sampling. The MT axis is now parallel to the *Y*-axis. (**c**) The *X*-axis has been altered by 90° to show the MT in cross section. The sampling thickness has been increased to 30 voxels to average several tubulin subunits along each PF. The center of the MT is identified by fitting the perimeter of a circle (blue) to the wall of the MT, simply by trial and error. Now the image can be moved so the center of the circle lies over the origin of Slicer's coordinate system (red cross). (**d**) After the tilt around *X* has been returned to the previously determined value, the axis of the MT is collinear with the *Y*-axis of the slicer window. One can now rotate around the *Y*-axis by altering the *Y* slider (red arrow). By inspection, one can usually find a PF that curls out from the MT axis. A graphic object can then be drawn on that PF, and its position/orientation recorded in the "Edit Angles" window (not shown). (**e**) If one now rotates ~27° either clockwise or counterclockwise around *Y*, one will usually encounter another PF. Sometimes this structure too is flared. The red arrow indicates the new *Y*-axis orientation. (**f**) A graphic representation of this PF can now be drawn and its position recorded with Edit Angles. When all visible PFs have been traced, the set of graphic objects gives a good representation of the MT end (**g**). This MT was in the process of shortening, following the addition of 20 mM $MgCl_2$ to the polymerization medium, which accounts for the large number of tubulin oligomers in the background of the image

3.3.4 Tracing Individual Protofilaments

1. In our experience, at least some of the PFs flare out from the MT axis, making them quite easy to see when the plane of sampling is oriented properly. To get such views, rotate the plane of sampling around *Y* by adjusting the *Y* slider in the slicer window (red arrow in Fig. 3d) with increments of 1° until a curving PF comes into view (*see* **Note 10**).
2. The tools for drawing models in IMOD are well described in the manual pages of that program and will not be repeated here. For drawing PFs, one wants the object type to be open (not forming closed loops), the point size to be slightly bigger than the line width, and the ends of the trace to be marked. The PFs are clearest if one uses the slicer window to display a thickness of image that includes a sufficient number of voxels to correspond to a little less than the thickness of a PF, which is 4 nm. Using these conditions, one can draw a trace that shows the position and orientation of the PF both while it is in the MT wall and as it flares out from the wall at the MT end.
3. The MT end under view will almost always have 13 PFs. To find additional PFs on the same MT end, look near the obvious one by rotating about the *Y*-axis by 360/13° (Fig. 3e). Given the possibility of distortions, we have been a little flexible about this angle and have looked over a range of 25–29° from the clear PF (Fig. 3f). If there is no flaring PF in that region, we have interpreted this structure as a non-flaring PF and have represented it by a single straight line that lies at that position in the MT wall (*see* **Notes 11**, **12**, and **13**).

3.3.5 Quantifying Structural Information from PF Traces

Each investigator who examines PF structure may wish to develop his or her own modes of analysis, but we have used two sets of programs that might be useful. In the IMOD suite there is a program called "howflared," written by David Mastronarde, which we have used for many purposes. It converts a manual trace of a straight and then curved object into points and calculates lengths, average curvatures, and related parameters; its use is described in the manual pages of IMOD. More specific study has required the development of additional programs that have taken the coordinates of traced points from the howflared program and analyzed the curvatures and lengths of the traces more fully. Those programs use the "point" output from howflared to calculate lengths and curvatures as a function of position, as well as average PF parameters; these are available from Dr. Nikita Gudimchuk (nikita_gb@mail.ru). The fruits of this kind of analysis can be seen in published work [26].

4 Notes

1. When axonemes are applied to an EM grid to be used as seeds, they should not be too numerous, or the MTs that grow from one axoneme cannot be distinguished from the many others that cover the grid. If they are too sparse, there are not enough MT ends to study. Identifying the right titer of axonemes is done by trial and error, using negative staining and conventional EM to avoid wasting time on a microscope with a field-emission gun for cryoEM.
2. The concentration of tubulin should be low enough that spontaneous initiation of MTs will be rare; otherwise there will be too many MTs on the grid to allow rapid finding of the polymers that were initiated by axonemal MTs.
3. We have found that 2–6 min of elongation is optimal, depending on the rate at which the tubulin is polymerizing. Under these conditions, the presence of growth is unambiguous, but the MTs have not elongated to the point that they fill the grid with MT walls, making the ends hard to find. For active tubulin at 10 μM, this is ~4 min, slightly longer than one might expect from a normal MT growth rate because the initial phase of elongating from an axonemal doublet is slower than the continued growth of a dynamic MT. With higher concentrations of tubulin, or in the presence of growth-enhancing or -inhibiting substances, the times must be shortened or lengthened appropriately.
4. Finding the best conditions for blotting is a matter of trial and error. We found that very light blotting worked well, and we often blotted from only the back side of the grid.
5. Methods for collection of tilt series images are described with detail in books on electron tomography. In brief, check the eucentricity of the stage at this part of the grid, make sure that the camera is properly adjusted (dark reference and gain reference up to date), and choose a magnification so the pixel size will deliver the desired resolution. This depends on the camera, but the pixels should be approximately one-third the desired resolution after appropriate binning (0.8–1.2 nm).
6. Several microscope manufacturers provide programs for aligning tilt series and computing tomograms. We have used the IMOD software package available from https://bio3d.colorado.edu//. Our tomograms have been most useful when filtered by nonlinear anisotropic diffusion [25], a routine that improves image signal-to-noise. The IMOD package includes an interface for doing these computations.

7. It is sometimes useful to acquire some images of the specimen that are not tomograms, but simply cryoEMs, taken with sufficient dose to provide a good signal-to-noise ratio, e.g., 30 $e^-/Å^2$.
8. The volume extracted should be big enough that its edges will not impinge on the view of the MT at high tilt. We have typically extracted volumes of 1000 × 800 × 300 voxels, each 0.9 nm on a side.
9. It usually takes a few tries to get the circle that represents the model point running smoothly around the MT wall. This operation can be difficult if the end of the MT is not a nice circle, which was quite common in our samples.
10. It is important NOT to change the position or orientation of the MT about either the *X*- or the *Z*-axis while rotating about *Y*; these positions have already been determined, and they should be left constant. In our experience, the entirety of a PF will commonly lie in a single plane that contains the MT axis, so such motions should be unnecessary. Occasionally, a PF will follow a slightly helical path, but this seems to be the exception, so we have made all our traces of PF shape on a single plane.
11. The 3dmod program used for tracing includes a useful set of windows and displays that can help with the tracing process. We have always used the "edit angles" feature that brings up a window (3dmod Slicer Angles) into which you can save the coordinates and orientations of each trace you make. This information is invaluable if you want to return to a given MT and check the traces you have made. We have also used the model view display to show the traces as they are made (Fig. 3g). This is both a good record of the shapes of the PFs seen and a good way to guide the person doing the traces, so the line segments that represent the PFs in the MT wall are positioned appropriately along the MT axis.
12. Given the issues of point spread along the beam axis and missing wedge artifacts in single-axis electron tomography, the quality of the image displayed in the slicer window will degrade as the angle about the *Y*-axis is increased above 45°. Although the PFs will not look sharp at higher angles, with practice they can be recognized with reasonable confidence. Moreover, in single-axis tomography, the resolution is affected by the variation "line spread" as one looks at linear objects that lie at different orientations relative to the axis of tilt: contrast is best for lines that run parallel to that axis. Thus, if the MT was approximately parallel to the tilt axis in the original tomogram, the PFs will have the best contrast when they are also parallel to that axis. A PF that curves will have variable line spread as one

follows it around its curve. With practice, the mind's eye can compensate for these kinds of variations, but it makes PF tracing more difficult than one would like.

13. Other labs have employed digital methods to trace PFs, using some of the excellent software available to the community, such as "tomoseg" from the Baylor College of Medicine [28]. The value of this approach is its consistency. Its limitation is that it is difficult, perhaps impossible, to train a computer program to deal with all of the expected variations in contrast and image noise that one encounters while trying to trace curving objects in 3D as reconstructed by electron tomography [18].

Acknowledgment

The development of these methods was supported in part by GM033787.

References

1. McDonald KL, Morphew M, Verkade P, Muller-Reichert T (2007) Recent advances in high-pressure freezing: equipment- and specimen-loading methods. Methods Mol Biol 369:143–173. https://doi.org/10.1007/978-1-59745-294-6_8
2. Dubochet J (2007) The physics of rapid cooling and its implications for cryoimmobilization of cells. Methods Cell Biol 79:7–21. https://doi.org/10.1016/S0091-679X(06)79001-X
3. Chretien D, Fuller SD, Karsenti E (1995) Structure of growing microtubule ends: two-dimensional sheets close into tubes at variable rates. J Cell Biol 129:1311–1328
4. Donohoe BS, Mogelsvang S, Staehelin LA (2006) Electron tomography of ER, Golgi and related membrane systems. Methods 39:154–162. https://doi.org/10.1016/j.ymeth.2006.05.013
5. Hoog JL, Huisman SM, Sebo-Lemke Z et al (2011) Electron tomography reveals a flared morphology on growing microtubule ends. J Cell Sci 124:693–698. https://doi.org/10.1242/jcs.072967
6. Kremer JR, Mastronarde DN, McIntosh JR (1996) Computer visualization of three-dimensional image data using IMOD. J Struct Biol 116:71–76
7. Baumeister W (2002) Electron tomography: towards visualizing the molecular organization of the cytoplasm. Curr Opin Struct Biol 12:679–684
8. VandenBeldt KJ, Barnard RM, Hergert PJ et al (2006) Kinetochores use a novel mechanism for coordinating the dynamics of individual microtubules. Curr Biol 16:1217–1223. https://doi.org/10.1016/j.cub.2006.04.046
9. Henderson LD, Beeby M (2018) High-throughput Electron Cryo-tomography of protein complexes and their assembly. Methods Mol Biol 1764:29–44. https://doi.org/10.1007/978-1-4939-7759-8_2
10. Oikonomou CM, Jensen GJ (2017) Cellular electron cryotomography: toward structural biology in situ. Annu Rev Biochem 86:873–896. https://doi.org/10.1146/annurev-biochem-061516-044741
11. Hyman A, Drechsel D, Kellogg D et al (1991) Preparation of modified tubulins. Methods Enzymol 196:478–485
12. Widlund PO, Podolski M, Reber S et al (2012) One-step purification of assembly-competent tubulin from diverse eukaryotic sources. Mol Biol Cell 23:4393–4401. https://doi.org/10.1091/mbc.E12-06-0444
13. Bergen LG, Kuriyama R, Borisy GG (1980) Polarity of microtubules nucleated by centrosomes and chromosomes of Chinese hamster ovary cells in vitro. J Cell Biol 84:151–159
14. Johnson KA, Borisy GG (1977) Kinetic analysis of microtubule self-assembly in vitro. J Mol Biol 117:1–31
15. Mitchison T, Kirschner M (1984) Dynamic instability of microtubule growth. Nature 312:237–242

16. Chretien D, Kenney JM, Fuller SD, Wade RH (1996) Determination of microtubule polarity by cryo-electron microscopy. Structure 4:1031–1040
17. Wade RH, Chrétien D, Job D (1990) Characterization of microtubule protofilament numbers. How does the surface lattice accommodate? J Mol Biol 212:775–786. https://doi.org/10.1016/0022-2836(90)90236-F
18. Atherton J, Jiang K, Stangier MM et al (2017) A structural model for microtubule minus-end recognition and protection by CAMSAP proteins. Nat Struct Mol Biol 24:931–943. https://doi.org/10.1038/nsmb.3483
19. Vemu A, Atherton J, Spector JO et al (2016) Structure and dynamics of single-isoform recombinant neuronal human tubulin. J Biol Chem 291:12907–12915. https://doi.org/10.1074/jbc.C116.731133
20. Sale WS, Gibbons IR (1979) Study of the mechanism of vanadate inhibition of the dynein cross-bridge cycle in sea urchin sperm flagella. J Cell Biol 82:291–298
21. Gibson TM, Asai DJ (2000) Isolation and characterization of 22S outer arm dynein from Tetrahymena cilia. Methods Cell Biol 62:433–440
22. Porter ME, Power J, Dutcher SK (1992) Extragenic suppressors of paralyzed flagellar mutations in Chlamydomonas reinhardtii identify loci that alter the inner dynein arms. J Cell Biol 118:1163–1176
23. Nicastro D, Schwartz C, Pierson J et al (2006) The molecular architecture of axonemes revealed by cryoelectron tomography. Science 313:944–948. https://doi.org/10.1126/science.1128618
24. Mastronarde DN (2005) Automated electron microscope tomography using robust prediction of specimen movements. J Struct Biol 152:36–51. https://doi.org/10.1016/j.jsb.2005.07.007
25. Frangakis AS, Hegerl R (2001) Noise reduction in electron tomographic reconstructions using nonlinear anisotropic diffusion. J Struct Biol 135:239–250. https://doi.org/10.1006/jsbi.2001.4406
26. McIntosh JR, O'Toole E, Morgan G et al (2018) Microtubules grow by the addition of bent guanosine triphosphate tubulin to the tips of curved protofilaments. J Cell Biol 217:2691–2708. https://doi.org/10.1083/jcb.201802138
27. Mastronarde DN (1997) Dual-axis tomography: an approach with alignment methods that preserve resolution. J Struct Biol 120:343–352. https://doi.org/10.1006/jsbi.1997.3919
28. Chen M, Dai W, Sun SY et al (2017) Convolutional neural networks for automated annotation of cellular cryo-electron tomograms. Nat Methods 14:983–985. https://doi.org/10.1038/nmeth.4405

Chapter 14

Optogenetic Control of Microtubule Dynamics

Jeffrey van Haren, Lauren S. Adachi, and Torsten Wittmann

Abstract

Light can be controlled with high spatial and temporal accuracy. Therefore, optogenetics is an attractive experimental approach to modulate intracellular cytoskeleton dynamics at much faster timescales than by genetic modification. For example, in mammalian cells, microtubules (MTs) grow tens of micrometers per minute and many intracellular MT functions are mediated by a complex of +TIP proteins that dynamically associate with growing MT plus ends. EB1 is a central component of this +TIP protein network, and we recently developed a photo-inactivated π-EB1 by inserting a blue light-sensitive LOV2/Zdk1 module between the EB1 MT-binding domain and the +TIP adaptor domain. Blue light-induced π-EB1 photodissociation results in disassembly of the +TIP complex and strongly attenuates MT growth in mammalian cells.

In this chapter, we discuss theoretical and practical aspects of how to perform high-resolution live-cell microscopy in combination with π-EB1 photodissociation. However, these techniques are broadly applicable to other LOV2-based and likely other blue light-sensitive optogenetics. In addition to being a tool to investigate +TIP functions acutely and with subcellular resolution, because of its dramatic and rapid change in intracellular localization, π-EB1 can serve as a powerful tool to test and characterize optogenetic illumination setups. We describe protocols on how to achieve micrometer-scale intracellular control of π-EB1 activity using patterned illumination, and we introduce a do-it-yourself LED cube design compatible with transmitted light microscopy in multiwell plates.

Key words Optogenetics, π-EB1, Photodissociation, Microtubules, EB1, +TIP, LOV2, LOVTRAP, Zdk1, Live-cell microscopy

1 Introduction

A challenge of the postgenomic era is that many intracellular processes occur at timescales that are orders of magnitude faster than the days or weeks required to deplete proteins by RNA interference or genetic deletion. Protein functions are often spatially and temporally restricted in cells, which makes mechanistic interpretation of such genetic knockout experiments challenging. This is particularly true for cytoskeleton dynamics that operate very locally and at second-to-minute timescales. For example, the interphase microtubule (MT) cytoskeleton reorganizes into a bipolar mitotic spindle

Helder Maiato (ed.), *Cytoskeleton Dynamics: Methods and Protocols*, Methods in Molecular Biology, vol. 2101,
https://doi.org/10.1007/978-1-0716-0219-5_14, © Springer Science+Business Media, LLC, part of Springer Nature 2020

in typically less than 30 min, and individual MT bundles dynamically connect to chromosomes to ensure accurate segregation of the genetic material. Similarly, both MTs, the F-actin cytoskeleton, and even much less dynamic intermediate filaments direct and respond to changes in interphase cell shape at much faster timescales than transcriptional or translational changes. Thus, experimental alteration of intracellular protein function by light is an attractive experimental approach because light exposure can be controlled with very high spatial and temporal precision.

Optogenetics originally involved expression of channelrhodopsins in neurons to elicit light-activated membrane currents and neuronal activity. However, the potential of light-modulated protein function to spatially and temporally control biochemistry in living cells is enormous, and optogenetics is no longer restricted to the neurosciences. An ever-increasing set of tools is being developed to manipulate protein interactions, localization, and activity by light [1, 2].

1.1 Optogenetics with LOV2 Domains

LOV domains were first identified as the light-sensing units of plant light receptors of the phototropin family [3, 4]. Phototropins are blue light-activated kinases, typically composed of two LOV domains (LOV1 and LOV2) in tandem followed by a serine/threonine kinase domain, and mediate plant growth toward light (phototropism), chloroplast migration, and stomata opening [5]. The light-absorbing unit of LOV2 domains is an obligatory flavin mononucleotide (FMN) cofactor. FMN is a riboflavin (vitamin B2) derivative, and a required cofactor in oxidoreductases in plants, animals, and bacteria. Thus, FMN is ubiquitously present in most life forms, which simplifies LOV2-based optogenetics in mammalian systems.

Blue light exposure of the LOV2-bound FMN leads to the formation of a flavin-cysteinyl adduct between the triplet-excited FMN isoalloxazine ring and a cysteine residue buried in the LOV2 core (C450 in *Avena sativa* LOV2) [6]. Adduct formation results in unwinding of the short A'α and the longer Jα helices at the LOV2 N- and C-termini [7–9]. Although the precise mechanism linking LOV2 photochemistry to rearrangements of protein structure is still under debate [10], the energy associated with this conformational change can be harnessed as "optogenetic work" [11]. Several recently developed optogenetics tools have in some way made use of the switch-like conformational changes of LOV2 domains to control protein activities in a light-dependent manner. For example, LOV2 domains appended to the termini of proteins or peptides can mask a protein activity in the dark [12, 13]. Similarly, LOV2 domains inserted into protein surface loops can allosterically control protein conformation and thus activity by light [14, 15]. Finally, several LOV2 domain-based optical dimerization systems have been developed in which unfolding of the LOV2 Jα helix results in association or dissociation of small proteins that interact specifically with either the light- or dark-state LOV2 conformations [16–18].

1.2 π-EB1

To control the function and dynamics of MT plus-end-associated protein complexes with high spatial and temporal accuracy, we recently developed a novel LOV2-based optogenetic tool to reversibly inactivate protein complexes that associate with growing MT ends. In mammalian cells, the small adaptor protein EB1 (EB3 in neurons) is required to assemble complexes of so-called +TIPs at growing MT ends [19, 20]. EB1 is a dimer with an N-terminal domain that recognizes the guanosine nucleotide state of growing MT ends, and a C-terminal domain that recruits +TIP proteins. We designed a light-sensitive variant of EB1 by connecting these two EB1 functional domains with the light-sensitive LOVTRAP module (Fig. 1) [16, 21]. This results in rapid and reversible blue light-induced dissociation of the C-terminal π-EB1 part and associated +TIPs from growing MT ends (Fig. 2b). In addition, we found that in the cell types we tested so far, π-EB1 photodissociation results in substantial inhibition of MT growth, and we propose that π-EB1 can thus be utilized to dissect the intracellular functions of +TIP complexes at high spatial and temporal accuracy. However, it is important to note that observing functional effects of π-EB1 photodissociation requires replacing endogenous EB1/3 activity with π-EB1. We recently published protocols on how this can be done using RNA interference or CRISPR/Cas9 genome editing [22], and we will not describe this here. Instead, we are focusing on the

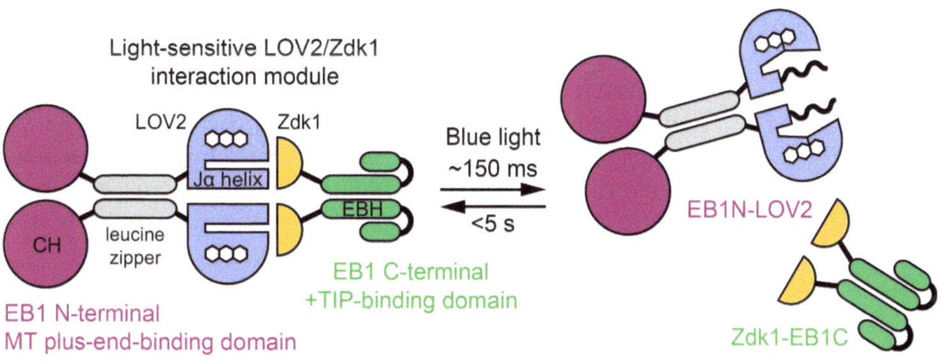

Fig. 1 Schematic of π-EB1, a split version of the MT plus-end tracking protein EB1, in which the N-terminal MT-plus-end binding (CH) domain is fused to the *A. sativa* phototropin 1 LOV2 domain. This N-terminal part also includes a short leucine zipper to maintain dimerization in the photodissociated state. The C-terminal +TIP adapter (EBH) domain is fused to Zdk1 (and a fluorescent protein). In the absence of blue light, Zdk1 and LOV2 domains bind with high affinity, thereby reconstituting a functional EB1 module that associates with growing MT plus ends, and is able to recruit other +TIP proteins to the MT end. In the presence of blue light, LOV2 and Zdk1 dissociate, thereby splitting the π-EB1 complex rapidly and reversibly due to the spontaneous dark-state recovery of the LOV2 domain. Adapted from [21]

technical aspects of how we perform optogenetics experiments on a microscope using π-EB1 as an example. We believe that these concepts and protocols can be adapted to other LOV2-based and likely other blue light-controlled optogenetic systems.

1.3 General Considerations Using LOV2 Domains

In addition to LOV2 domains, many other plant and bacterial light sensors have been developed into optogenetic tools. While no single system is perfect, we find that the small size and high solubility in combination with fast-switching kinetics and the ubiquitous FMN cofactor make LOV2 domains ideal to achieve subcellular spatial control in mammalian cells. Next, we discuss some of the characteristics of LOV2 domains that need to be considered when planning an experiment with π-EB1 or other LOV2-based systems.

1.3.1 Spectral Response

The spectral response of LOV2 domains is determined by the FMN spectrum, which also makes it unlikely that red-shifted LOV domain variants are possible. Although FMN peak absorption is around 440–470 nm, the excitation spectrum reaches into the UV and above 500 nm (Fig. 2a). In live fluorescence imaging experiments, this limits the fluorescent proteins (FPs) that can be used in combination with LOV2 optogenetics to red and far-red FPs with excitation wavelengths above 550 nm, or red and far-red SNAP or HALO tag-compatible fluorescent dyes [23]. Blue, cyan, and green fluorescent proteins are essentially incompatible as their excitation wavelengths very efficiently photoactivate LOV2. In our hands, even 515 nm yellow FP excitation resulted in π-EB1 photodissociation albeit less efficiently and should probably not be used in combination with LOV2 optogenetics (Fig. 2b). To be safe, we use a red filter in the transmitted lightpath for extended transmitted light imaging although at least wild-type LOV2 does not seem to be activated efficiently by the blue light component in typically used transmitted light intensities. However, the broad LOV2 spectral response has the advantage that at least temporal photoactivation control can be easily implemented on any microscope that has a GFP excitation channel.

1.3.2 Kinetics

The photochemistry underlying LOV2 photoactivation is very fast with formation of the flavin-cysteinyl adduct in the microsecond range [8, 24, 25]. While recent work showed that initial unfolding of the Jα helix happens concurrently with adduct formation, full unfolding occurs in the millisecond range in isolated LOV2 domains [8, 26]. However, subsequent conformational changes and association/dissociation of effector proteins are likely substantially slower, as they depend on the binding kinetics and diffusion rates of the molecules involved. In the case of π-EB1, we measured a blue light-induced photodissociation half-life of <200 ms. However, reassociation of π-EB1 after blue light exposure is much

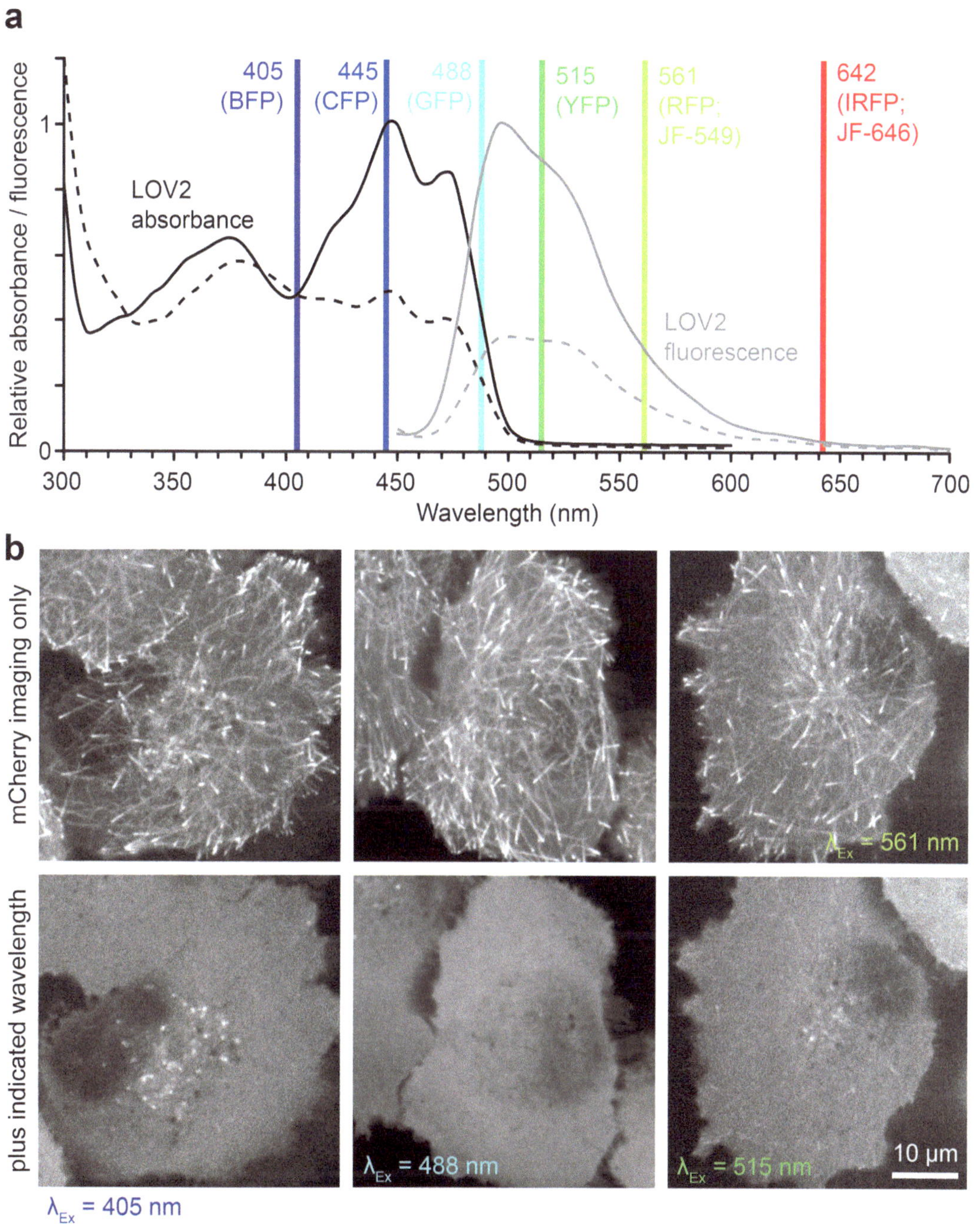

Fig. 2 LOV2 domain spectral response and compatibility with fluorescence microscopy. (**a**) Absorbance (black lines) and fluorescence spectra (gray lines; $\lambda_{Ex} = 405$ nm) of purified GST-tagged *A. sativa* phototropin 1 LOV2 domain, before (solid lines) and after (dashed lines) 10 s of blue light illumination with a 36 W 460 nm LED lamp. The flavin-cysteinyl adduct no longer absorbs blue light and thus quenches LOV2 fluorescence. Laser lines commonly used in fluorescence microscopy are indicated. (**b**) π-EB1 photodissociation by different wavelengths in cells expressing untagged EB1N-LOV2 and mCherry-Zdk1-EB1C. mCherry-Zdk1-EB1C was imaged before (top) and after (bottom) switching on the other wavelength. Note that even 515 nm light at the

slower as this depends on both the dark-state recovery of LOV2 (*see* **Note 1**) as well as the two π-EB1 halves finding and binding to each other. In our original π-EB1 design, we used a LOV2 domain with a Jα helix-stabilizing LAAA N terminus [18] for which we measured a half-life of π-EB1 dark reassociation of around 10 s. It is important to be aware of these kinetics as they affect both spatial and temporal resolution of subcellular optogenetics experiments. In addition, in most cases we use pulsed blue light illumination to control light exposure dose, minimize phototoxicity, and avoid cross talk with imaging channels. With this original π-EB1 design, a pulse frequency of 1 Hz is enough to keep average π-EB1 activity below 3%. However, one needs to be aware that the relative amount of dissociated π-EB1 is not constant, and gradually decreases until the next blue light exposure depending on pulse frequency and π-EB1 reassociation rate (Fig. 3c). Using a LOV2 variant with a faster dark-state recovery rate, we were able to reduce the dark recovery half-life by a factor of two. However, one should bear in mind that keeping a fast LOV2 domain activated will require more frequent light exposure. In addition, dark recovery of other LOV2-based optogenetics systems may be much faster especially if they only involve single-molecule conformational changes, and on and off kinetics should always be characterized to determine the optimal photoactivation illumination sequences.

1.3.3 Temperature

Although the photochemistry underlying LOV2 photoactivation is largely temperature insensitive within the biological range, the thermal decay of the flavin-cysteinyl adduct in the dark is not and occurs substantially slower at lower temperature (Fig. 3a) [6, 27]. However, somewhat surprisingly, we also found that while π-EB1 photodissociation occurred very rapidly in mammalian cells at 37 °C, it was much slower and incomplete at lower temperatures (Fig. 3b). We do not fully understand why this is the case, but it may be related to a decreased LOV2/Zdk1 off-rate at decreased temperature. This is further supported by our finding that photodissociation of π-EB1 with a wild-type LOV2 domain with LATT N-terminus that was originally used to develop LOVTRAP and likely has a higher affinity for Zdk1 is even less efficient at room temperature. It is possible that this temperature sensitivity is specific to the LOVTRAP system in which thermal energy facilitates LOV2/Zdk1 photodissociation. From a practical point of view, this is important when these systems are adapted to nonmammalian experimental models with lower temperature optimum such as

Fig. 2 (continued) tail end of the LOV2 absorbance spectrum (only ~3% of the peak absorbance) still results in π-EB1 photodissociation although much less efficiently than shorter wavelengths. This makes YFP a risky choice for live-cell imaging in combination with LOV2 optogenetics, especially at higher acquisition frame rates

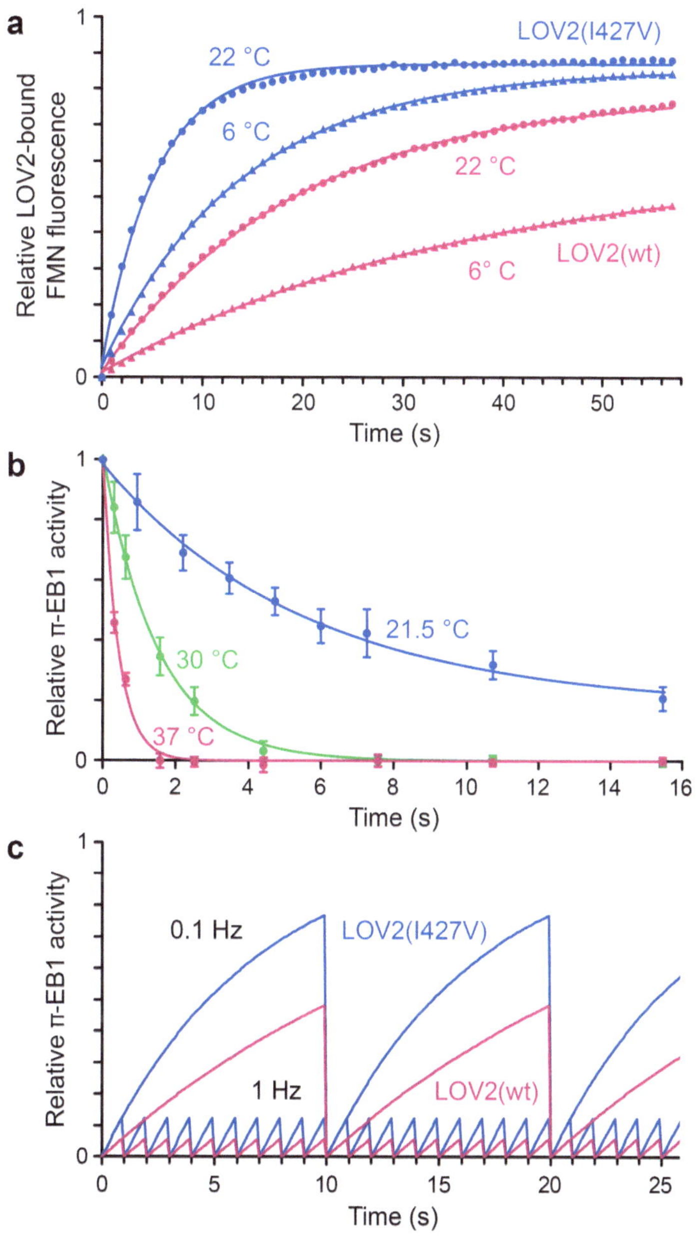

Fig. 3 Factors influencing π-EB1 photodissociation kinetics. (**a**) LOV2 reversion to the dark state is temperature sensitive and tunable. Dark-state recovery was analyzed by measuring flavin fluorescence ($\lambda_{Ex} = 475$ nm, $\lambda_{Em} = 537/35$ nm), which recovers upon flavin-cysteinyl adduct decay, at the indicated temperatures in *E. coli* colonies expressing a wild-type LOV2 domain or a fast-cycling variant LOV2(I427V) after photoactivation with higher intensity 475 nm light. The LOV2 dark-state recovery rate is a limiting factor in optogenetics experiments, because it impacts how much light is required to maintain a certain level of photoactivation. In addition, intracellular diffusion of photoactivated molecules limits spatial resolution. (**b**) Temperature sensitivity of π-EB1 photodissociation.

fruit flies, nematodes, or zebra fish. To verify that an optogenetic system works as anticipated in a given experimental system, it is therefore highly recommended to have a direct microscopy-based assay in place—such as blue light-induced dissociation of the π-EB1 C-terminal half from MT ends (Fig. 2b).

1.3.4 Phototoxicity

Phototoxicity is always a concern in live imaging experiments, especially with higher energy blue light [28]. In general, we find that the amount of light energy required for π-EB1 photodissociation is substantially lower than what is typically used for fluorescence imaging. For example, a single 100-ms exposure at 0.1–1 W mm^{-2} irradiance, which is a typical light intensity range in both epifluorescence or spinning disk confocal microscopy [29], amounts to a radiant exposure of 10–100 mJ mm^{-2} for each image taken. In contrast, we find that π-EB1 can be efficiently maintained in the dissociated state by short blue light pulses with energies below 0.2 mJ mm^{-2} [21]. However, to maintain π-EB1 photodissociation these blue light pulses must be administered at a frequency of 1 Hz or higher (Fig. 3c). Thus, in short-term experiments or experiments utilizing spatially patterned blue light, radiant exposure is quite low and likely negligible, but in longer term experiments the accumulated radiant energy becomes substantial (i.e., 3600 pulses per hour equals 720 mJ mm^{-2}). Slower, more light-sensitive LOV2 variants requiring less frequent blue light exposure may therefore be better suited for long-term experiments [30, 31]. It should also be noted that different cell types may have very different sensitivities to blue light. In addition, one should be aware that common cell culture media ingredients such as riboflavin are strong blue light photosensitizers. Strikingly, riboflavin-mediated generation of reactive oxygen species and hydrogen peroxide in blue light is further exacerbated in the presence of HEPES [32–34], which is frequently used to buffer cell culture media without CO_2 during microscopy experiments. Blue light-mediated reactive oxygen generation may be particularly

Fig. 3 (continued) π-EB1 activity was measured by monitoring the ratio of EGFP-Zdk1-EB1C on MT plus ends versus cytoplasm in time-lapse experiments at the indicated temperatures with identical blue light exposure settings. π-EB1 photodissociation is much slower and inefficient at room temperature compared with 37 °C. (**c**) Comparison of the predicted π-EB1 activity levels at different saturating blue light pulse frequencies based on measured exponential recovery rates. For example, at a 1 Hz blue light pulse frequency wild-type LOV2 recovers to ~5% before the next blue light pulse hits. Because of the faster dark recovery, the fast-cycling LOV2(I427V) variant would require an approximately twofold higher blue light pulse frequency to maintain the same average activation state. 0.1 Hz is insufficient to maintain a high average π-EB1 photodissociation (>50% recovery before the next blue light exposure) problematic in experiments using bulk blue light illumination of entire dishes or wells that would generate significant amounts of phototoxic compounds throughout the culture medium volume. For such experiments it is likely indicated to use special media formulations that reduce the concentration of photosensitive compounds [35]. In any case, control experiments with cells not expressing the light-sensitive construct at identical blue light exposure conditions must always be included.

2 Materials

It is difficult to present a comprehensive list of materials for live-cell microscopy experiments and the following should be seen only as guidelines that will need to be adapted for specific experimental questions.

2.1 Plasmids

1. π-EB1 Plasmids

pEB1N-LZ-LOV2(wt)	(Addgene #107614)	LAAA N-terminus
pEB1N-LZ-LOV2(fast)	(Addgene #107690)	
pmCherry-Zdk1-EB1C	(Addgene #107695)	
pEGFP-Zdk1-EB1C	(Addgene #107696)	

2. EB1/3 shRNA and CRISPR Plasmids

pLKO.1EB1 shRNA #3	(Addgene #37927)
pSpCas9_BB_2A_GFP_MAPRE1-gRNAs	(Addgene #107726, 107727, and 107728)
pSpCas9_BB_2A_GFP_MAPRE3-gRNAs	(Addgene #107729, 107730, and 107731)

3. Original LOVTRAP Plasmids

pTriEX-NTOM20-LOV2	(Addgene #81009)	LATT N-terminus
pTriEX-mCherry-Zdk1	(Addgene #81057)	

2.2 Cell Culture

1. H1299 lung cancer cells (ATCC CRL-5803), or other mammalian cell line.
2. RPMI-1640 medium with 10% FBS, 1× penicillin/streptomycin, 1× nonessential amino acids (NEAA), or medium as appropriate for different cell lines.
3. Transfection reagent (Fugene-6 or as required for other cell lines).
4. Reduced serum transfection media (e.g., Opti-MEM).

2.3 Microscopy

1. Inverted spinning disk confocal microscope. For a more detailed description of our system with a Borealis-modified Yokogawa CSU-X1, Photometrics cMyo CCD camera, ASI MS-2000 motorized stage, and an environmental enclosure on a Nikon Ti-E stand, *see* [36].
2. Digital micromirror device (DMD) pattern illuminator (e.g., Mightex Polygon400) with a blue 470 nm light source.
3. Imaging software that can control microscope and DMD device.
4. Light power meter (e.g., X-Cite XR2100).
5. Microscopy medium with reduced autofluorescence (e.g., FluoroBrite DMEM) supplemented with 10–20 mM HEPES buffer if CO_2 control is not available on the microscope.
6. Glass-bottom dishes (e.g., MatTek P35G-1.5-20-C).
7. Glass-bottom 12-well plates with black plastic frame to prevent cross-illumination between wells (e.g., Cellvis P12-1.5H-N).
8. Dow Corning high-vacuum grease in a 10 mL plastic syringe.
9. Yellow highlighter marker (e.g., Stabilo Boss) to visualize DMD patterns by fluorescence.
10. Black marker (e.g., Sharpie) to help with focusing on cover glass surface.

2.4 Miscellaneous Materials for LED Cube and Triggering

1. Arduino UNO microcontroller with external 9 V power supply.
2. TTL output from imaging computer (we use a National Instruments USB-6501 digital I/O board that is also used to control lasers and other shutters on our system).
3. Prototyping Breadboard and jumper wires.
4. 3 mm 60° viewing angle 470 nm LEDs (> 1 Cd brightness), for example Visual Communications Company VAOL-3LSBY4.
5. 24 V DC (500 mA–1 A) power supply to power the LED circuits.
6. Resistors (100–1000 Ω, 1/4 W).
7. Low-power general-purpose NPN switching transistor (e.g., 2N2222A).
8. 3D printer (often accessible in Maker lab spaces).
9. 15 mm Round cover glasses (e.g., Warner Instruments).

3 Methods

Since every fluorescence microscope will have a blue light excitation channel for imaging of GFP or similar fluorophores, the most basic implementation of LOV2 photoactivation with temporal control can be accomplished relatively easily on most microscopes. Spatial targeting of subcellular regions of interest however requires additional equipment to project blue light patterns onto the image plane of the microscope. Finally, controlling larger regions beyond the objective field of view also requires a different blue light illumination setup to allow simultaneous blue light exposure of larger areas or multiple wells while being compatible with specific imaging requirements. Here we focus on the technical aspects of how we perform these three different types of optogenetics experiments.

3.1 Temporal π-EB1 Photoinactivation Using Microscope Illumination

Any inverted fluorescence microscope will work for this. We use spinning disk confocal microscopy to clearly visualize MT plus-end-associated π-EB1, but it should be possible to see EB1 plus-end tracking with a conventional epifluorescence microscope equipped with high NA ×60 or ×100 oil immersion objectives. To test π-EB1 photodissociation cells can simply be transfected with plasmids encoding the N- and C-terminal π-EB1 halves, which we describe here. However, experiments assessing the function of EB1 +TIP complexes require removing endogenous EB1 (and possibly EB3). H1299 lung cancer cells in which endogenous EB1 has been depleted by either RNA interference or genetic knockout and replaced with π-EB1 are available on request. For procedures to establish these π-EB1 cell lines *see* [22]. All required plasmids are available on Addgene (*see* Subheading 2).

1. Seed cells on a 3.5 cm glass-bottom dish at ~10% confluency, and place in 37 °C, 5% CO_2, incubator overnight.
2. The next day co-transfect with plasmids encoding π-EB1 halves. Prepare transfection mix in a 1.5 mL Eppendorf tube (one per dish): Add 3 μL Fugene 6–100 μL Opti-MEM. Mix well. Add 1.5 μg EB1 N-LOV2 and 0.5 μg mCherry-Zdk1-EB1C (*see* **Note 2**). Mix well. Incubate for 15 min at room temperature. Add transfection mix dropwise to cell culture dish and return to incubator. Note that different types of cells will require different types of transfection procedures.
3. The next day, prepare the dish for microscopy by replacing the medium with low-fluorescence FluoroBrite DMEM. To avoid medium evaporation, we also often seal the lid of the dish onto the bottom by squeezing a bead of vacuum grease around the inside edge of the lid before closing.
4. Place dish in environmental chamber on the microscope stage and let equilibrate for 10 min.
5. Using the optical configuration for imaging red fluorescent proteins ($\lambda_{Ex} > 550$ nm), look for cells that show the typical EB1 comet pattern in cells (*see* Fig. 2b). MT plus-end localization of the mCherry-tagged C-terminal half indicates that the untagged N-terminal half is expressed at sufficient levels (*see* **Note 3**).

6. Set up a time-lapse experiment with two phases with one frame per second acquisition rate using filter sets for multiband GFP and RFP excitation (i.e., $\lambda_{Ex} = 488$ and 561 nm) and RFP emission ($\lambda_{Em} > 580$ nm). Each phase should be 20–30 s long with 200–400-ms exposure times:

   Phase 1: Only 561 nm excitation is on for imaging mCherry-Zdk1-EB1C; 488 nm light source is off.

   Phase 2: Both 561 nm and 488 nm light sources are on for LOV2 photoactivation (*see* **Note 4**).

7. Run imaging experiment.

The expected outcome is that mCherry-Zdk1-EB1C tracks growing MT ends in phase 1, but rapidly dissociates from MT ends in phase 2 when the 488 nm illumination source has been turned on (*see* Fig. 2b). The 488 nm light source should be adjusted to the minimal intensity that shows complete mCherry-Zdk1-EB1C dissociation in 2–3 images. On our Borealis-modified spinning disk setup this amounts to 10% power of the 100 mW 488 nm laser (~2 mW mm^{-2} in the image plane) with <200-ms exposure times [36].

3.2 Spatial π-EB1 Photoinactivation Using a Digital Micromirror Device (DMD)

Spatial control requires projecting light patterns of interest onto the specimen plane. While this can be accomplished using a point scanning system, we find that especially for the low light intensities required for LOV2 photoactivation, a digital micromirror device (DMD)-based illuminator is the most versatile and practical solution. Most commonly a DMD pattern illuminator is coupled into the microscope through a secondary epifluorescence port. In Nikon microscopes this is realized through two filter turrets stacked on top of each other. However, our system is a little unconventional in that we use a secondary camera port in the eyepiece lightpath to connect an older model Polygon400 pattern illuminator with a built-in 1 W 470 nm LED (Fig. 4b). Using an 80/20 beamsplitter in the emission lightpath, this has the advantage that the blue light pattern can be imaged directly with the spinning disk camera (Fig. 4c), which makes it very easy to focus the DMD pattern on the image plane, but it does require manual switching between eyepieces and DMD port before starting an experiment. Another disadvantage is that only 20% of the blue light pattern intensity reaches the specimen, but as outlined below this is still more than enough for LOV2 photoactivation.

3.2.1 Visualizing and Focusing the DMD Blue Light Pattern

In our setup, the blue light pattern can be directly imaged through the spinning disk microscope. However, if the DMD device is coupled through the epifluorescence excitation lightpath, it will likely be very difficult to directly image the blue light pattern reflection because the UV-selective dichroic mirror, which directs

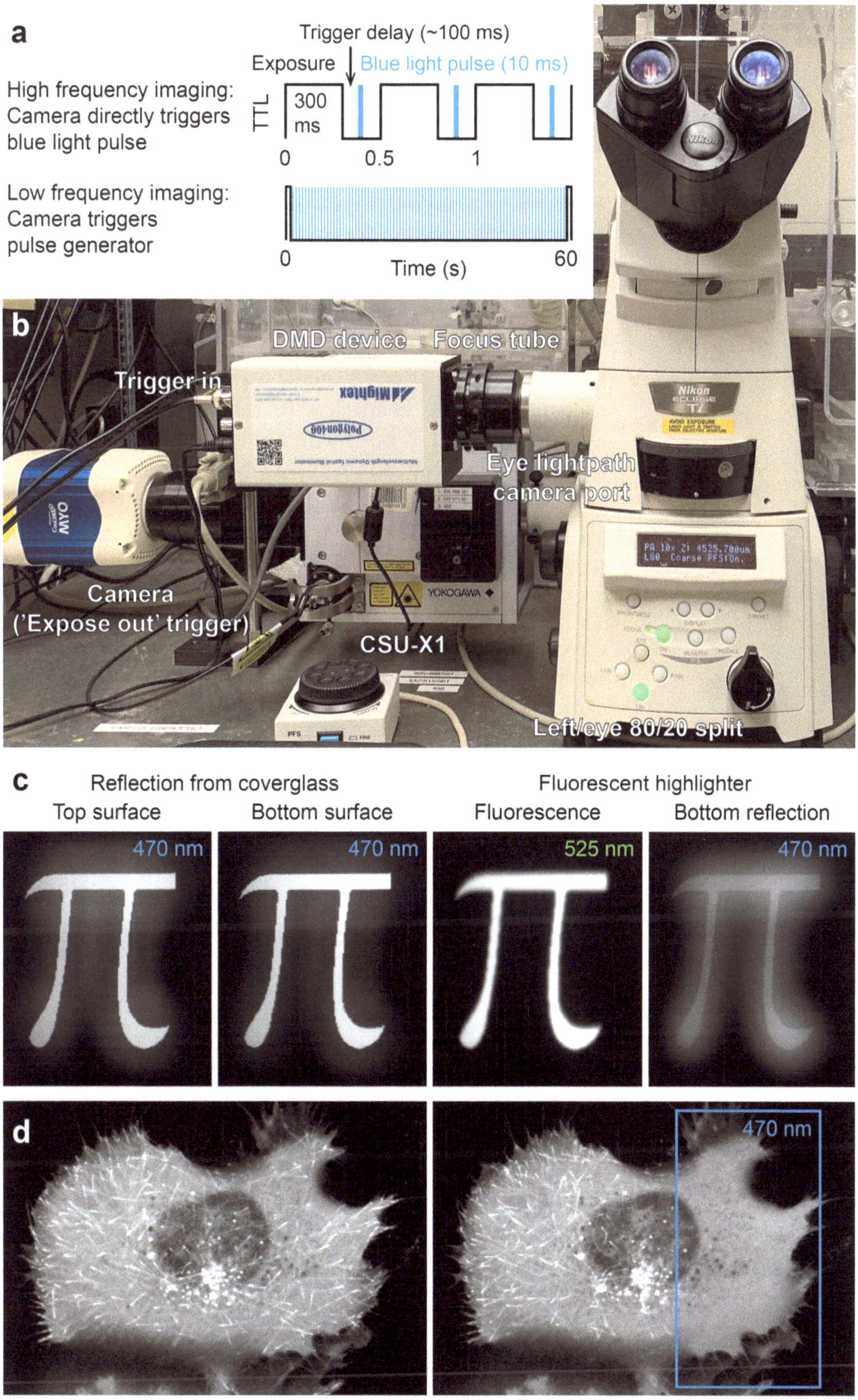

Fig. 4 Spatial control with a DMD device. (**a**) Schematic of the two methods triggering the DMD blue light pattern exposure depending on image acquisition rate. (**b**) Setup of the Polygon400 pattern illuminator on our

the excitation light from the DMD device to the specimen, will also prevent most light reflected back from the cover glass from reaching the camera. In this case fluorescence from a yellow/green highlighter marker can be used to visualize the pattern.

1. Mark a glass-bottom dish with fluorescent highlighter. Omit if directly imaging the blue light reflection. Place the dish on the microscope stage.
2. Set up imaging in the GFP channel but turn off the 488 nm excitation laser. If directly visualizing the blue light reflection, also remove the GFP emission filter from the lightpath. Start live imaging.
3. Focus on the top cover glass surface using the camera and transmitted light. A mark with a black Sharpie marker helps with focusing. It is not terribly important if this is the top or bottom surface reflection. It is only important that the camera chip and the DMD chip end up parfocal.
4. Turn off transmitted light. Turn on blue light pattern (*see* **Note 5**).
5. Refocus on the blue light pattern as much as possible.
6. Carefully slide the DMD device in the focus tube to obtain the sharpest possible image. Also turn sideways to align with the camera angle as much as possible. Best results likely require a few iterations of Steps 5 and 6. Once satisfied, tighten set screws on the DMD focus tube.

3.2.2 Measuring Irradiance at the Specimen Plane

To minimize phototoxicity and achieve maximal spatial control, it is important to determine the optimal pulse frequency and light exposure, and thus to characterize the illumination setup as accurately as possible. The light power at the specimen plane can be measured with a light meter. However, this is only informative if we also know the area of the exposed surface. The amount of light power received per surface area is defined as irradiance (in W per unit area).

1. Using the procedure described above, take an image of the reflection (or fluorescence) of a rectangular blue light pattern.

Fig. 4 (continued) spinning disk confocal microscope system through an additional camera port in the eyepiece lightpath. (**c**) Visualization of the blue light pattern. The left two panels show the reflection of the pattern (π symbol) on the top and bottom surfaces of a cover glass. Note the slight haze around the π symbol resulting from the out-of-focus reflection from the other side of the cover glass. The two panels on the right show fluorescence of a yellow highlighter mark and the much less bright direct blue light reflection from the bottom cover glass surface. (**d**) Spatially restricted π-EB1 photodissociation. Cell expressing untagged EB1 N-LOV2(I427V) and mCherry-Zdk1-EB1C. Images were acquired at one frame per second. 20-ms pulses of the DMD blue light pattern were triggered directly by the camera in between image acquisitions

2. Calculate the area of the blue light pattern in μm^{-2}. This requires knowledge of the calibrated pixel size. If there is no intermediate magnification in the lightpath, this should be very close to the pixel size of the camera divided by the objective magnification.
3. Remove cover glass or dish. If an oil immersion lens was used, carefully wipe the oil off the objective. Do not put immersion oil on the light sensor!
4. Place the sensor of the microscope light meter on the microscope stage with the sensor window facing toward the objective. Set the light meter to the correct wavelength (i.e., 470 nm). While the sensor of the light meter should not be too far from the focal plane, it does not have to be perfectly in focus. The photon flux in and out of focus is the same, only spread over a different area.
5. Turn on the same photoactivation pattern in continuous mode, not pulsed, to measure light power.
6. Calculate irradiance by dividing the power reading by the area of the blue light pattern. Example: The light meter reads 5 mW. The pattern is a square with 500 μm sides. Thus, the irradiance is 5 mW divided by 0.25 mm^2, or 20 mW mm^{-2}. Beware of the squares of metric prefixes.

While we can measure the irradiance of different photoactivation setups, it is important to remember that irradiance characterizes the brightness at the specimen plane but does not by itself inform how much light energy is received by the specimen as that also depends on the amount of time of irradiation. The amount of light energy received is defined as radiant exposure (J per unit area), which is the irradiance multiplied by the exposure time.

We typically use the DMD LED at the lowest power setting. With a ×60 1.49 NA oil immersion lens, this gives us an irradiance at the specimen plane of just below 20 mW mm^{-2}. Thus, a single 10-ms blue light pulse delivers a radiant exposure of almost 0.2 mJ mm^{-2}, and one hundred 10-ms pulses will deliver a cumulative 20 mJ mm^{-2} (*see* **Note 6**). Both irradiance and radiant exposure should be reported in optogenetics experiments (*see* **Note 7**). Statements in experimental procedures of percent laser power used or even light power coming out of the objective are essentially meaningless, contain no information about the amount of light energy a sample was exposed to, and should be avoided.

3.2.3 Blue Light Triggering

Most light sources including LEDs have only a limited and nonlinear range of intensity control. Thus, the best way to control the amount of blue light exposure is by modulating exposure time at a fixed intensity. In addition, to prevent interference of the blue light pattern with image acquisition (we observe some scattered blue light in the RFP channel, and 470 nm light excites RFP fluorescence to some extent), it is best to time patterned blue light exposures such that they occur in between image acquisitions (Fig. 4a). Depending on the experiment, different hardware solutions may have to be implemented to trigger blue light pattern pulse sequences:

Fast time lapse: For short experiments with relatively high acquisition frequency (e.g., MT dynamics at two frames per second image acquisition rate), the blue light pattern can be directly triggered by a high- to low-edge TTL output when the camera has finished image exposure. Most scientific cameras provide such a TTL signal, but the name varies by manufacturer (Photometrics calls it "Expose out"). In this case, the blue light pulse frequency is determined by the image acquisition frequency and needs to be sufficiently high to maintain π-EB1 photodissociation (i.e., >1 Hz for π-EB1; *see* Fig. 3c). It is also important to make sure that there is enough time in between images for the blue light pulse (i.e., a 300-ms image exposure at two frames per second leaves 200 ms for the blue light pulse). In addition, a trigger delay can be set to temporally shift the blue light pulse away from the image acquisition window.

Slow time lapse: For longer experiments with lower acquisition frequency (e.g., cell migration at one frame per minute or less), it is necessary to utilize an external pulse generator at at least 1 Hz if π-EB1 photodissociation is to be maintained. Such a pulse generator can be quite easily implemented using an Arduino microcontroller similar to what we describe below (*see* Fig. 5c; the output pin 5 of the Arduino board would then directly connect to the trigger-in of the Polygon400). However, because the Polygon400 is unresponsive to external triggers once a blue light pulse has started, the timing of blue light pulses in relation to the imaging frame rate must be carefully planned to avoid overlap with the next image acquisition. In addition, the pulse generator should be synchronized after each image exposure; otherwise small timing inaccuracies will eventually result in overlap.

3.3 LED Cube Design for Multiwell Plate Photoactivation

Because blue light exposure through the objective only allows for photoactivation of one field of view at a time, especially long-term experiments observing cell behavior over several hours become very low throughput. Photoactivation of larger areas or multiple wells (to allow for control and experimental conditions in parallel) therefore requires an external light source. To further permit transmitted light imaging in combination with π-EB1 photodissociation, we designed a 3D printed LED cube that fits over two wells of a 12-well glass-bottom plate (Fig. 5a,b). This design in which an observation window is surrounded by six 3 mm LEDs allows simultaneous and more or less evenly distributed blue light exposure and transmitted light imaging of two complete wells. The

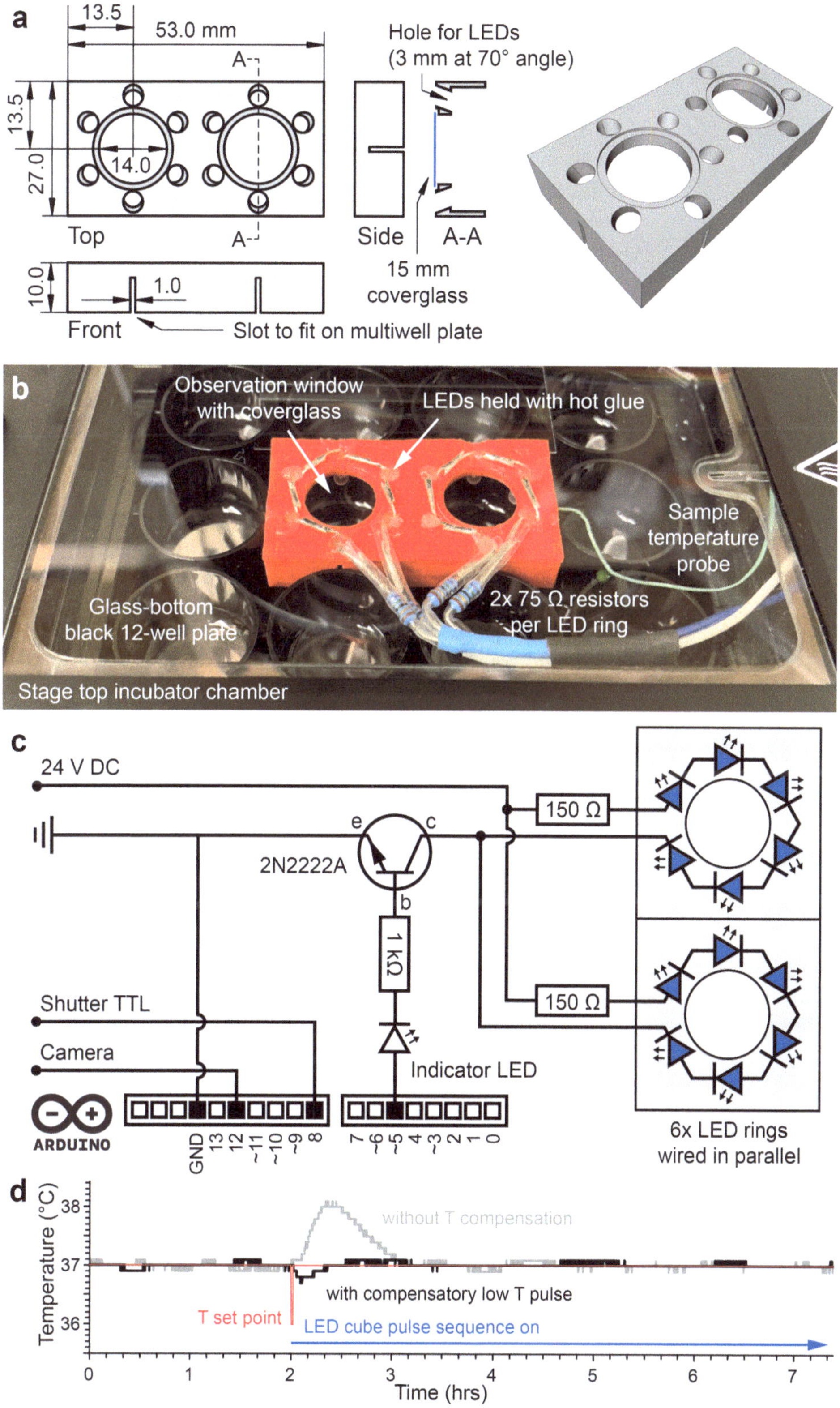

Fig. 5 DIY LED cube design for multiwell π-EB1 photodissociation. (**a**) Technical drawing of the 3D printed plastic part of the LED cube (FreeCAD 3D file available upon request). (**b**) Image of the LED cube on a 12-well

observation window can further be covered with a 15 mm cover glass to minimize media evaporation, and the LED cube fits inside an Okolab stage top incubator. Pulse frequency and duration are controlled by an Arduino UNO microcontroller that also takes a TTL input from a shutter or the imaging software to switch photo-activation on and off, and a TTL input from the camera to turn off the blue LEDs when an image is taken (Fig. 5c; *see* **Notes 8** and **9** for a more detailed explanation of the control circuit and corresponding Arduino code, respectively).

Because the blue light is spread over a much larger area the irradiance in this LED cube setup is substantially lower than what is achieved with the DMD device. Even though we are using some of the brightest available LEDs with a wide viewing angle, the power measured with a light sensor just below the glass bottom is <15 mW. The diameter of the EXFO light sensor is 1 cm (i.e., the area is 78.5 mm^2), which gives an irradiance of a little less than 0.2 mW mm^{-2}. This is approximately 100 times less than what we typically use with the DMD. Thus, to achieve a similar radiant exposure per blue light pulse, pulse length with this LED cube has to be much longer. Empirically, we find that 400-ms blue light pulses are enough to achieve sustained π-EB1 photodissociation.

Finally, it is important to note that even these low-power LEDs dissipate a significant amount of heat resulting in an increase in sample temperature. Because we directly measure sample temperature with a thermocouple submersed in one of the blue light-exposed wells (adjacent wells will not reflect temperature in the illuminated wells), feedback to the heating system eventually returns temperature to the set point. We were also able to largely compensate for this transient temperature increase by temporarily turning off the heating of the stage top incubator at the same time as the blue light exposure is turned on (Fig. 5d).

4 Notes

1. Many LOV2 mutations have been identified that affect the dark-state recovery rate, with half-lives from 2 s to over 30 min [30, 31, 37]. One must be aware though that most of these LOV2 mutants have not been characterized well in the

Fig. 5 (continued) plate inside an Okolab stage top incubator. The LED power wires are threaded through holes in the side of the incubator chamber. Six LEDs are hot glued into the 3 mm holes and leads from adjacent LEDs are soldered together in series on top of the plastic part. (**c**) Circuit diagram of the Arduino-based LED pulse generator. (**d**) Temperature trace inside a blue light exposed well before and after LEDs are turned on. A short low-temperature pulse (1 min to 36 °C at the start of blue light exposure) effectively compensates for the transient temperature increase

context of optogenetic switches and that slower LOV2 variants will be much more sensitive to ambient light and extended transmitted light imaging. In this case, a red filter is strongly recommended.

2. Transient transfection with untagged EB1N-LOV2 and mCherry-Zdk1-EB1C is a simple and useful way to characterize the photoactivation setup. However, for making cell lines in which endogenous EB1 is replaced by π-EB1 to optically control MT dynamics, we have used EGFP-Zdk1-EB1C in combination with an untagged version of EB1 N-LOV2 instead. While this does not allow observation of EGFP-Zdk1-EB1C MT plus-end tracking, it is possible to acquire single images in the GFP channel to evaluate MT end localization and ensure that π-EB1 photodissociation works as expected. A GFP tag also greatly facilitates FACS sorting and clonal selection. Most importantly, labeling π-EB1 with GFP leaves the mCherry channel available for other markers of interest. In π-EB1 cells without endogenous EB1 (and EB3), other mCherry-tagged +TIPs should disappear from growing MT ends as a consequence of π-EB1 photodissociation. In addition, for analyzing MT organization and growth dynamics, we transiently transfect mCherry-tubulin, or an mCherry-tagged version of the π-EB1 N-terminal half, which remains bound to MT ends in the presence of blue light.
3. No MT end localization of mCherry-Zdk1-EB1C after a few minutes of no blue light exposure indicates that EB1 N-LOV2 is not expressed. mCherry-Zdk1-EB1C localization along MTs indicates too high expression levels.
4. In Nikon NIS Elements, the software we use to control our microscopes, this can be done by defining two consecutive time phases in the ND acquisition dialog. In the advanced settings, a macro is executed that changes 488 nm laser power from 0% to 10% between the two time phases.
5. We are using a first-generation Polygon400 that can only be controlled by the (now obsolete) Mightex PolyLite software. This requires that a transparent DMD ROI window is aligned and overlaid with the camera window in the imaging software to draw and target patterns. There are surely much more elegant ways to do this, but this is the cheapest and requires no software integration. To turn on the pattern continuously in PolyLite, we use autotrigger mode and set trigger period and pulse duration to the same value.
6. When measuring irradiance on the microscope, one must keep in mind that photon flux density scales with the square of objective magnification. Because of the difference in area, with the same light source, the irradiance with a ×10 objective

is 36-fold smaller than with a ×60 objective. In other words, a 10-ms blue light pulse at ×60 will deliver the same amount of energy per unit area as a 360-ms pulse at ×10.

7. Confusingly, for pulsed light exposure irradiance is sometimes multiplied by the pulse duty cycle and expressed as "average irradiance"; that is, a 1 Hz sequence of 10-ms pulses has a 1% duty cycle and at 100 mW mm^{-2} would have a 1 mW mm^{-2} "average irradiance." However, this becomes very confusing when one wants to figure out the light energy received by a specimen. It seems more meaningful for both regular microscopy experiments and optogenetics to report the radiant exposure received by the sample.

8. Control of a low-power LED array is implemented as a low-side transistor switch using low-cost components and can be easily implemented on a prototyping breadboard. In this example, the LED pulse sequence is controlled by an Arduino microcontroller, but any other TTL pulse generator should work equally well. The value of the current limiting resistor in the LED series depends on the power ratings of the specific LEDs. We used six 3 mm through hole 470 nm LEDs from Visual Communications (VCC) with luminous intensity 1200 mcd, forward current $I_f = 20$ mA, and forward voltage $V_f = 3.5$ V. The value of the current limiting resistor is calculated as $R = (V_{supply} - n_{LEDs} \cdot V_f)/I_f = (24\ V - 6 \cdot 3.5\ V)/0.02\ A = 150\ \Omega$. We use 1/4 W resistors to be on the safe side. These LEDs have a 60° viewing angle to cover the bottom of the dish as evenly as possible. The 2N2222A NPN transistor is rated up to 40 V and 600 mA, so it should be able to handle many of these LED circuits in parallel.

9. Arduino code to control LEDs using the circuit shown in Fig. 5c is listed below. Shutter is a software-controlled TTL output. In this case, the Arduino board does not have to be connected to the imaging computer and can stand alone. However, if a TTL output is not available this could also be implemented directly as a serial command to the Arduino board. Multiple different pulse sequences can be programmed onto different output pins in parallel. Time measurements in milliseconds are not entirely precise on Arduino, which counts time in 1.024-ms increments. To avoid drift over longer time periods, the LED pulse cycle is reset after each camera exposure.

```
class Blues
{
 int ledPin; // the number of the LED output pin
 long OnTime; // Time LED is turned on in milliseconds
```

```
long OffTime; // Time LED is turned off in milliseconds
int trigState;
int camCurr;
int camLast;
int shuttState;
unsigned long previous; // last time LED was updated

public:
Blues(int pin, long on, long off)
{
ledPin = pin;
pinMode(ledPin, OUTPUT);

OnTime = on; // transfer and initialize variables
OffTime = off;
trigState = LOW;
camCurr = LOW;
camLast = LOW;
shuttState = LOW;
previous = 0;
}

void Update()
{
unsigned long current = millis();
if ((trigState == HIGH) && (current - previous >= OnTime))
{
trigState = LOW; // time to turn LED off
previous = current; // remember the time
}
if ((trigState == LOW) && (current - previous >= OffTime))
{
trigState = HIGH; // time to turn LED on
previous = current; // remember the time
}

shuttState = digitalRead(8); // read external shutter state
from pin 8
camCurr = digitalRead(12); // read camera exposure state from
pin 12

if ((camCurr == LOW) && (shuttState == HIGH) && (trigState ==
HIGH))
{
digitalWrite(ledPin, HIGH);
} // turn on LED only if shutter is on and camera is not
exposing
{
```

```
 digitalWrite(ledPin, LOW);
 }

if ((camLast == HIGH) && (camCurr == LOW))
 {
 previous = 0;
 } // reset time after camera exposure

camLast = camCurr; // remember previous camera state
 }
};

Blues led5(5, 400, 600); // define output channel pin 5 at 1 Hz
400 ms on

// define additional output channels
// Blues ledX(pin number, time on in ms, time off in ms)

void setup()
{
 pinMode(8, INPUT); // set shutter input pin
 pinMode(12, INPUT); // set camera input pin
}

void loop()
{
 led5.Update(); // update LED state

// update additional output channels
// ledX.Update();
}
```

Acknowledgments

This work was supported by NIH grants R01 GM094819, R01 NS107480, and R21 CA224194 to T.W. We would also like to thank Dylan Romero and Jenny Tai from The Makers Lab at UCSF for valuable advice and assistance with 3D printing.

References

1. Losi A, Gardner KH, Möglich A (2018) Blue-light receptors for optogenetics. Chem Rev 118:10659–10709. https://doi.org/10.1021/acs.chemrev.8b00163
2. Goglia AG, Toettcher JE (2019) A bright future: optogenetics to dissect the spatiotemporal control of cell behavior. Curr Opin Chem Biol 48:106–113. https://doi.org/10.1016/j.cbpa.2018.11.010

3. Huala E, Oeller PW, Liscum E et al (1997) Arabidopsis NPH1: a protein kinase with a putative redox-sensing domain. Science 278:2120–2123. https://doi.org/10.1126/science.278.5346.2120
4. Christie JM, Salomon M, Nozue K et al (1999) LOV (light, oxygen, or voltage) domains of the blue-light photoreceptor phototropin (nph1): binding sites for the chromophore flavin mononucleotide. Proc Natl Acad Sci U S A 96:8779–8783. https://doi.org/10.1073/pnas.96.15.8779
5. Christie JM (2007) Phototropin blue-light receptors. Annu Rev Plant Biol 58:21–45. https://doi.org/10.1146/annurev.arplant.58.032806.103951
6. Salomon M, Christie JM, Knieb E et al (2000) Photochemical and mutational analysis of the FMN-binding domains of the plant blue light receptor, phototropin. Biochemistry 39:9401–9410
7. Harper SM, Neil LC, Gardner KH (2003) Structural basis of a phototropin light switch. Science 301:1541–1544. https://doi.org/10.1126/science.1086810
8. Konold PE, Mathes T, Weienborn J et al (2016) Unfolding of the C-terminal Jα Helix in the LOV2 photoreceptor domain observed by time-resolved vibrational spectroscopy. J Phys Chem Lett 7:3472–3476. https://doi.org/10.1021/acs.jpclett.6b01484
9. Zayner JP, Antoniou C, Sosnick TR (2012) The amino-terminal helix modulates light-activated conformational changes in AsLOV2. J Mol Biol 419:61–74. https://doi.org/10.1016/j.jmb.2012.02.037
10. Kottke T, Xie A, Larsen DS, Hoff WD (2018) Photoreceptors take charge: emerging principles for light sensing. Annu Rev Biophys 47:291–313. https://doi.org/10.1146/annurev-biophys-070317-033047
11. Yao X, Rosen MK, Gardner KH (2008) Estimation of the available free energy in a LOV2-Jα photoswitch. Nat Chem Biol 4:491–497. https://doi.org/10.1038/nchembio.99
12. Wu YI, Frey D, Lungu OI et al (2009) A genetically encoded photoactivatable Rac controls the motility of living cells. Nature 461:104–108. https://doi.org/10.1038/nature08241
13. Lungu OI, Hallett RA, Choi EJ et al (2012) Designing Photoswitchable peptides using the AsLOV2 domain. Chem Biol 19:507–517. https://doi.org/10.1016/j.chembiol.2012.02.006
14. Dagliyan O, Tarnawski M, Chu PH et al (2016) Engineering extrinsic disorder to control protein activity in living cells. Science 354:1441–1444. https://doi.org/10.1126/science.aah3404
15. Reynolds KA, McLaughlin RN, Ranganathan R (2011) Hot spots for allosteric regulation on protein surfaces. Cell 147:1564–1575. https://doi.org/10.1016/j.cell.2011.10.049
16. Wang H, Vilela M, Winkler A et al (2016) LOVTRAP: an optogenetic system for photoinduced protein dissociation. Nat Methods 13:755–758. https://doi.org/10.1038/nmeth.3926
17. Guntas G, Hallett RA, Zimmerman SP et al (2015) Engineering an improved light-induced dimer (iLID) for controlling the localization and activity of signaling proteins. Proc Natl Acad Sci U S A 112:112–117. https://doi.org/10.1073/pnas.1417910112
18. Strickland D, Lin Y, Wagner E et al (2012) TULIPs: tunable, light-controlled interacting protein tags for cell biology. Nat Methods 9:379–384. https://doi.org/10.1038/nmeth.1904
19. Kumar P, Wittmann T (2012) +TIPs: SxIPping along microtubule ends. Trends Cell Biol 22:418–428
20. van Haren J, Wittmann T (2019) Microtubule plus end dynamics—do we know how microtubules grow? BioEssays 41:e1800194. https://doi.org/10.1002/bies.201800194
21. van Haren J, Charafeddine RA, Ettinger A et al (2018) Local control of intracellular microtubule dynamics by EB1 photodissociation. Nat Cell Biol 20:252–261. https://doi.org/10.1038/s41556-017-0028-5
22. Wittmann T, van Haren J (2018) Generation of cell lines with light-controlled microtubule dynamics. Protoc Exchange. https://doi.org/10.1038/protex.2017.155
23. Grimm JB, Muthusamy AK, Liang Y et al (2017) A general method to fine-tune fluorophores for live-cell and in vivo imaging. Nat Methods 14:987–994. https://doi.org/10.1038/nmeth.4403
24. Swartz TE, Corchnoy SB, Christie JM et al (2001) The photocycle of a flavin-binding domain of the blue light photoreceptor phototropin. J Biol Chem 276:36493–36500. https://doi.org/10.1074/jbc.M103114200
25. Kennis JTM, Crosson S, Gauden M et al (2003) Primary reactions of the LOV2 domain of phototropin, a plant blue-light photoreceptor. Biochemistry 42:3385–3392. https://doi.org/10.1021/bi034022k
26. Eitoku T, Nakasone Y, Matsuoka D et al (2005) Conformational dynamics of phototropin 2 LOV2 domain with the linker upon

photoexcitation. J Am Chem Soc 127:13238–13244. https://doi.org/10.1021/ja052523i

27. Kottke T, Heberle J, Hehn D et al (2003) Phot-LOV1: photocycle of a blue-light receptor domain from the green alga Chlamydomonas reinhardtii. Biophys J 84:1192–1201. https://doi.org/10.1016/S0006-3495(03)74933-9
28. Laissue PP, Alghamdi RA, Tomancak P et al (2017) Assessing phototoxicity in live fluorescence imaging. Nat Methods 14:657–661. https://doi.org/10.1038/nmeth.4344
29. Ettinger A, Wittmann T (2014) Fluorescence live cell imaging. Methods Cell Biol 123:77–94. https://doi.org/10.1016/B978-0-12-420138-5.00005-7
30. Zayner JP, Sosnick TR (2014) Factors that control the chemistry of the LOV domain photocycle. PLoS One 9:e87074. https://doi.org/10.1371/journal.pone.0087074
31. Kawano F, Aono Y, Suzuki H, Sato M (2013) Fluorescence imaging-based high-throughput screening of fast- and slow-cycling LOV proteins. PLoS One 8:e82693. https://doi.org/10.1371/journal.pone.0082693
32. Grzelak A, Rychlik B, Bartosz G (2001) Light-dependent generation of reactive oxygen species in cell culture media. Free Radic Biol Med 30:1418–1425. https://doi.org/10.1016/S0891-5849(01)00545-7
33. Zigler JS, Lepe-Zuniga JL, Vistica B, Gery I (1985) Analysis of the cytotoxic effects of light-exposed HEPES-containing culture medium. In Vitro Cell Dev Biol 21:282–287. https://doi.org/10.1007/BF02620943
34. Halliwell B, Butt VS (1972) Flavin mononucleotide-sensitized photo-oxidation of glyoxylate in Good's buffers. (Short Communications). Biochem J 129:1157–1158. https://doi.org/10.1042/bj1291157
35. Stockley JH, Evans K, Matthey M et al (2017) Surpassing light-induced cell damage in vitro with novel cell culture media. Sci Rep 7:849. https://doi.org/10.1038/s41598-017-00829-x
36. Stehbens S, Pemble H, Murrow L, Wittmann T (2012) Imaging intracellular protein dynamics by spinning disk confocal microscopy. Methods Enzymol 504:293–313. https://doi.org/10.1016/B978-0-12-391857-4.00015-X
37. Christie JM, Corchnoy SB, Swartz TE et al (2007) Steric interactions stabilize the signaling state of the LOV2 domain of phototropin 1. Biochemistry 46:9310–9319. https://doi.org/10.1021/bi700852w

Chapter 15

Measurement of Microtubule Half-Life and Poleward Flux in the Mitotic Spindle by Photoactivation of Fluorescent Tubulin

Hugo Girão and Helder Maiato

Abstract

The study of microtubule dynamics is of utmost importance for the understanding of the mechanisms underlying mitotic fidelity. During mitosis, the microtubular cytoskeleton reorganizes to assemble a mitotic spindle necessary for chromosome segregation. Several methods, such as controlled exposure to cold, high pressure, high calcium concentration, or microtubule depolymerizing drugs, have been widely used to evaluate the dynamic properties of specific spindle microtubule populations. However, while these methods offer a qualitative approach that is sufficient to discern differences among specific spindle microtubule populations, they fall short in providing a robust quantitative picture that is sensitive enough to highlight minor differences, for example when comparing spindle microtubule dynamics in different genetic backgrounds. In this chapter we describe a detailed methodology to measure spindle microtubule dynamics using photoactivation of fluorescently tagged tubulin in living cells. This methodology allows the quantitative discrimination of the turnover of specific microtubule populations (e.g., kinetochore vs. non-kinetochore microtubules), as well as determination of microtubule poleward flux rates. These two conspicuous features of metazoan spindles must be tightly regulated to allow, on the one hand, efficient error correction, and on the other hand the satisfaction of the spindle assembly checkpoint that controls mitotic fidelity.

Key words Photoactivation, Fluorescence microscopy, Microtubule dynamics, Mitosis, Mitotic spindle, Flux

1 Introduction

Cytoskeleton microtubules consist of highly dynamic polymeric protofilaments of α- and β-tubulin that continuously switch between phases of growth and shrinkage, a process known as dynamic instability [1]. Among other vital cellular processes, microtubules play a key role during mitosis, since they establish the mitotic spindle required for chromosome segregation. During this process, centrosomal and non-centrosomal microtubules eventually interact with the kinetochore region of chromosomes

Helder Maiato (ed.), *Cytoskeleton Dynamics: Methods and Protocols*, Methods in Molecular Biology, vol. 2101, https://doi.org/10.1007/978-1-0716-0219-5_15, © Springer Science+Business Media, LLC, part of Springer Nature 2020

establishing kinetochore fibers (k-fibers), while other microtubules from opposing poles interact to form interpolar microtubules, or interact with the cell cortex to orient the spindle along the cell division axis. Although the mitotic spindle at steady state appears as a rigid structure, microtubules are continuously turning over due to dynamic instability and poleward flux of tubulin on stable microtubule populations [2]. In fact, the highest dynamic state of microtubules peaks during mitosis [3, 4]. Due to the stochastic nature of microtubule-kinetochore interactions during mitosis, erroneous attachments often occur, and their correction requires microtubule turnover at kinetochores to avoid chromosome missegregation that might result in aneuploidy [5, 6]. In contrast, the satisfaction of the spindle assembly checkpoint (SAC) that controls the metaphase-anaphase transition requires the formation of stable kinetochore-microtubule attachments that are competent to segregate chromosomes during anaphase [7]. Therefore, obtaining a quantitative picture of the mechanisms and principles underlying the regulation of microtubule dynamics in the complex context of the mitotic spindle will be critical for our fundamental understanding of mitosis and the respective implications for human diseases, such as cancer [8].

The use of quantitative fluorescence microscopy techniques in the study of microtubule dynamics during mitosis has become the gold standard for understanding the structure, assembly, and function of the mitotic spindle [2, 9, 10]. In this regard, the development of photoactivatable derivatives of fluorescent proteins [2, 11, 12] has played a critical role in the study of microtubule dynamics by live-cell imaging. Accordingly, by being exposed to a specific wavelength beam, nonfluorescent probes convert to a fluorescence-emitting form [13], and this spectral shift can be reversible or irreversible [14]. By fusing such probes with tubulin, followed by expression in living cells, they incorporate into microtubules of the mitotic spindle due to their dynamic nature. Then, by using spatially and temporally confined irradiation it is possible to locally stimulate a defined region of interest in the mitotic spindle and simultaneously follow the movement and the dissipation of the fluorescent signal over time [15]. This allows the determination of both spindle microtubule turnover and microtubule poleward flux velocity [7, 12]. This method shows advantages when compared with photobleaching techniques, since it has a better signal-to noise ratio and, most importantly, allows the photochemical switch without side reactions that lead to the production of heat or reactive oxygen species (ROS) [16–18].

In this chapter we describe detailed methodology to measure microtubule turnover and determine the stability of distinct spindle microtubule populations during mitosis by fluorescence dissipation after photoactivation (FDAPA). This technique uses a 405 nm laser to activate photoactivatable (PA)-green fluorescent protein

(GFP)-α-tubulin and measure the respective fluorescence dissipation over time. By fitting the fluorescence dissipation curve after photoactivation to a double exponential, it is possible to determine the half-life of two spindle microtubule subpopulations with slow and fast recovery kinetics [7]. These two microtubule subpopulations have been interpreted to correspond to kinetochore and non-kinetochore microtubules, respectively. Additionally we show the utility of photoactivation for the determination of poleward microtubule flux velocity in the mitotic spindle by kymograph analysis [19].

2 Materials

2.1 Cell Culture

1. Cell line stably expressing PA-GFP-α-tubulin (optional: if possible, additional expression of tubulin with a red fluorescent tag, for example mCherry or mRFP, might be useful to define the entire spindle).
2. T25 and T75 cell culture flasks.
3. Dulbecco's modified Eagle medium (DMEM), 4.5 g/L glucose, L-glutamate, sodium pyruvate.
4. Fetal bovine serum (FBS), heat inactivated.
5. 35 mm MatTek dish #1.5 coverslip, 14 mm glass diameter.
6. 9.5 cm^2, 34.8 mm diameter 6-well plates.
7. Squared glass coverslips 22 × 22 mm, #1.5.
8. Cell culture incubator with CO_2 supply.

2.2 Imaging

1. Leibovitz's L-15 medium phenol-red-free, L-glutamate.
2. DMEM phenol-red-free, 4.5 g/L glucose, 25 mM of 4-(2-hydroxyethyl)-1-piperazineethanesulfonic acid (HEPES).
3. MG-132 proteasome inhibitor.
4. Imaging chambers (e.g., CYTOO chamber).

2.3 Microscope

1. Inverted microscope (e.g., Nikon TE2000U).
2. CSU-X1 spinning disk confocal head (Yokogawa Corporation of America).
3. ×100 1.4 NA Plan-Apochromatic differential interference contrast (DIC) objective.
4. Heated chamber (37 °C) (optional CO_2 supplier).
5. Two laser lines 488 and 561 nm.
6. iXonEM+ EM-CCD camera (Andor Technology).
7. Mosaic DMD-based patterning system (Andor) equipped with a 405 nm diode laser.

2.4 Software

1. NIS Elements Advance Research (Nikon).
2. Fiji—ImageJ.
3. MatLab (Mathworks).
4. Custom-written routine LAPSO software [19].

3 Methods

3.1 Sample Preparation

1. Culture cells in T25 of T75 flasks with DMEM supplemented with 10% FBS and grown at 37 °C in a 5% CO_2, humidified atmosphere.
2. Two to three days prior to imaging, seed cells stably expressing PA-GFP-α-tubulin in 6-well plates containing square glass coverslips in a total of 2 mL of DMEM with 10% FBS in each well (*see* **Notes 1** and **2**). Alternatively, cells may be seeded directly into 3.5 mm MatTek glass-bottom dishes. Incubate the samples at 37 °C in a 5% CO_2, humidified atmosphere.
3. One day before imaging replace the medium with 2 mL of DMEM without phenol red and with 25 mM HEPES supplemented with 10% FBS (*see* **Note 3**). Incubate the samples at 37 °C in a 5% CO_2, humidified atmosphere.
4. To prepare the samples for microscope imaging, carefully take out the coverslip from the well with the help of some tweezers and mount it in a CYTOO chamber (*see* **Note 4**). Use 1 mL of conditioned medium from the well in the chamber and use a glass top cover (*see* **Note 5**). If using MatTek chambers, use directly the dish in the microscope.

3.2 Microscope Preparation for Imaging

1. In the temperature controller set the temperature to 37 °C and if the microscope has CO_2 supply, select levels to 5%.
2. Turn on all parts of the microscope and the computer and open the NIS Elements Advance Research application.
3. To calibrate the 405 nm diode laser, select the ×100 oil-immersion 1.4 NA objective and focus a scratched glass mirror slide in DIC.
4. Open the Mosaic submenu and set the laser intensity above 90% and exposure time to 500 ms.
5. Select "Calibration," align the cursor for the three calibration points, and double-click to set the point of calibration. In the end select "Calibrate" (*see* **Note 6**).
6. Set the intensity of the 488 nm laser to 75% and exposure time of 25 ms.
7. Set the imaging settings to 1 μm separated *z*-planes captured every 10-15 s for a total of 4.5 min (*see* **Note 7**).

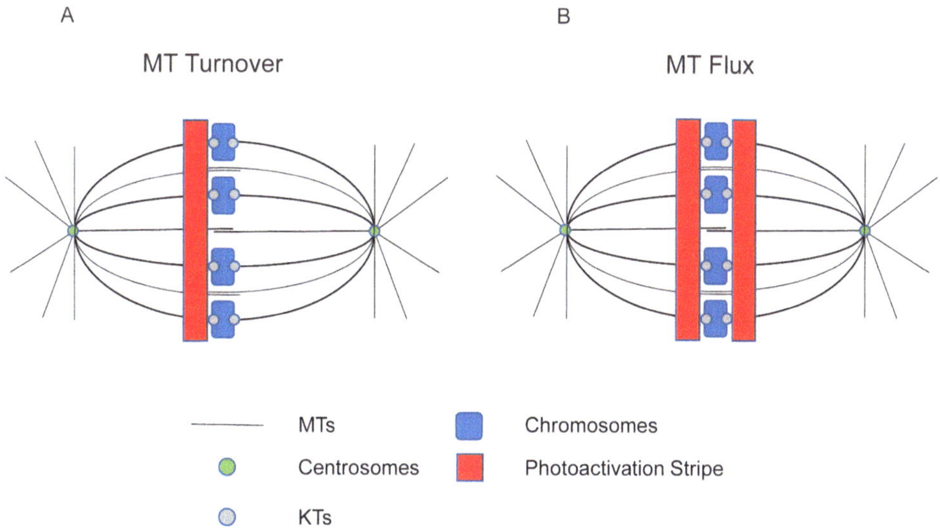

Fig. 1 Drawing the excitation stripes for photoactivation of the mitotic spindle. (**a**) To measure microtubule turnover define a stripe in one half-spindle, near the interface between microtubules and chromosomes. (**b**) To determine microtubule flux define two stripes in both half-spindles near the microtubule-chromosome interface

3.3 Photoactivation

1. Identify a prometaphase or a metaphase cell of interest using DIC, checking the alignment of the chromosomes.
2. Capture a DIC image of the cell. This will be important to classify the cell according to mitotic stage, by observing whether the chromosomes are fully aligned or not.
3. Determine the center of the spindle, by focusing the spindle with the 561 nm laser (*see* **Note 8**). Make sure that the spindle poles are near the same focal plane and avoid tilted spindles (*see* **Note 9**).
4. To determine microtubule turnover rates, define a stripe in one half-spindle near the microtubule-chromosome interface (Fig. 1a) by using the "Define Region" option in the Mosaic menu (*see* **Note 10**). Do not define a region that is too close to the spindle pole (*see* **Note 11**). If the goal is to determine spindle microtubule poleward flux, define two stripes at the spindle-chromosome interface, one in each half-spindle (Fig. 1b).
5. Capture a first z-stack image of pre-photoactivation with the 488 nm laser, and immediately after trigger the 405 nm laser to excite the previously defined stripe(s).
6. Afterwards, capture z-stack images in the 488 nm channel using 10-15-s intervals until it automatically finishes at 4.5 min (*see* **Notes 12** and **13**).

3.4 Determining Spindle Microtubule Turnover

1. Open your images in Fiji/ImageJ. Select the channel of the photoactivation signal and do a sum projection of all stacks.
2. If using LAPSO extension in MatLab, save the sum projection as an image sequence in TIFF format, choosing a numerical name.
3. In the sum projection, rotate the image in order to align the spindle poles horizontally (Fig. 2a). If using LAPSO, open the sum projection file by using as prefix the numerical name given, starting from time point 1 to 19 and selecting the appropriate pixel size. Then, rotate the image of the spindle in order to align the poles horizontally.
4. Generate a kymograph that includes all time points of the time-lapse movie (Fig. 2b). In LAPSO, use the "Kymo" option to define a region of interest containing the spindle, and then choose sum collapse to generate the kymograph for all time points [19].
5. Quantify the fluorescence intensities for the photoactivated region for all time points individually and normalize all values to the intensity of the first time point. In LAPSO, choose "Path/Turnover" option and draw a region that follows and contains all the signal along the time points (Fig. 2b). The program normalizes the intensities to the first time point after photoactivation.
6. Next, quantify background values in the non-photoactivated region in the other half-spindle and subtract these to the previously determined values. In LAPSO, draw a region in the non-photoactivated region in the other half-spindle covering all time points (Fig. 2b), and the program will subtract that to the previous values.
7. Quantify the fluorescence loss of the whole cell, by measuring the signal intensity in all cell (including the cytoplasm) for all the time points. In LAPSO, return to the "Kymo" option, but now select a region that covers all the cell and sum collapse and then choose "Path/Turnover," selecting here a region that covers all the signal until the visible borders of the cell (Fig. 2b).
8. Correct values for photobleaching by normalizing to the values obtained in **step** 7. This method allows for the precise measurement of photobleaching for each cell at the individual level and avoids any potential unspecific Taxol-associated phenotypes. Under these conditions, the photoactivated region dissipates yet the photoactivated molecules are retained in the cytoplasm (*see* **Note 14**).
9. To calculate microtubule turnover, fit the corrected normalized intensities at each time point to a double-exponential curve

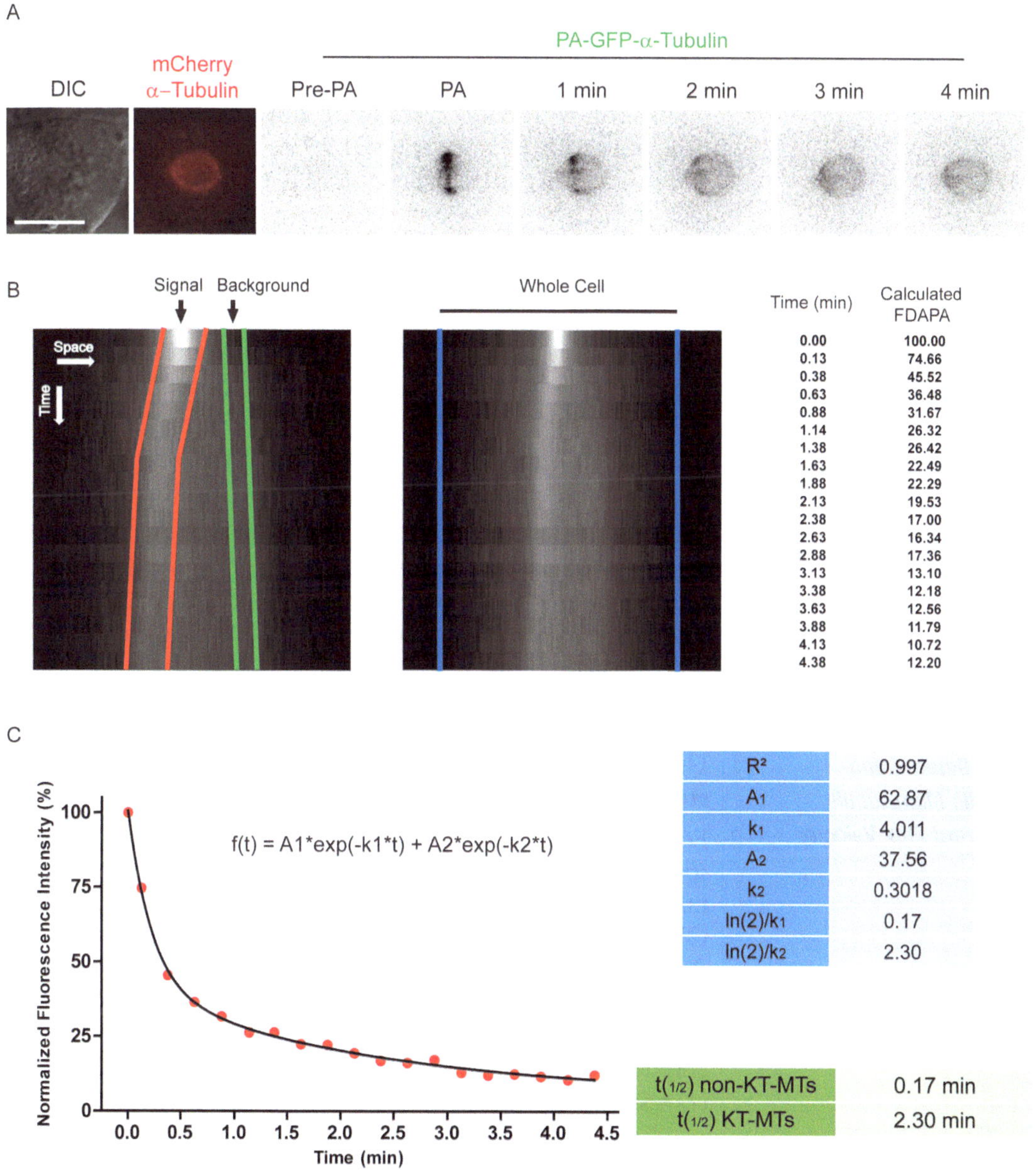

Fig. 2 Determination of spindle microtubule turnover rates. (**a**) DIC and time-lapse fluorescence images of a representative U2OS cell in metaphase expressing PA-GFP-α-tubulin and mCherry-α-tubulin. The mitotic spindle is visible with mCherry; the dissipation of PA-GFP over time is inverted for better visualization; scale bar is 5 μm. (**b**) Sum-projected kymograph for the mitotic spindle. Red lines show the region used to quantify the signal generated from photoactivation. Green lines show the regions used to determine background levels; sum-projected kymograph for whole cell (between blue lines) used for photobleaching correction; calculated FDAPA values for all time points. (**c**) Fitting of the normalized fluorescence intensities in a double-exponential curve; the values obtained in the curve fitting allow the calculation of the half-life values for non-kinetochore microtubules and kinetochore microtubules

[$A1 * \exp.(-k1 * t) + A2*\exp(-k2*t$], in which t is time, $A1$ represents the less stable (non-kinetochore microtubules) population, and $A2$ is the more stable (kinetochore microtubules) population with decay rates of $k1$ and $k2$, respectively. Avoid using cells with R^2 value below 0.99 for the best fitting possible and to improve the quality of MT turnover quantification (Fig. 2c).

10. After obtaining the curves, use the rate constants $k1$ corresponding to the fast turnover microtubules, typically representing the non-kinetochore microtubules and $k2$ corresponding to the slow turnover microtubules, typically representing the kinetochore microtubules to calculate half-life ($t_{1/2}$) values (*see* **Note 15**).
11. To calculate half-life ($t_{1/2}$) for non-kinetochore microtubules use $\ln(2)/k1$ and for kinetochore microtubules $\ln(2)/k2$ (Fig. 2c).
12. To perform statistical analysis use Student's two-tailed t-test or Mann-Whitney U-test with 95% confidence, depending on the normality of the data distribution. Make sure that you do separate analysis for prometaphase and metaphase cells, since they have different microtubule stability properties.

3.5 Determining Spindle Microtubule Poleward Flux Velocity

1. Use the same methodology described in Subheading 3.4, from **steps 1** to **4**.
2. Follow one of the signals from the photoactivation of one half-spindle throughout all time points, and draw a line that better fits the slope of the progression of the fluorescence (Fig. 3b). Obtain the values of time (Δt in minutes) and the distance of the movement (Δx in μm) and calculate the velocity ($\Delta x/\Delta t$ in μm/min) (Fig. 3b). If using LAPSO, just draw a line in the kymograph that better fits the progression of the fluorescence, and the program will give directly the value of the slope corresponding to the average velocity of flux (*see* **Note 16**).
3. Repeat **step 2**, now for the photoactivation signal in the other half-spindle.
4. To determine the microtubule flux velocity calculate the arithmetic average of the absolute values of velocity obtained in the previous points (Fig. 3b).
5. To perform statistical analysis use Student's two-tailed t-test or Mann-Whitney U-test with 95% confidence, depending on the normality of the values distribution.

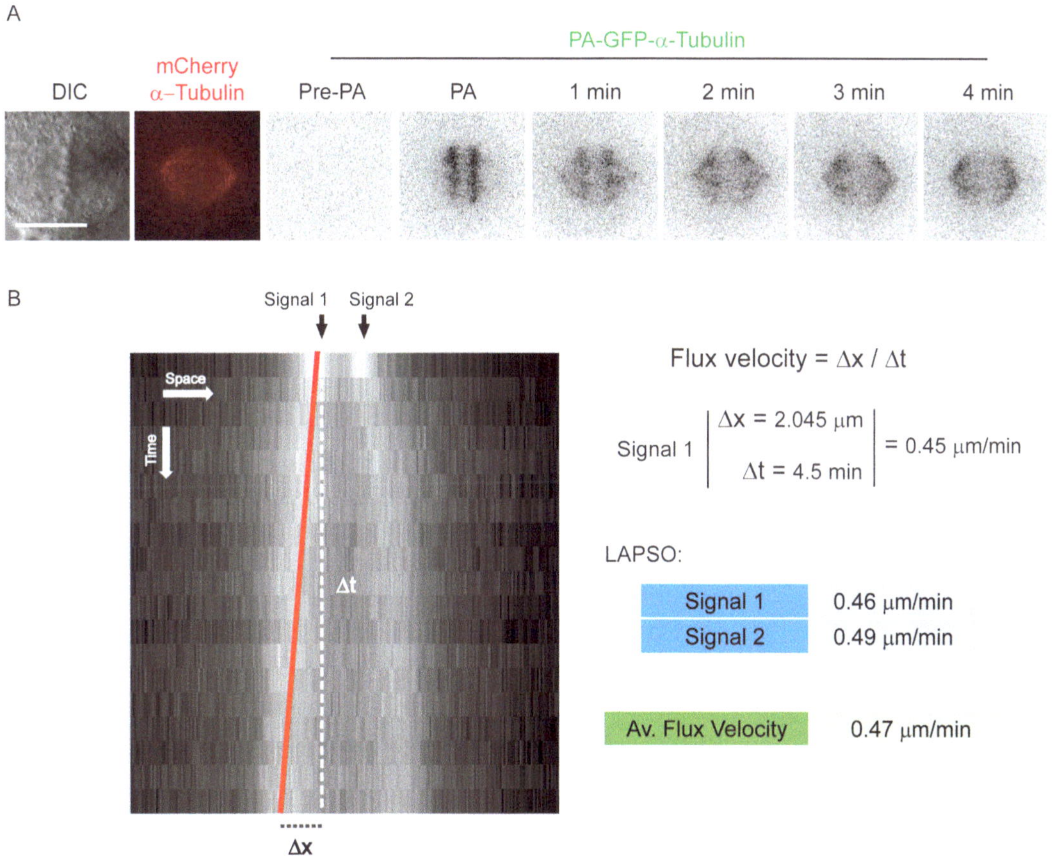

Fig. 3 Determination of flux rates in mitotic spindle. (**a**) DIC and time-lapse fluorescence images of a representative metaphase U2OS cell photoactivated in both half-spindles, expressing PA-GFP-α-tubulin and mCherry-α-tubulin. The mitotic spindle is visible with mCherry; the dissipation of PA-GFP over time is inverted for better visualization; scale bar is 5 μm. (**b**) Sum-projected kymograph for the mitotic spindle, showing in red the line slope adjusted to the progression of the fluoresce signal, in dashed white the Δt, and in dashed black Δx; calculation of flux velocity for one half-spindle and two half-spindles using LAPSO software

4 Notes

1. Seed enough cells so that in the moment of imaging they are at approximately 80–90% confluence. For U2OS cells this corresponds approximately to 0.2×10^6 cells in each well or MatTek.
2. You can also use coated coverslips or MatTek chambers with fibronectin or poly-L-lysine.
3. When using DMEM without phenol red it is preferable to add the medium to the cells one day before the experiment. However, if using L-15, it is better to replace the medium immediately before preparing the sample for microscope visualization.

If the microscope has CO_2 supply, the use of a CO_2-independent medium is not required, and regular growth medium can be used without phenol red indicator to avoid autofluorescence and without HEPES buffer.

4. It is extremely important to check for medium leakage, since it can do serious damage to the objective. Carefully confirm if the silicon gasket is perfectly fitted in the main body of the chamber before closing it. A simple way to confirm the correct sealing is by putting the chamber on top of a piece of absorbing paper and check for leakage for few minutes in the cell culture incubator.
5. When studying microtubule dynamics in prometaphase and metaphase, it is advisable to prevent mitotic exit. For that purpose we use 5 μM of the proteasome inhibitor MG-132. This prevents cyclin B1 degradation and consequently arrests cells in metaphase. MG-132 is ideal for this purpose since it does not affect microtubule dynamics [20, 21]. However, pay attention for prolonged incubations, since it can lead to cohesion fatigue [22].
6. It is highly advisable to check if the calibration was correct, by using the "Test Calibration" option and drawing a random image. Try to get the maximum amplitude of the field to confirm if the stimulation is similar in the entire field.
7. If using tubulin tagged with a red fluorescent protein such as mRFP or mCherry, it is advisable to capture an image of the spindle in the first time point z-stack for reference.
8. In the cases where it is not possible to visualize the spindle due to lack of a fluorescent marker (e.g., tubulin tagged with mRFP or mCherry), it is possible (although not very precise) to infer the center by checking the localization of the chromosomes in the DIC channel.
9. Using tilted spindles can lead to the excitation of one half-spindle in one set of z-stacks and the other half-spindle in another set of z-stacks, which creates two photoactivated signals moving through opposite poles, making it impossible to quantify microtubule turnover.
10. In the cases where the spindle is not possible to visualize (*see* **Note 8**), one can define a stripe that approximately matches the half-spindle by using the DIC image of the aligned chromosomes. This is easy for metaphase or late prometaphase cells. However, for early-prometaphase cells, it is almost impossible to infer the orientation of the spindle.
11. Some experimental conditions (e.g., depletion of certain proteins) can result in shorter mitotic spindles, which compromises the definition of a stripe for photoactivation that is far

enough from the spindle poles. In such cases, try to draw a region that best matches the interface between microtubules and chromosomes.

12. If not using MG-132, make sure that the cell does not enter anaphase, since the dynamic state of microtubules changes, making it impossible to accurately determine microtubule turnover or flux velocities.
13. If the cell moved significantly in the *x*, *y*, or *z* planes, do not consider it for analysis, since it makes it very difficult to generate a kymograph under these conditions.
14. As an alternative, it is possible to correct the values normalizing to averages obtained in quantifications for microtubule turnover in cells treated with 1–5 μM of Taxol, which stabilizes microtubules [21].
15. Note that the general equation of a double-exponential curve is $[a * \exp(b*x) + c * \exp(d * x)]$, so the equivalent to our equation is $a = A1$, $b = -k1$, $c = A2$, and $d = -k2$. To correctly calculate half-lives, the values of b and d must be negative and the percentage of $A1$ and $A2$ populations must be below 2%. If this is not verified, it means that the obtained curve fitting is not valid and does not reliably represent our model, so do not consider those cells for analysis.
16. If doing experiments without MG-132, and the goal is to determine microtubule poleward flux velocities in the metaphase spindle, do not consider the time points from which the cell has entered anaphase.

Acknowledgments

Work in the laboratory of H.M. is funded by the European Research Council (ERC) under the European Union's Horizon 2020 research and innovation program (grant agreement No. 681443) and FLAD Life Science 2020. H.G. holds a Ph.D. fellowship from Fundação para a Ciência e a Tecnologia (SFRH/BD/141066/2018).

References

1. Mitchison T, Kirschner M (1984) Dynamic instability of microtubule growth. Nature 312 (5991):237–242
2. Mitchison T (1989) Polewards microtubule flux in the mitotic spindle: evidence from photoactivation of fluorescence. J Cell Biol 109(2):637–652
3. Belmont L, Hyman A, Sawin K, Mitchison T (1990) Real-time visualization of cell cycle-dependent changes in microtubule dynamics in cytoplasmic extracts. Cell 62(3):579–589
4. Piehl M, Cassimeris L (2003) Organization and dynamics of growing microtubule plus ends during early mitosis. Mol Biol Cell 14 (3):916–925

5. Bakhoum SF, Compton DA (2012) Kinetochores and disease: keeping microtubule dynamics in check. Curr Opin Cell Biol 24 (1):64–70
6. Matos I, Maiato H (2011) Prevention and correction mechanisms behind anaphase synchrony: implications for the genesis of aneuploidy. Cytogenet Genome Res 133 (2–4):243–253
7. Zhai Y, Kronebusch PJ, Borisy GG (1995) Kinetochore microtubule dynamics and the metaphase-anaphase transition. J Cell Biol 131(3):721–734
8. Godek KM, Kabeche L, Compton DA (2015) Regulation of kinetochore–microtubule attachments through homeostatic control during mitosis. Nat Rev Mol Cell Biol 16(1):57
9. Waterman-Storer CM, Desai A, Bulinski JC, Salmon E (1998) Fluorescent speckle microscopy, a method to visualize the dynamics of protein assemblies in living cells. Curr Biol 8 (22):1227–S1221
10. Carminati JL, Stearns T (1997) Microtubules orient the mitotic spindle in yeast through dynein-dependent interactions with the cell cortex. J Cell Biol 138(3):629–641
11. Fernández-Suárez M, Ting AY (2008) Fluorescent probes for super-resolution imaging in living cells. Nat Rev Mol Cell Biol 9(12):929
12. Patterson GH, Lippincott-Schwartz J (2002) A photoactivatable GFP for selective photolabeling of proteins and cells. Science 297 (5588):1873–1877
13. Sample V, Newman RH, Zhang J (2009) The structure and function of fluorescent proteins. Chem Soc Rev 38(10):2852–2864
14. Nienhaus K, Nienhaus GU (2014) Fluorescent proteins for live-cell imaging with super-resolution. Chem Soc Rev 43(4):1088–1106
15. Renz M, Lippincott-Schwartz J (2014) Optical highlighters: applications to cell biology. In: Day RN, Davidson MW (eds) The fluorescent protein revolution, 1st edn. CRC Press Taylor & Francis, pp 197–224
16. Wadsworth P, Salmon E (1986) Analysis of the treadmilling model during metaphase of mitosis using fluorescence redistribution after photobleaching. J Cell Biol 102 (3):1032–1038
17. Krafft G, Cummings R, Dizio J, Furukawa R, Brvenik L, Sutton W, Ware B (1986) Fluorescence photoactivation and dissipation (FPD). In: Nucleocytoplasmic transport. Springer, Berlin, pp 35–52
18. Vigers G, Coue M, McIntosh J (1988) Fluorescent microtubules break up under illumination. J Cell Biol 107(3):1011–1024
19. Pereira AJ, Maiato H (2010) Improved kymography tools and its applications to mitosis. Methods 51(2):214–219
20. Kabeche L, Compton DA (2012) Checkpoint-independent stabilization of kinetochore-microtubule attachments by Mad2 in human cells. Curr Biol 22(7):638–644
21. Orr B, Talje L, Liu Z, Kwok BH, Compton DA (2016) Adaptive resistance to an inhibitor of chromosomal instability in human cancer cells. Cell Rep 17(7):1755–1763
22. Gorbsky GJ (2013) Cohesion fatigue. Curr Biol 23(22):R986–R988. https://doi.org/10.1016/j.cub.2013.08.017

Chapter 16

Functional Dissection of Mitosis Using Immortalized Fibroblasts from the Indian Muntjac, a Placental Mammal with Only Three Chromosomes

Ana C. Almeida, Danica Drpic, Naoyuki Okada, Joana Bravo, Marta Madureira, and Helder Maiato

Abstract

During cell division in eukaryotes a microtubule-based network undergoes drastic changes and remodeling to assemble a mitotic spindle competent to segregate chromosomes. Several model systems have been widely used to dissect the molecular and structural mechanisms behind mitotic spindle assembly and function. These include budding and fission yeasts, which are ideal for genetic and molecular approaches, but show limitations in high-resolution live-cell imaging, while being evolutionarily distant from humans. On the other hand, systems that were historically used for their exceptional properties for live-cell imaging of mitosis (e.g., newt lung cells and *Haemanthus* endosperm cells) lack the necessary genomic tools for molecular studies. In a CRISPR-Cas9 era, human cultured cells have conquered the privilege to be positioned among the most powerful genetically manipulatable systems, but their high chromosome number remains a significant bottleneck for the molecular dissection of mitosis in mammals. We believe that we can significantly broaden this scenario by establishing a unique placental mammal model system that combines the powerful genetic tools and low chromosome number of fission yeast and *Drosophila melanogaster*, with the exceptional cytological features of a rat kangaroo cell. This system is based on hTERT-immortalized fibroblasts from a female Indian muntjac, a placental mammal with the lowest known chromosome number ($n = 3$). Here we describe a series of methodologies established in our laboratory for the study of mitosis in Indian muntjac. These include standard techniques such as immunofluorescence, western blotting, and FISH, but also several state-of-the-art methodologies, including live-cell imaging, cell confinement, RNAi, super-resolution STED microscopy, and laser microsurgery.

Key words Indian muntjac, Mitosis, Kinetochore, Spindle, Microtubule, Laser microsurgery, STED, RNAi

1 Introduction

In every cell cycle the genomic content must be replicated in order to be equally distributed into two identical daughter cells in the

---

Ana C. Almeida, Danica Drpic and Naoyuki Okada contributed equally to this work.

Helder Maiato (ed.), *Cytoskeleton Dynamics: Methods and Protocols*, Methods in Molecular Biology, vol. 2101, https://doi.org/10.1007/978-1-0716-0219-5_16,

form of chromosomes. Faithful chromosome segregation relies on the action of cytoskeletal microtubules, which organize as a bipolar spindle during mitosis [1]. Arguably, to understand the molecular and structural principles behind mitotic spindle assembly and function in mammals, the simpler the system, the better. Here we describe a series of methodologies that take advantage of the unique cytological features of the Indian muntjac, a small deer whose females have the lowest known chromosome number in mammals.

Diploid chromosome number among mammalian species is generally well restrained, typically ranging between 36 to 60 [2–4]. However, rare exceptions do exist, such as the red viscacha rat (*Tympanoctomys barrerae*), whose genome is distributed among 102 chromosomes [5], and on the other extreme the Indian muntjac (*M. muntjak*) with only six or seven chromosomes in females or males, respectively [6]. Asian muntjacs have drawn attention of many geneticists and biologists because they exhibit the greatest chromosomal diversity within related species. The genus *Muntiacus* underwent an extreme karyotype diversification with chromosome numbers extending from the thought last common ancestor of all *Cervidae* with $2n = 70$ [7] to $2n = 46$ (*M. reveesi*) [8], $2n$ = 13♀/14♂ (*M. feae*) [9], $2n$ = 8♀/9♂ (*M. crinifrons*) [10], and $2n$ = 6♀/7♂ (*M. muntjak*) [6]. The latter is thought to result from a repeated series of tandem and centric fusions [1112], giving rise to large and morphologically distinct chromosomes, with one pair of acrocentric chromosomes (chromosome 3 + X) containing an unusually large compound kinetochore (~2 μm linear length) [1314]. This represents a unique advantage for micromanipulation and high-resolution live-cell studies of mitosis, if combined with state-of-the-art molecular manipulation.

In the past, primary Indian muntjac fibroblasts were used extensively to study kinetochore structure [13, 15–18], sister chromatid exchange [19], as well as chromosomal structure [11]. However, the use of the unique features of these cells for the study of mitosis has been limited by the fact that they undergo replicative senescence, preventing the establishment of stable lines expressing useful fluorescent markers. Moreover, molecular studies were hampered by the lack of a sequenced genome. More recently, by using hTERT-immortalized fibroblasts from a female Indian muntjac and a draft genome sequence we obtained proof of principle that the Indian muntjac is a powerful system to study different molecular and physiological aspects of mitosis, unveiling the relevance of kinetochore size for biased chromosome congression and segregation in mammals [14]. Here we extend the breadth of experimental approaches using immortalized Indian muntjac fibroblasts in culture, to establish this system as a powerfully simple model to study mitosis in mammals.

2 Materials

2.1 Cell Culture

1. Disposable 25 cm^2 tissue culture flasks and/or 6-well plates.
2. Minimum essential medium (MEM) with L-glutamine.
3. Heat-inactivated fetal bovine serum (FBS).
4. TrypL Express Enzyme, phenol red.
5. Sterile phosphate-buffered solution (PBS), pH 7.4.
6. Disposable sterile pipettes (1, 5, and 10 mL).

2.2 Cell Immortalization

1. Indian muntjac fibroblasts growing in 6-well plates.
2. MEM supplemented with 10% FBS.
3. MEM supplemented with 5% FBS.
4. Lipofectamine 2000 (Invitrogen).
5. Immortalization plasmid (pBabe-puro-hTERT, *kind gift from Dr. Jerry W Shay, University of Texas*).
6. Puromycin DIHCL.

2.3 Transient Transfection and Establishment of Stable Cell Lines

1. Indian muntjac fibroblasts growing in 6-well plates.
2. Opti-MEM Reduced Serum Media.
3. Lipofectamine 2000.
4. DNA expression vector:
   (a) H2B-GFP *from Geoff Wahl lab* (Addgene plasmid #11680).
   (b) pIRESneo3-EGFP-α-tubulin *(kind gift from Dr. Patrick Meraldi, University of Geneva).*
5. Complete growth medium.
6. Selection antibiotics (Geneticin® Selective Antibiotic—G418 Sulfate).

2.4 Viral Transduction

1. Human embryonic kidney (HEK) 293T cells growing in a 10 cm tissue culture dish.
2. Dulbecco's modified Eagle medium (DMEM) supplemented with 10% FBS.
3. Indian muntjac cells growing in 6-well plates.
4. MEM supplemented with 10% FBS.
5. Lipofectamine 2000.
6. Opti-MEM Reduced Serum Media.
7. pMD2.G (Addgene plasmid #12259).
8. psPAX2 (Addgene plasmid #12260).
9. Viral plasmid: pRRL-CMV-EGFP-α-tubulin.
10. 0.45 μm Syringe filters.

11. 50 mL Syringe.
12. Polybrene.

2.5 Cell Sorting

1. Indian muntjac cells stably expressing H2B-GFP or GFP-α-tubulin.
2. Indian muntjac immortalized parental cells.
3. Complete growth medium.
4. Universal tabletop centrifuge: 4× CytoRotor.
5. Basic sorting buffer: 1× PBS, 1 mM EDTA, 25 mM HEPES, and 1% heat-inactivated FBS.
6. Cell strainer 40 μm.
7. FACS tubes.

2.6 RNAi

1. Indian muntjac cells growing in 6-well plates.
2. MEM supplemented with 5% FBS.
3. Lipofectamine RNAi Max.
4. Opti-MEM Reduced Serum Media.
5. siRNA oligos.

2.7 Western Blotting

1. Indian muntjac cells growing in disposable 25 cm^2 flasks or 6-well plates.
2. Complete growth medium.
3. TrypL Express Enzyme.
4. PBS 1×.
5. Microcentrifuge.
6. Lysis buffer: 20 mM HEPES/KOH pH 7.9, 1 mM EDTA, 1 mM EGTA, 150 mM NaCl, 0.5% of NP-40, 10% glycerol, 2 mM DDT, supplemented with 1:50 protease inhibitor cocktail and 1:100 phenylmethylsulfonyl fluoride (PMSF).
7. Liquid nitrogen or ice.
8. Coomassie Plus Protein Assay Kit (Thermo-Scientific).
9. Laemmli sample buffer: 50 mM Tris–HCl pH 6.8, 2% 422 SDS, 10% glycerol, 1% β-mercaptoethanol, 12.5 mM EDTA, and 0.02% bromophenol 423 blue.
10. Nitrocellulose membrane and transfer device: Trans-Blot® Turbo™ (Bio-Rad).
11. PBS containing 0.1% Tween with 5% nonfat dry milk.
12. Primary antibodies and HRP-conjugated secondary antibodies.
13. Clarity Western ECL Substrate (Bio-Rad).
14. Detection device (e.g., Bio-Rad ChemiDoc XRS system).

2.8 Immuno-fluorescence

1. Indian muntjac cells growing in 6-well plates.
2. MEM supplemented with 10% FBS.
3. Cover glass 22 × 22 mm, thickness 1½.
4. Fibronectin bovine plasma.
5. Pre-cold methanol.
6. Paraformaldehyde 20% solution (Electron Microscopy Sciences) diluted in cytoskeleton buffer (CB): 27a mM NaCl, 10 mM KCl, 2.2 mM Na_2HPO_4, 0.8 mM KH_2PO_4, 4 mM $MgCl_2$, 10 mM Pipes, 10 nM glucose, pH 6.1.
7. Blocking solution: CB supplemented with 10% FBS.
8. Tween 20.
9. Triton X-100.
10. 4′6′-Diamidino-2-phenylindole (DAPI, Sigma-Aldrich).
11. Glass slides.
12. Mounting medium: 20 mM Tris-HCl pH 8, 0.5% *N*-propyl gallate, and 90% glycerol.

2.9 STED Microscopy

1. *See* Section 2, Subheading 2.8.1–10.
2. Glutaraldehyde EM Grade 25% (Electron Microscopy Sciences).
3. STAR 580, donkey anti-human IgG, 500 μg (Abberior).
4. STAR-Red Goat anti-rat IgG 500 μg (Abberior).
5. STAR 635p anti-mouse IgG 500 μg (Abberior).
6. Microscope setup: Abberior Instruments "Expert Line" gated-STED coupled to a Nikon Ti microscope. Oil-immersion 60× 1.4NA Plan-Apo objective (Nikon, Lambda Series) and pin-hole size of 0.8.

2.10 Live-Cell Imaging

1. Disposable 6-well plates.
2. Cover glass 22 × 22 mm, thickness 1½ coated with fibronectin.
3. Indian muntjac fibroblasts exponentially growing in 6-well plates.
4. MEM supplemented with 10% FBS.
5. Leibovitz's L-15 medium without phenol red supplemented with 20% FBS.
6. SiR-Tubulin (Spirochrome) to label spindle microtubules or SiR-DNA (Spirochrome) to stain chromosomes.
7. Magnetic imaging chamber for 22 × 22 coverslips (e.g., Chamlide).

2.11 Laser Microsurgery

1. Indian muntjac fibroblasts growing in fibronectin-coated coverslips.
2. Complete growth medium.
3. L-15 medium without phenol red supplemented with 20% FBS.
4. SiR-Tubulin.
5. Magnetic imaging chamber for 22 × 22 coverslips.

2.12 Cell Confinement

1. Glass plates.
2. Round coverslips (18 mm), thickness 1½.
3. PDMS (A) (TECHSiL).
4. PDMS (B) (TECHSiL).
5. Metal mold (custom-designed layout to fit a 35 mm dish [20]).
6. Harris Uni-Core puncher with 0.75 mm.
7. Fluorodish (World Precision Instrument).
8. Indian muntjac cells stably expressing H2B-GFP growing in a 6-well plate.
9. L-15 supplemented with 20% FBS.
10. SiR-Tubulin.

2.13 Fish

1. Indian muntjac cells growing in 6-well plates.
2. MEM supplemented with 10% FBS.
3. 3.3 mM Nocodazole.
4. TrypL Express Enzyme.
5. PBS 1×.
6. Hypotonic solution (medium:water 1:1 and 3.3 μM nocodazole).
7. Carnoy's fixative (75% methanol, 25% acetic acid).
8. 1% Pepsin in 10 mM HCl.
9. 70%, 90%, and 100% ethanol.
10. Preheated water bath at 65 °C.
11. Fluorescent probes.
12. 10× Taq DNA polymerase buffer.
13. 20 μM Primer 6 MW (CCGACTCGAGNNNNNNATGTGG).
14. 2.5 mM dNTPs.
15. 50 mM $MgCl_2$.
16. Taq DNA polymerase.
17. dH_2O.
18. Biotin/FITC/Cy3-labeled UTPs.
19. 2.5 mM ½ dNTPs (50% of TTPs).

20. PCR machine.
21. 1 μL Secondary PCR product labeled with fluorochrome (paint).
22. 10 μL of hybridization buffer (LSI/WCP HyBuffer, Abbott Molecular).
23. Ice-cold methanol.
24. Coverslips 22 × 22 mm, thickness 1½.
25. Superfrost Plus microscope slides.
26. Fixo Gum (Pattex).
27. Preheated water bath (45 °C).
28. 2× Saline sodium citrate (SSC): 3 M Sodium chloride and 300 mM trisodium citrate dihydrate, pH 7.0.
29. Denaturing fluid: 70% Formamide with 30% of 2× SSC.
30. 4×T solution: 250 mL of 4× SSC containing 125 μL of Nonidet.
31. Displacement pipettes.
32. 2× Stringency wash solution: 50% Formamide with 50% of 1× SSC.

3 Methods

3.1 Cell Culture

Indian muntjac fibroblasts are cultured in minimum essential medium (MEM) supplemented with 10% heat-inactivated FBS at 37 °C and 5% CO_2. To enrich mitotic index (e.g., before imaging), supplement the medium with 20% FBS. All cell manipulation procedures should be performed in a sterile laminar flow hood (*see* **Notes 1–3**).

3.2 Cell Immortalization

1. On day 0, seed cells at 60–70% confluence using MEM supplemented with 10% FBS.
2. One hour before transfection replace the medium to MEM supplemented with 5% FBS.
3. Transfect cells with pBabepuro hTERT plasmid using Lipofectamine 2000 or viral transfection [21].
4. 4–5 h later change the transfection media to MEM supplemented with 10% FBS and keep cells in culture for 48 h before selection with puromycin.

3.3 Transient Transfection and Establishment of Stable Cell Lines

1. On the day prior to transfection, plate Indian muntjac parental cells at 60–70% confluence in MEM supplemented with 10% FBS.
2. Rinse the cells three times with PBS and incubate with Opti-MEM Reduced Serum Medium.

3. Prepare the following solutions:
   (a) Solution A: 250 μL of Opti-MEM reduced serum medium with 5 μL Lipofectamine 2000.
   (b) Solution B: 250 μL of Opti-MEM reduced serum medium with 3 μg of the DNA plasmid.
4. Wait for 5 min at room temperature.
5. Mix A and B, vortex, and incubate the mixture for 20 min at room temperature.
6. Add the transfection mixture to the cells dropwise.
7. Wait for 4 h and replace the medium to MEM supplemented with 10% FBS.
8. 48 h after transfection, select the positive cells by adding the appropriate selection marker (e.g., G418) to the growth medium.

3.4 Viral Transduction

3.4.1 Lentivirus Production

1. On day 0, plate human embryonic kidney (HEK) 293T cells in a 10 cm tissue culture dish and wait until they reach 60–80% confluence (around 6.0×10^6 cells). Grow cells in complete DMEM medium at 37 °C and 5% CO_2.
2. On the day of transfection prepare the following mixtures:
   (a) Solution A: 5.6 μg of VSV-G/pMD2.G, 16.6 μg of Pax2/psPAX2, 22.3 μg of the plasmid of interest in 2 mL of Opti-MEM Reduced Serum Media.
   (b) Solution B: 30 μL Lipofectamine 2000 in 2 mL of Opti-MEM Reduced Serum Media.
3. Mix thoroughly and incubate solutions A and B separately for 5 min at room temperature.
4. Mix A to B and incubate for 30 min at room temperature to allow the DNA–lipid complexes to form.
5. Remove the medium and wash gently with 10 mL PBS (*see* **Note 4**).
6. Add the DNA–lipid complex solution dropwise and incubate the cells at 37 °C, 5% CO_2, for 5 h.
7. Replace the medium with 10 mL complete growth medium and incubate at 37 °C, 5% CO_2, for 72 h.
8. Three days after, remove the medium to a 15 mL Falcon tube and centrifuge at $200 \times g$ for 10 min.
9. Filter the harvested supernatants using a 0.45 μm cellulose acetate filter in order to eliminate cellular debris (*see* **Note 5**).
10. Aliquot the viral supernatants in 500 μL aliquots and store at −80 °C.

3.4.2 Lentivirus Transduction in Target Cells

1. Seed Indian muntjac parental fibroblasts at 0.5×10^6 confluence in a 6-well plate. Grow them in 2000 μL MEM supplemented with 10% FBS at 37 °C, 5% CO_2, until they reach 90% confluence.
2. On the day after remove the growth medium and add 500 μL of viral particles and 10 μg/mL of polybrene (*see* **Note 6**). Mix by gentle swirling.
3. Incubate for 24–48 h at 37 °C, 5% CO_2.
4. Replace the medium to fresh MEM supplemented with 10% FBS and incubate for additional 24–48 h.

3.5 Cell Sorting

1. Grow Indian muntjac fibroblasts stably expressing GFP-α-tubulin or H2B-GFP in 10 mL complete growth medium supplemented with G-418 in a 75 cm^2 flask.
2. Harvest cells after trypsinization.
3. Centrifuge the cells for 5 min at 200 × *g*.
4. Discard the supernatant and resuspend the cells in 1 mL of pre-warmed basic sorting buffer.
5. Filter the cells using a cell strainer into a FACS tube.
6. After cell sorting, collect the positive clones to 15 mL Falcon (*see* **Note 7**).
7. Centrifuge the cells at 200 × *g* for 5 min at room temperature.
8. Discard the supernatant and resuspend the pellet in MEM supplemented with 10% FBS and antibiotics.

3.6 RNAi

3.6.1 Design of siRNAs

1. Design the siRNA sequence using the online application BLOCK-ITTM RNAi Designer (Thermo Fisher Scientific). Provide the nucleotide sequence of the genes of interest, select an ideal GC percentage between 35 and 55%, and choose the default motif pattern for the RNAi design. Select the sequence ranked with higher probability of knockdown.
2. Purchase siRNA oligonucleotides specific for the mRNA encoding for the genes of interest (e.g., CLASP1, Fig. 1).

3.6.2 siRNA Transfection

1. Culture Indian muntjac fibroblasts at 60–70% confluence in 6-well plate/35 mm dishes.
2. On the following day, replace the medium to MEM supplemented with 5% FBS.
3. Simultaneously prepare the following solutions:
   (a) Solution A: 250 μL of Opti-MEM Reduced Serum Medium with 5 μL Lipofectamine RNAi Max.
   (b) Solution B: 250 μL of Opti-MEM Reduced Serum Medium with 5 μL of the oligonucleotide (50 nM).

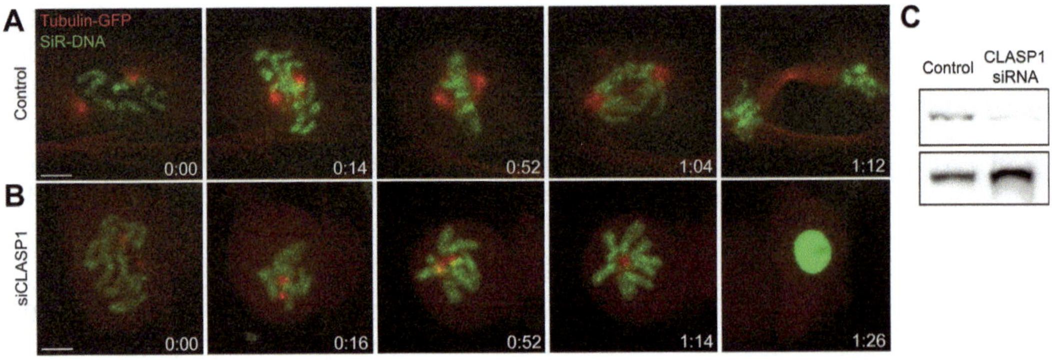

Fig. 1 Example of RNAi-mediated depletion of CLASP1 in Indian muntjac cells. (**a**) Live-cell imaging of control Indian muntjac fibroblasts stably expressing GFP-α-tubulin to visualize the spindle microtubules (red) and treated with 75 nM SiR-DNA to label chromosomes (green). (**b**) Similar conditions but after CLASP1 knockdown by RNAi. Scale bar, 5 μm. Time, h:min. (**c**) Monitoring RNAi efficiency by western blotting with an antibody against CLASP1. GAPDH was used as loading control

4. Mix thoroughly and incubate separately for 10–15 min at room temperature.
5. Mix solution A with solution B. Vortex and incubate for 15–30 min at room temperature.
6. Add the oligo-lipid solution to the wells dropwise and incubate for 6–8 h.
7. Change medium to complete growth medium and incubate for 24, 48, or 72 h.
8. Collect cell extracts to confirm the transfection efficiency by western blot or fix cells for immunofluorescence analysis (*see* Subheadings 3.7 and 3.8).

3.7 Western Blotting

3.7.1 Preparation of Cell Lysates

1. Grow cells until 90% confluence in complete growth medium.
2. Collect the cells by centrifugation at 200 × *g* for 5 min.
3. Wash pellets once 500 μL with warm PBS.
4. Centrifuge at 200 × *g* for 5 min.
5. Resuspend the pellets in ice-cold lysis buffer freshly supplemented with protease inhibitor cocktail (1:50) and PMSF (1:100). Incubate on ice for 30 min (*see* **Note 8**).
6. Clarify the lysate by centrifugation at 20,800 × *g* for 15 min at 4 °C.

3.7.2 Sample Preparation

1. Determine total protein concentration using the Bradford assay.
2. Denature protein samples in Laemmli sample buffer at 95 °C for 5 min.

3.7.3 Electrophoresis

1. Separate 25–50 μg of total proteins by 7.5/10/15% (v/v) SDS-PAGE gel electrophoresis at 80 V through the stacking gel and increase to 120 V for the resolving migration.
2. Transfer proteins to a nitrocellulose membrane using a semidry blotting system when transferring low/medium-molecular-weight proteins (*see* **Note 9**).
3. To evaluate transfer efficiency, incubate the membrane with Ponceau S solution for 2–3 min.

3.7.4 Immunodetection

1. Block nonspecific binding sites by incubating membranes in blocking solution (PBST supplemented with 5% nonfat dry milk) for 1.5 h at RT with gentle agitation.
2. Add the primary antibody (diluted in PBST supplemented with 1% nonfat dry milk) to the membrane and incubate for 1 h at RT or overnight at 4 °C.
3. Wash the membrane 3× in PBST for 10 min.
4. Prepare fresh the HRP-conjugated secondary antibody (diluted in PBST supplemented with 1% nonfat dry milk) to the membrane and incubate for 1 h at RT (gentle agitation).
5. Wash the membrane 3× in PBST for 10 min.
6. Incubate the membrane for 1–2 min in enhanced chemiluminescence (ECL) mixture (prepared following the manufacturer's instructions).
7. Detect and acquire the chemiluminescence signal using an imaging system (e.g., Chemidoc).

3.8 Immunofluorescence

In this section all the solutions must be prepared fresh.

3.8.1 Preparation of Fibronectin-Coated Coverslips

1. Prepare 50 μg/mL fibronectin solution in sterile PBS.
2. Incubate 22 × 22 mm glass coverslips in fibronectin solution for 1 h with gentle agitation.
3. Wash coverslips with sterile PBS for 5 min in order to remove the fibronectin in excess.
4. Allow to dry and UV sterilize for 15 min.

3.8.2 Fixation with Paraformaldehyde

1. Prepare PFA 4% by diluting commercial PFA 20% in CB (*see* **Note 10**).
2. Trash cell medium and add 2 mL of PFA 4% for 10 min (*see* **Note 11**).

3.8.3 Fixation with Cold Methanol

1. Prepare a 6-well plate containing 2 mL of ice-cold methanol per well.
2. Transfer coverslips to the cold methanol and incubate for 4 min at −20 °C (*see* **Note 12**).

3. Prepare the plate for rehydration: Add 3–4 mL of CB to a new 6-well plate.

3.8.4 Cell Fixation and Staining

1. Seed Indian muntjac parental fibroblasts at 2×10^5 confluence in a 6-well plate. Grow them in 2 mL complete medium, at 37 °C and 5% CO_2.
2. Discard cell medium and add 2 mL of PFA 4% for 10 min. Alternatively add 2 mL of ice-cold methanol for 4 min and skip **step 3**.
3. Extract with 0.1–0.5% Triton X-100 in CB for 10 min.
4. Wash three times with CB + 0.05% Tween-20 with gentle swirling.
5. Perform a blocking step for 30 min using CB supplemented with 10% FBS.
6. Dilute the primary antibodies in the same solution as in **step 5**.
7. Incubate the samples with 50–100 μL of the primary antibody mixture for 1 h at room temperature. Optional: This step can be done overnight at 4 °C (*see* **Note 13**).
8. Wash three times with CB + 0.05% Tween-20 with gentle mixing.
9. Prepare the secondary antibody solution (1:1000) and incubate at room temperature for 45 min.
10. Wash three times with CB + 0.05% Tween-20 with gentle mixing.
11. Add 1 μg/mL DAPI for 5 min.
12. Rinse with PBS and seal on glass slides using mounting media.

3.9 STED Microscopy

1. Seed 0.5×10^6 Indian muntjac parental fibroblasts in a 6-well plate containing fibronectin-coated coverslips in 2 mL MEM supplemented with 10% FBS. Grow cells 24 h.
2. Remove the media.
3. Fix the cells with 4% PFA + 0.25% glutaraldehyde (diluted in CB) to visualize microtubules. Incubate for 10 min at room temperature (*see* **Notes 14** and **15**).
4. Remove the fixative solution and add 0.1% solution of sodium borohydride (diluted in PBS) for quenching. Incubate for 7 min.
5. Extract the cells with CB 0.1–0.5% Triton. Incubate for 10 min.
6. Wash three times with 2 mL 0.05% CB with gentle mixing.
7. Prepare the blocking solution: 10% FBS diluted in the washing solution. Incubate for 30 min.

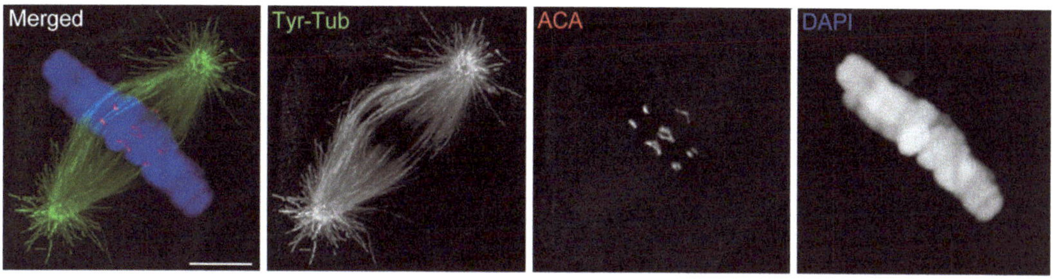

Fig. 2 Maximum intensity projection of a metaphase Indian muntjac fibroblast acquired with CH-STED. Microtubules (tyrosinated tubulin—green), chromosomes (DAPI—blue), and kinetochores (ACA—red) are indicated. Cells were arrested in metaphase with 3 μM of MG-132. Scale bar, 5 μm

8. Prepare the primary antibody mixture. Dilute 1:200 ACA (human) and 1:100 anti-tyrosinated tubulin (rat)—or alternatively 1:200 anti-α-tubulin—in the blocking solution and incubate for 1 h at room temperature.
9. Rinse three times with 2 mL 0.05% T-CB with gentle mixing.
10. Prepare the secondary antibody mixture: Mix 1 μL anti-Rat STAR-Red and 1 μL anti-Human STAR-580 in 100 μL 0.05% PBST for one coverslip.
11. Wash three times with 2 mL 0.05% T-CB with gentle mixing.
12. Add 1 μg/mL DAPI for 5 min.
13. Wash with PBS and seal the coverslips on glass slides using mounting media.
14. For image acquisition we use a pulsed gated STED microscope with 2D-STED or CH-STED modes [22] (Fig. 2).

3.10 Live-Cell Imaging

1. Seed Indian muntjac fibroblasts stably expressing H2B-GFP or GFP-α-tubulin at 0.5×10^6 confluence in a 6-well plate containing fibronectin-coated coverslips. Grow the cells in MEM supplemented with 10% FBS for at least 24 h.
2. 2 h before imaging, change medium to L-15 supplemented with 20%FBS and add 20–50 nM SiR-Tubulin to visualize microtubules or 75 nM of SiR-DNA to label the chromosomes (*see* **Note 16**).
3. Assemble the magnetic imaging chamber and supply with 1 mL of the imaging medium containing the respective dye. Start imaging (Fig. 1).

3.11 Laser Microsurgery

1. Seed cells on fibronectin-coated coverslips 2 days before preforming laser microsurgery. The confluence should be around 60%.
2. On the day of experiment change MEM medium supplemented with 10% FBS to L-15 medium with 20% FBS.

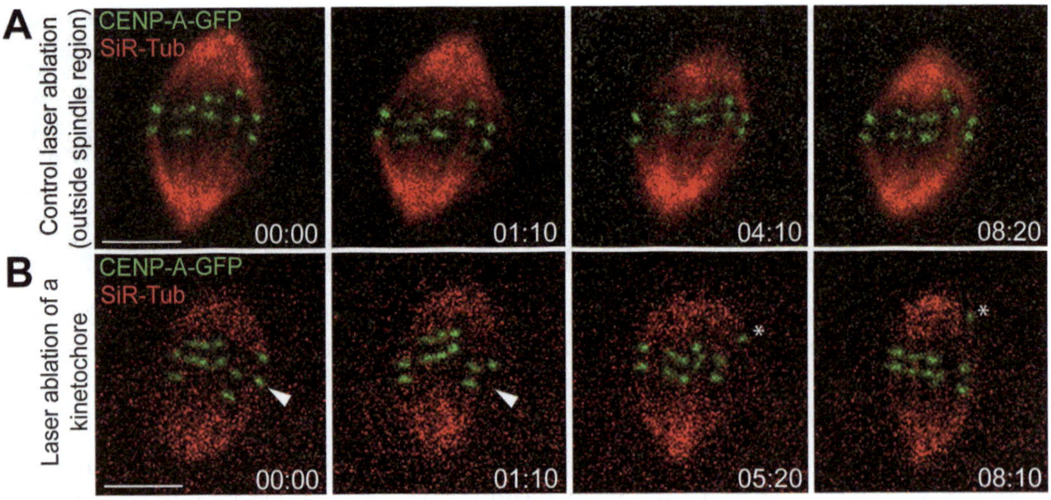

Fig. 3 Example of laser microsurgery in Indian muntjac fibroblasts stably expressing CENP-A-GFP to visualize the kinetochores (green) and treated with 20 nM SiR-Tubulin to label spindle microtubules (red). (**a**) Control laser microsurgery outside the mitotic spindle region. (**b**) Laser microsurgery of a kinetochore in a metaphase spindle of an Indian muntjac cell. White arrowhead indicates the ablated kinetochore. Asterisks indicate the poleward movement of the chromosome with a single kinetochore. Scale bar, 5 μm. Time, min:sec

3. Place cells in magnetic chambers adapted for live-cell imaging.
4. Laser microsurgery using Nikon confocal spinning disk microscope.
5. First align the laser on the glass coverslip using higher laser power.
6. Perform the laser alignment in the fluorescence channel on the structures that will be severed, using lower laser power. A detailed description of laser microsurgery has been published [23].
7. Perform laser microsurgery followed by live-cell imaging acquiring images every 10 s, 0.5 μm single Z-stack for 5–10 min (Fig. 3).

3.12 Cell Confinement

3.12.1 Making the PDMS Pillar

1. Mix 10 mL PDMS (A) and 1 mL PDMS (B). Centrifuge at 3900 × *g* for 10 min to avoid air bubbles.
2. Glue two round coverslips (18 mm in diameter) to a glass plate surface, using a drop of PDMS while heating for 3 min at 95 °C.
3. Place the metal mold on the coverslips and add PDMS (*see* **Note 17**).
4. Heat for 30 min at 85 °C on a hot plate.
5. Wash with isopropanol to detach PDMS pillar from the mold and the glass.

6. Puncture the side wall of PDMS pillar using a Harris Uni-Core puncher with 0.75 mm.
7. Wash the PDMS pillar with water and 70% ethanol (*see* **Note 18**).
8. Attach the round coverslip to the center of the dry PDMS pillar by gently pressing them together.
9. After attachment of the coverslip, connect it to the pressure and vacuum generator.

3.12.2 Preparation of Cells

1. Seed Indian muntjac cells stably expressing H2B-GFP at 0.5×10^6 cells in a 6-well plate.
2. Grow them for 48 h.
3. Change the growth medium to L-15 supplemented with 20% FBS and add 50 nM SiR-Tubulin to visualize microtubules.
4. Right before imaging, harvest cells after trypsinization. Centrifuge for 5 min at $200 \times g$ and remove supernatant. Resuspend the pellet into 1 mL L-15 supplemented with 20% FBS containing 50 nM SiR-Tubulin.

3.12.3 At the Microscope

1. Place 200 μL of trypsinized cell on the FluoroDish (*see* **Note 19**).
2. Connect the PDMS pillar to the vacuum generator with a thin wire and tube, through the hole on the side wall of the pillar. Gently place it on top of cell culture dish.
3. Turn on the vacuum and set at −30 mbar (*see* **Note 20**).
4. Make sure that the cell confiner does not move. In order to squeeze the cells, slowly increase vacuum pressure to −100 mbar (−10 kPa) (*see* **Notes 20** and **21**) (Fig. 4).

3.13 Fish

3.13.1 Degenerate Oligonucleotide-Primed (DOP) PCR

1. For reamplification reaction, prepare per tube 2.5 μL of 10× buffer, 2.5 μL of 20 μM DOP primer, 2.5 μL 2.5 mM dNTPs, 1.25 μL 50 mM $MgCl_2$, 0.25 μL 5 U/μL Taq, and 15 μL dH_2O.
2. For labeling reaction, prepare per tube 2.5 μL of 10× buffer, 2.5 μL of 20 μM DOP primer, 2.5 μL biotin/FITC/CY3-labeled UTPs, 2.5 μL ½ dNTPs (50% of dTTP), 1.25 μL 50 mM $MgCl_2$, 0.25 μL 5 U/μL Taq, 1 μL primary DOP-PCR product, and 12.5 μL dH_2O.
3. Vortex and spin down the tubes.
4. Define the following PCR program: 94 °C—5 min; 94 °C—1.30 min, 62 °C–1.30 min, 72 °C—3 min (27 cycles); 72 °C—8 min; 4 °C—pause.
5. Store the DOP product at −20 °C (*see* **Note 22**). A detailed description of FISH has been published [24, 25].

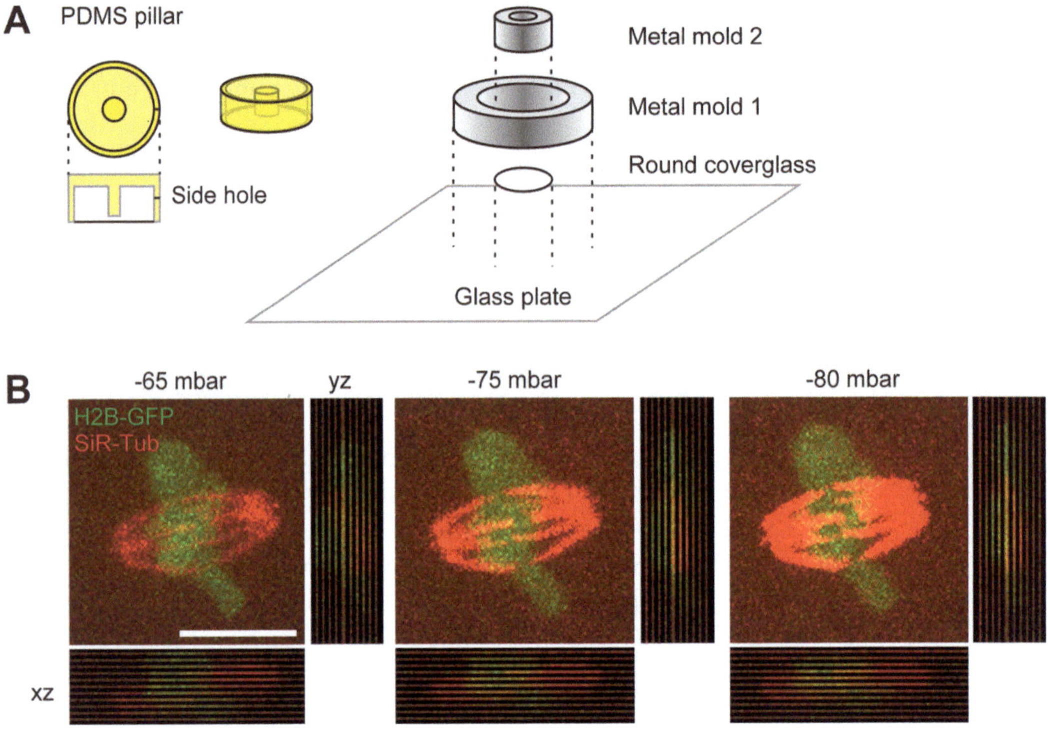

Fig. 4 Live-cell imaging of Indian muntjac fibroblasts under cell confinement. (**a**) Schematics representing the PMDS pillar assembly and experimental setup: two ring-shape metal molds assembled on a glass plate with a coverslip glued on top. (**b**) Prometaphase cell treated with 50 nM SiR-Tubulin confined using −65, −75, and −80 mbar. Scale bar, 10 μm

3.13.2 Chromosome Spreads

1. Seed Indian muntjac fibroblasts in disposable 6-well plates.
2. Add 3.3 μM nocodazole to the cells and incubate for 7–8 h (or overnight) at 37 °C, 5% CO_2.
3. Rinse the well with 500 μL of 1× trypsin and transfer the cells to a 15 mL Falcon tube.
4. Centrifuge for 5 min at 200 × *g*.
5. Resuspend the pellet in 500 μL of the supernatant and add dropwise the hypotonic solution, until the final volume of 5 mL.
6. Incubate the solution for 20 min at 37 °C.
7. Centrifuge for 5 min at 200 × *g*.
8. Discard the supernatant and fix the cells with cold Carnoy's solution (−20 °C, overnight).
9. In the following day repeat the fixation with Carnoy's solution.
10. Spread the solution drop by drop onto a glass slide.

3.13.3 Slide Preparation and Probe Labeling

1. Rinse the slides with 100% ethanol for 5 min and wait for them to air-dry.
2. Incubate the slides in 1% pepsin in 10 mM HCl for 3 min.
3. Wash two times in 2× SSC and one time in deionized water.
4. Dehydrate the slides through ethanol series (2× 70% ethanol, 2× 90% ethanol, 2 min per wash, and finally 4 min in 100% ethanol).
5. Denature the slides by incubating 70% of formamide in 2× SSC, at 67 °C for 2 min (*see* **Note 24**).
6. Cool down the samples in ice-cold 70% ethanol and dehydrate through a 70, 90, and 100% ethanol series (*see* **Note 23**).
7. Denature the FISH probes (5% probe diluted in HyBuffer) for 10 min at 75 °C and incubate at 37 °C during the pre-annealing stage.
8. Apply the denatured probes onto glass slides.
9. Cover with a coverslip and seal using Fixo Gum. Allow the hybridization overnight at 37 ° C in a wet chamber.
10. On the day after hybridization, remove the coverslips and wash the slides twice with 50% formamide/50% 1× SSC for 5 min at 45 °C; twice in 1× SSC for 5 min at 45 °C; and one time in 4× T for 4 min at 45 °C (*see* **Note 24**).
11. Label two chromosomes simultaneously using probes with different labels and signal colors:
    (a) 1:1000 Cy3-streptavidin for chromosome 1 (red)
    (b) 1:200 anti-digoxigenin-fluorescein, fab fragments, for chromosome X + 3 (green)
12. After incubation, wash the slides three times in 4× SSC/NP-40 for 3 min at 45 °C.
13. Incubate with DAPI.
14. Seal the coverslips with mounting medium (Fig. 5).

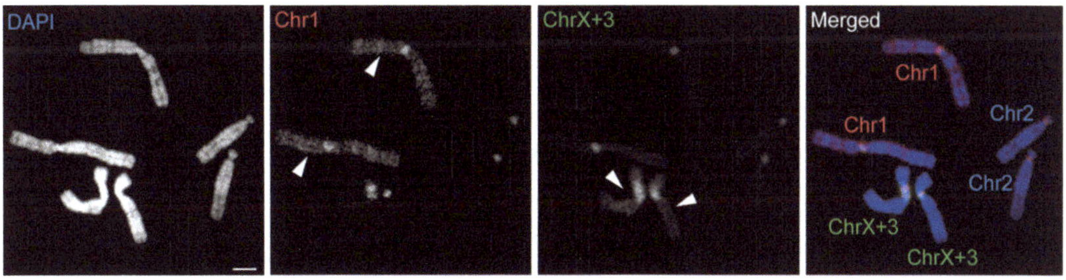

Fig. 5 Chromosome painting with probes against chromosomes 1 and 3+X. Chromosome 1-specific probe (Cy3) is labeled in red and chromosome 3+X-specific probe (anti-digoxigenin FITC conjugate) is labeled in green. DAPI staining was used as a DNA marker for all chromosomes. By exclusion in the merged image, chromosomes 2 will be stained blue. Scale bar, 5 μm

4 Notes

1. Ideally tissue culture should be performed without the use of antibiotics, which may select resistant subclones. Nevertheless, if it is extremely necessary the use of antibiotic-antimycotic 100× is recommended.
2. Indian muntjac fibroblasts usually take around 4 days to achieve total confluence and they grow better if kept concentrated in culture.
3. hTERT-immortalized Indian muntjac fibroblasts tend to become polyploid at higher passages. For this reason, it is very important to keep low culture passages. Coating the flasks/plates with fibronectin might be an option to slow the polyploid appearance rates.
4. HEK293T cells are very easy to detach from tissue culture dish. The washing step should be performed carefully.
5. The filter membrane might clot with cellular debris. In case of feeling pressure, change to a new one.
6. Polybrene is used to improve transduction efficiency, as it is a polycation that decreases charge repulsion between the virus and the cellular membrane.
7. For FACS: Prepare an extra tube with unstained cells and mono-labels, in case you have a mix of fluorochromes in your sample to be sorted.
8. Optional: Freeze by immersion in liquid nitrogen.
9. Semidry transfer devices provide rapid transfers with high efficiency, but when blotting proteins of high molecular weight, you might consider using a wet transfer system. In this case, use a constant voltage of 100 V for 1 h. Transfers are typically performed with an ice pack and with ice-cold transfer buffer to mitigate the heat produce.
10. On the day of fixation, CB should be supplemented with 10 nM of filtered glucose in order to maintain cell iso-osmotic pressure which also helps preserving internal cell structures.
11. Keep cells at 37 °C to avoid temperature fluctuations.
12. Perform this step as gently as possible to prevent cell dislodgement from the coverslip.
13. For primary and secondary antibody incubation, place the coverslips on top of parafilm layer and use a wet chamber, in order to minimize the evaporation events, thus reducing the volume of antibody per staining.
14. The sodium borohydride will bubble vigorously and may cause coverslips to float and flip occasionally.

15. Glutaraldehyde concentration might need to be troubleshoot, depending on the proteins of interest.
16. SiR-Tubulin is based on the microtubule-binding drug docetaxel. Be aware that at higher concentrations microtubule dynamics might be affected. For Indian muntjac cells, concentrations of 20–50 nM do not impair the normal progression of mitosis [14].
17. Be careful not to form any air bubbles, as this will affect confinement performance.
18. Clean the bottom with Magic Scotch Tape (3M). Do not leave confiner with ethanol for a long time as it will absorb it and become toxic.
19. Use the minimum volume as possible; otherwise it can damage the vacuum generator. Up to 500 μL should be enough.
20. A pressure of −30 mbar ensures that the device is held in place.
21. Above 5 kPa cells should be already confined.
22. Always reamplify 50–100 μL of each chromosome probe, and work with the reamplified product, and not from the DOP PCR product.
23. Ethanol solutions used for rinsing slides should be kept at room temperature. This prevents exposing the slides to large temperature variations.
24. Adding the slides into the water bath during denaturation and washing steps changes the temperature of solution for approximately 1 °C per slide which can affect the efficiency of the hybridization. Therefore, avoid incubating more than four slides at the time and consider temperature variations before incubating slides (set few degrees higher temperature of the water bath).

Acknowledgment

This project has received funding from the European Research Council (ERC) under the European Union's Horizon 2020 Research and Innovation Programme (grant agreement no. 681443) and FLAD Life Science 2020 (261/2014).

References

1. Heald R, Khodjakov A (2015) Thirty years of search and capture: the complex simplicity of mitotic spindle assembly. J Cell Biol 211 (6):1103–1111. https://doi.org/10.1083/jcb.201510015
2. Scherthan H, Cremer T, Arnason U, Weier HU, Lima-de-Faria A, Fronicke L (1994) Comparative chromosome painting discloses homologous segments in distantly related mammals. Nat Genet 6(4):342–347. https://doi.org/10.1038/ng0494-342
3. Matthey R (1973) The chromosome formulae of eutherian mammals. In: ABCE C (ed)

Cytotaxonomy and vertebral evolution, vol 531–616. Academic Press, New York

4. Hartmann N, Scherthan H (2004) Characterization of ancestral chromosome fusion points in the Indian muntjac deer. Chromosoma 112 (5):213–220. https://doi.org/10.1007/s00412-003-0262-4
5. Contreras LC, Torres-Mura JC, Spotorno AE (1990) The largest known chromosome number for a mammal, in a South American desert rodent. Experientia 46(5):506–508
6. Wurster DH, Benirschke K (1970) Indian muntjac, Muntiacus muntjak: a deer with a low diploid chromosome number. Science (New York, NY) 168(3937):1364–1366
7. Bogenberger JM, Neitzel H, Fittler F (1987) A highly repetitive DNA component common to all Cervidae: its organization and chromosomal distribution during evolution. Chromosoma 95(2):154–161
8. Yang F, O'Brien PC, Wienberg J, Neitzel H, Lin CC, Ferguson-Smith MA (1997) Chromosomal evolution of the Chinese muntjac (Muntiacus reevesi). Chromosoma 106(1):37–43
9. Soma H, Kada H, Mtayoshi K, Suzuki Y, Meckvichal C, Mahannop A, Vatanaromya B (1983) The chromosomes of Muntiacus feae. Cytogenet Cell Genet 35(2):156–158
10. Ma SL, Wang YX, Shi LM (1990) A new species of the genus Muntiacus from Yunnan, China. Zool Res 11(1):47–53
11. Hsu TC, Pathak S, Chen TR (1975) The possibility of latent centromeres and a proposed nomenclature system for total chromosome and whole arm translocations. Cytogenet Cell Genet 15(1):41–49
12. Chi JX, Huang L, Nie W, Wang J, Su B, Yang F (2005) Defining the orientation of the tandem fusions that occurred during the evolution of Indian muntjac chromosomes by BAC mapping. Chromosoma 114(3):167–172. https://doi.org/10.1007/s00412-005-0004-x
13. Rattner JB, Bazett-Jones DP (1989) Kinetochore structure: electron spectroscopic imaging of the kinetochore. J Cell Biol 108 (4):1209–1219. https://doi.org/10.1083/jcb.108.4.1209
14. Drpic D, Almeida AC, Aguiar P, Renda F, Damas J, Lewin HA, Larkin DM, Khodjakov A, Maiato H (2018) Chromosome segregation is biased by kinetochore size. Curr Biol 28(9):1344–1356. e1345. https://doi.org/10.1016/j.cub.2018.03.023
15. Comings DE, Okada TA (1971) Fine structure of kinetochore in Indian muntjac. Exp Cell Res 67(1):97–110
16. Brown P, Loughman W (1980) Visible light observations on the kinetochore of the Indian muntjac, Muntiacus muntjac, Z. Cytogenet Genome Res 27(2–3):123–128
17. Brinkley BR, Valdivia MM, Tousson A, Brenner SL (1984) Compound kinetochores of the Indian muntjac. Evolution by linear fusion of unit kinetochores. Chromosoma 91(1):1–11
18. Brinkley BR, Zinkowski RP, Mollon WL, Davis FM, Pisegna MA, Pershouse M, Rao PN (1988) Movement and segregation of kinetochores experimentally detached from mammalian chromosomes. Nature 336 (6196):251–254. https://doi.org/10.1038/336251a0
19. Carrano A, Wolff S (1975) Distribution of sister chromatid exchanges in the euchromatin and heterochromatin of the Indian muntjac. Chromosoma 53(4):361–369
20. Le Berre M, Aubertin J, Piel M (2012) Fine control of nuclear confinement identifies a threshold deformation leading to lamina rupture and induction of specific genes. Integr Biol 4(11):1406–1414
21. Zou Y, Yi X, Wright WE, Shay JW (2002) Human telomerase can immortalize Indian muntjac cells. Exp Cell Res 281(1):63–76
22. Pereira A, Sousa M, Almeida AC, Ferreira LT, Costa AR, Novais-Cruz M, Ferras C, Sousa MM, Sampaio P, Belsley M, Maiato H (2019) Coherent-hybrid STED: high contrast sub-diffraction imaging using a bi-vortex depletion beam. Opt Express 27 (6):8092–8111. https://doi.org/10.1364/OE.27.008092
23. Pereira AJ, Matos I, Lince-Faria M, Maiato H (2009) Dissecting mitosis with laser microsurgery and RNAi in Drosophila cells. Methods Mol Biol 545:145–164
24. Yang F, Trifonov V, Ng BL, Kosyakova N, Carter NP (2009) Generation of paint probes by flow-sorted and microdissected chromosomes. In: Fluorescence in situ hybridization (FISH)—application guide. Springer, Cham, pp 35–52
25. Yang F, Graphodatsky AS (2009) Animal probes and ZOO-FISH. In: Fluorescence in situ hybridization (FISH)—application guide. Springer, Cham, pp 323–346

Chapter 17

Visualization and Functional Analysis of Spindle Actin and Chromosome Segregation in Mammalian Oocytes

Binyam Mogessie

Abstract

Chromosome segregation is conserved throughout eukaryotes. In most systems, it is solely driven by a spindle machinery that is assembled from microtubules. We have recently discovered that actin filaments that are embedded inside meiotic spindles (spindle actin) are needed for accurate chromosome segregation in mammalian oocytes. To understand the function of spindle actin in oocyte meiosis, we have developed high-resolution and super-resolution live and immunofluorescence microscopy assays that are described in this chapter.

Key words Actin, Microtubules, Chromosomes, Meiosis, Oocytes, Eggs, Spindle, Fertility, High-resolution live microscopy

1 Introduction

Every mammalian life begins with the fertilization of an egg by a sperm cell. For the resulting genetically unique zygote to grow into a healthy offspring, both the egg and sperm should first contain the correct number of chromosomes. However, for reasons we are only starting to understand, eggs are often likely to have additional or missing copies of certain chromosomes before fertilization—they are aneuploid [1]. Embryos formed from fertilization of aneuploid eggs frequently die, leading to pregnancy failures, or result in offspring with genetic disorders such as Down's syndrome. Importantly, the rate of egg aneuploidy increases dramatically with advancing maternal age [2]. This phenomenon, often referred to as "the maternal age effect," is highly attributed to errors in meiosis, the specialized form of cell division that generates eggs from oocytes [3]. Indeed, a number of factors including oocyte

The original version of this chapter was revised. The correction to this chapter is available at https://doi.org/10.1007/978-1-0716-0219-5_22

Helder Maiato (ed.), *Cytoskeleton Dynamics: Methods and Protocols*, Methods in Molecular Biology, vol. 2101,
https://doi.org/10.1007/978-1-0716-0219-5_17,

chromosome cohesion [4] and microtubule dynamics [5] are known to deteriorate with increasing maternal age.

Meiotic chromosome segregation is driven by a spindle machinery that is assembled from microtubules and separates the chromosomes in two rounds of cell division [3]. Interestingly, we recently discovered that the actin cytoskeleton plays a vital role inside the meiotic spindle—actin filaments embedded inside the spindle help to organize microtubules into functional bundles that can accurately separate the chromosomes [6]. This finding constitutes an important safety mechanism in mammalian meiosis that prevents aneuploidy in oocytes and eggs. The association of actin filaments with meiotic spindles of mouse oocytes has long been known [7, 8], but its functional analyses had been thwarted due to lack of high spatial and temporal resolution microscopy assays. This chapter discusses in detail the live and immunofluorescence microscopy assays that have for the first time enabled highly resolved visualization and functional analysis of spindle actin in mammalian oocyte meiosis.

2 Materials

2.1 Ovary Dissection and Oocyte Isolation

1. N6,2′-O-dibutyryladenosine 3′,5′-cyclic monophosphate sodium salt (dbcAMP).
2. 35 mm Tissue culture dishes.
3. Pastettes (e.g., Alpha Laboratories, LW4206).
4. Paraffin (mineral oil).
5. 37 °C incubator.
6. 1.5 mL microcentrifuge tubes.
7. Mouth aspirator tube assemblies for calibrated microcapillary pipettes (e.g., Sigma, A5177-5EA).
8. 100 μL micropipettes (e.g. PIP3022, Scientific Laboratory Supplies).
9. 0.22 μm filters.
10. 2× Dumont Tweezers #2, 12 cm, Straight, 0.34 × 0.14 mm Tips (e.g., Electron Microscopy Sciences, 0103-2-PO).
11. 2× Dumont Tweezers #2, 8.3 cm, Straight, 0.06 × 0.1 mm Tips (e.g., Electron Microscopy Sciences, 0302-M5S-PS).
12. 1 mL BD Plastipak precision syringe (e.g., VWR, 613-3908).
13. 50 mL BD Plastipak luer-lock concentric syringe (e.g., VWR, 613-3925).
14. 13 mm gauge needle.
15. Mouse ovaries from 8- to 12-week-old CD1, 129 s, FVBN, or C57BL/6 mice.
16. Zeiss Discovery V8 stereomicroscope.

2.2 Stock Solutions for Oocyte Culturing and In Vitro Maturation M2 Medium

1. BSA stock: 10 mg/mL BSA in embryo transfer water.
2. Stock A: 55.5 mg/mL NaCl, 3.8 mg/mL KCl, 1.74 mg/mL KH_2PO_4, 3.14 mg/mL $MgSO_4$, 10.76 mg/mL glucose, 0.67 mg/mL penicillin sodium salt, 0.54 mg/mL streptomycin sulfate, 2.1% w/w sodium lactate in embryo transfer water.
3. Stock B: 3.4 mg/mL $NaHCO_3$ in embryo transfer water.
4. Stock C: 3.36 mM sodium pyruvate in embryo transfer water.
5. Stock D: 2.6 mg/mL $CaCl_2$ in embryo transfer water.
6. Stock E: 0.02 mg/mL phenol red in embryo transfer water.
7. Stock F: 175.2 mM HEPES.
8. Store all stock solutions at −20 °C until use.
9. Thoroughly defrost aliquots of 10 mg/mL BSA and M2 stock solutions in a beaker filled with double-distilled or Milli-Q water or by placing on a rotator at room temperature. Make sure that there are no precipitates in any of the stock solutions before proceeding.
10. Constitute M2 medium by adding 5 mL each of stock solutions A, B,C, D, E, and F to 20 mL of 10 mg/mL BSA solution in a 50 mL tube and mixing by inversion. Refreeze unused stocks for future preparation of M2 medium.
11. Add 20 μL of 1 M NaOH to the M2 medium and transfer 2 mL of it into a 14 mL tube for pH measurement.
12. Measure the pH and adjust to pH 7.2–7.4 by adding 1 M NaOH in 2 μL increments to the 50 mL tube of M2.
13. Once the desired pH range is reached, load M2 medium into a 50 mL syringe and filter it through a 0.22 μm filter into a new 50 ml tube.
14. Divide filtered M2 medium into 8 mL fractions in 15 mL tubes and store at 4 °C for a maximum of 2 weeks.

2.3 mRNA Synthesis and Preparation for Microinjection

1. mRNA synthesis kit (e.g., Thermo Fisher, AM1344).
2. Nuclease-free water.
3. Sterile microtubes.
4. Special glass capillary tube (100 mm × 0.8 mm outer diameter) (e.g., Jaytec Glass, CAP-SPEC-100-8).
5. Silicone grease, medium viscosity, Bayer (e.g. VWR, ACRO386110010).
6. Microscope slide.
7. 15 cm Petri dish
8. Dimethylpolysiloxane (capillary oil).

2.4 Microinjection

1. 22 × 22 mm Coverslips
2. Hydrochloric acid solution.
3. Microwave.
4. Absolute ethanol.
5. Tungsten carbide scriber (glass cutter) (e.g., Micro-Mark, 50299).
6. Microinjection chamber constructed from plastic according to dimensions specified in Subheading 3.2, **step 2**.
7. Scotch double-sided tape, 12.7 × 7.9 mm.
8. P-1000 micropipette puller fitted with a box filament (World Precision Instruments).
9. Zeiss AxioVert A1 microscope equipped with Narishige MN-4, MO-202 U and NZ-19-2 micromanipulators, and a CellTram 4r Oil hydraulic manual microinjector (Eppendorf).
10. Mercury.

2.5 High-Resolution and Super-Resolution Live Imaging, and Drug Addition Experiments

1. For high-resolution live imaging, a Zeiss LSM 710, LSM 780, LSM 800, LSM 880, or Leica SP8 microscope equipped with sensitive gallium arsenide phosphide (GaAsP) detectors and a piezo stage.
2. For super-resolution live imaging, a Zeiss LSM 800 or LSM 880 microscope equipped with Airyscan technology.
3. Glass-bottom imaging dishes (e.g., MatTek Corporation, P35G-0-14-C).
4. Dimethyl sulfoxide (DMSO).
5. 0.6 mM Stock SNAP-Cell-647-SiR solution by dissolving compound in DMSO (New England Biolabs).
6. 1 mg/mL Cytochalasin D stock solution prepared from powder.
7. 5 mM Nocodazole stock solution prepared from powder.
8. 1 mM SiR-Actin stock solution prepared by dissolving compound in DMSO (Cytoskeleton Inc.).

2.6 Immuno-fluorescence Microscopy of Fixed Oocytes

1. 60-Well Terasaki plates (Greiner Bio-One).
2. Soft tissue paper such as Kimwipes.
3. BSA.
4. HEPES.
5. EGTA.
6. $MgSO_4 \cdot 7H_2O$.
7. Formaldehyde, 10%, methanol free, ultrapure.
8. Triton X-100 solution.

9. A 37 °C incubator designated for storage of fixative solution-containing plates.
10. Fixative stock solutions dissolved or diluted in double-distilled or Milli-Q water (1 M HEPES, 0.25 M EGTA, 1 M $MgSO_4 \cdot 7H_2O$, 10% v/v TritonX-100).
11. Fixative solution in double-distilled or Milli-Q water (0.1 M HEPES, 0.05 M EGTA, 0.01 M $MgSO_4$, 2% formaldehyde, 0.2% TritonX-100).
12. Extraction buffer PBT (PBS + 0.1% v/v Triton X-100).
13. Wash buffer (3% v/v PBT + BSA).
14. Anti-tubulin primary antibody.
15. Stock labeling solutions from Thermo Fisher prepared according to the manufacturer's instructions (Phalloidin A488, A12379; Rhodamine Phalloidin, R415; Hoechst, 62249).

3 Methods

3.1 Oocyte Isolation, Culturing, and In Vitro Maturation

Mouse ovaries are typically collected from 6- to 12-week-old females (e.g., 129 s, CD1, C57BL/6). Depending on the strain used, typically 30–50 oocytes can be obtained from each animal. This is normally sufficient for one live imaging or immunofluorescence experiment. Ovaries should be dissected and processed with as little delay as possible to maximize the number of meiosis-competent oocytes that can be isolated. All culturing medium and dishes should be prepared and stored at 37 °C for prompt transfer of dissected ovaries. Oocytes should be cultured in dbcAMP (a non-hydrolyzable analog of cAMP) to keep them arrested in prophase during micromanipulation.

3.1.1 Preparation of Oocyte Transfer and Washing Dishes

1. Defrost an aliquot of 100 mM stock dbcAMP by briefly shaking at room temperature or 37 °C.
2. Dilute dbcAMP to 250 μM in M2 medium. For 129 s strains, we find that three times higher concentration of dbcAMP is necessary to maintain prophase arrest (*see* **Note 1**).
3. Prepare two 35 mm culture dishes with 1 mL and 2 mL of M2 + dbcAMP and two 35 mm culture dishes each with nine droplets of M2 + dbcAMP (Fig. 1).
4. Overlay the droplets with mineral oil using a pastette as in Fig. 1 and prewarm dishes at 37 °C for a minimum of 5 min.
5. Prepare a sterile microcentrifuge tube with 500 μL of M2 + dbcAMP for ovary collection and prewarm at 37 °C for 5 min (*see* **Note 2**).

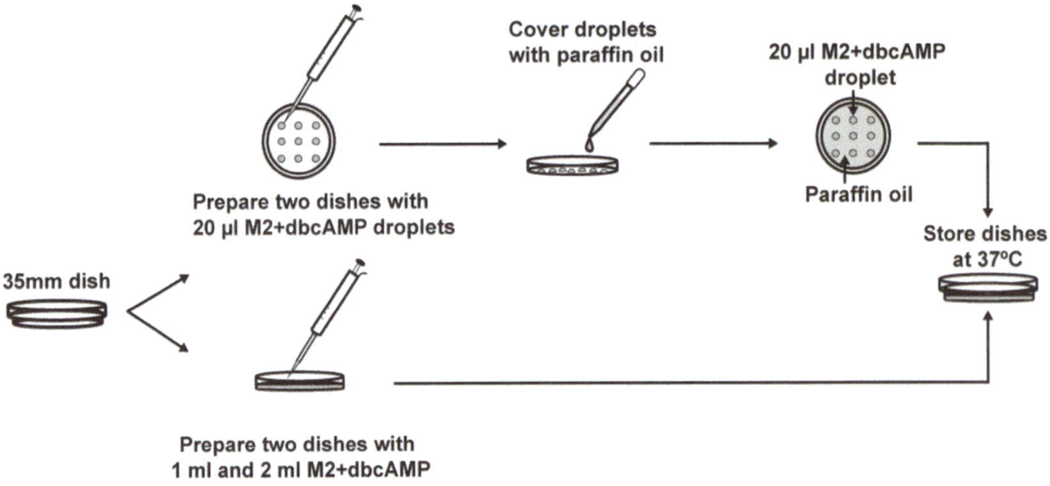

Fig. 1 Preparation of oocyte culturing and in vitro maturation dishes. Dishes should be prepared well ahead of oocyte isolation and microinjection experiments and stored at 37 °C

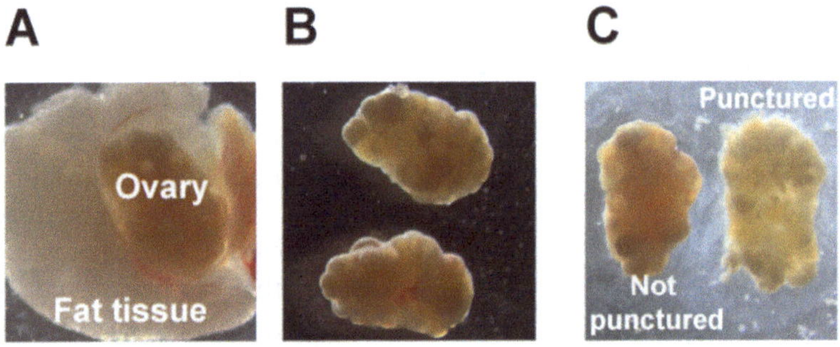

Fig. 2 Ovary dissection and puncturing. (**a**) A mouse ovary surrounded by fat tissue. (**b**) Mouse ovaries after extraction from surrounding fat tissue. (**c**) Intact (left) and punctured (right) mouse ovaries

3.1.2 Isolation and Preparation of Oocytes for Microinjection

1. Transfer collected ovaries to the 1 ml M2 + dbcAMP culture dish above (Fig. 1).
2. Separate ovaries from associated fat and other tissues using fine tweezers while taking extra care not to tear them (Fig. 2a, b).
3. Transfer ovaries to the 2 mL dbcAMP culture dish above (Fig. 1).
4. Puncture one ovary at a time with a 13 mm gauge needle fitted to a 1 mL syringe to expel oocytes from follicles (Fig. 2c). For efficient puncturing, hold the ovary down with one needle and puncture with the other (*see* **Note 3**).
5. When finished puncturing one side of the ovary, flip it over and continue to puncture until it appears flat (Fig. 2c).
6. Exchange syringes between hands and puncture the second ovary similarly with the unused needle (*see* **Note 4**).

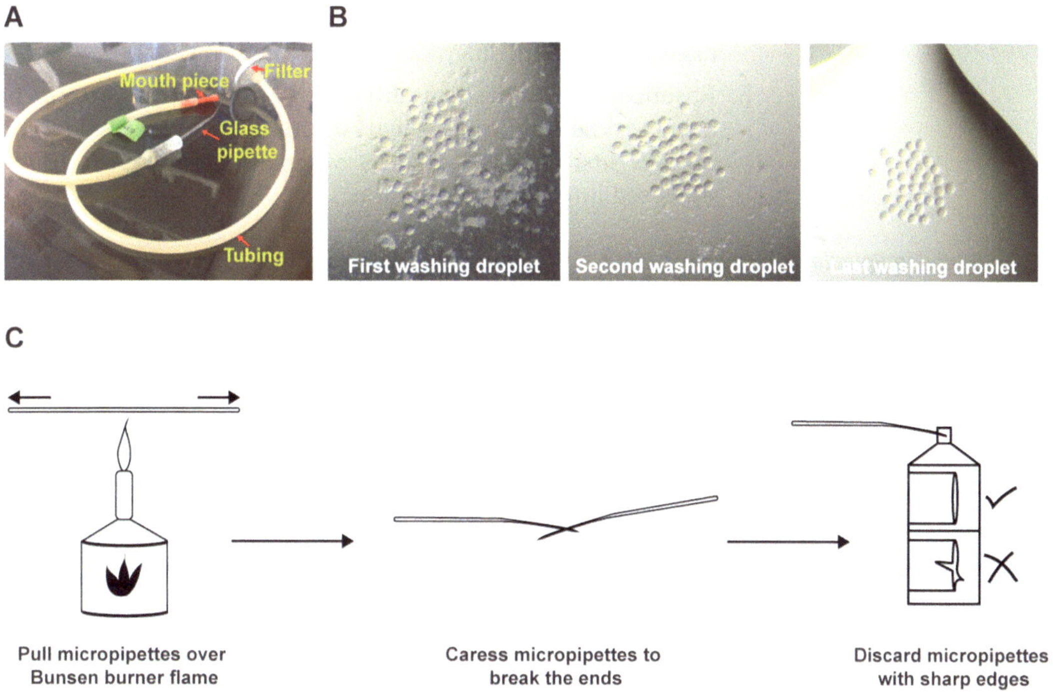

Fig. 3 Mouth pipette production and assembly for aspiration of mouse oocytes. (**a**) A fully assembled mouth pipette containing mouth piece-fitted tubing and a pulled glass micropipette separated with a 0.22 μm filter. (**b**) The process of cleaning isolated oocytes from debris by washing through nine droplets of M2 + dbcAMP. (**c**) Production of glass micropipettes for mouth pipetting using fire-assisted manual pulling techniques. Micropipettes with sharp or broken edges should be discarded to avoid oocyte death during mouth pipetting

7. Transfer oocytes from the puncture dish into a dish with M2 + dbcAMP droplets (Fig. 1) using a mouth pipette assembly (Fig. 3a) and micropipettes produced as described in Subheading 3.1, **step 3** (*see* **Note 5**).
8. Remove debris and sort meiosis-competent oocytes by washing oocytes that have centrally positioned nuclei through M2 + dbcAMP droplets prepared (Fig. 3b).
9. Leave cleaned oocytes in the last droplet and keep at 37 °C while preparing mRNAs for microinjection.

3.1.3 Production of Capillaries for Mouth Pipetting

1. Using both hands, hold a micropipette over a Bunsen burner so that the middle of the glass is exposed to the flame (Fig. 3c).
2. When the glass is red hot, pull both ends of the micropipette apart with force. This will produce two pieces of micropipettes with bent ends where the glass surface was heated (Fig. 3c).
3. While still holding non-heated ends of the two pieces of micropipettes, gently caress the bent end of one piece with that of the other until it breaks and repeat the same for the second piece (Fig. 3c).

4. Observe pulled micropipettes under a stereomicroscope and discard those with sharp or broken edges (Fig. 3c).
5. Ensure that produced mouth pipettes are at least 80 μm in diameter before using them routinely by pipetting some oocytes first. Discard pipettes where oocytes appear to be squeezed into during aspiration.

3.2 Studying Spindle Actin Function Using High-Resolution and Super-Resolution Live Microscopy

Microinjection of oocytes is a principal method for studying mammalian meiosis by high-resolution live imaging. We routinely perform quantitative microinjection of oocytes using a modification of a method that was described in great detail by Jaffe and Terasaki [9]. In this section, technical details in the context of mouse oocyte microinjection are provided along with visual aids.

3.2.1 Preparation of mRNAs for Microinjection

Visualization of spindle actin, microtubules, and chromosomes in live oocytes is achieved by microinjection of in vitro-transcribed mRNAs that encode their fluorescently labeled markers. We routinely use the calponin-homology domain of utrophin (UtrCH) to mark actin filaments, the microtubule-binding domain of MAP4 (MAP4-MTBD) to mark meiotic spindles, and histone H2B to mark chromosomes. To avoid degradation of mRNAs, they should be prepared for microinjection as follows only after isolation of oocytes from ovaries. In addition, all mRNA samples should be handled on RNase-free bench with gloved hands and using pipettes thoroughly cleaned with 70% ethanol.

1. Defrost mRNAs quickly under RNase-free conditions and store on ice.
2. If two separate mRNAs are to be microinjected, prepare a new RNase-free tube and mix them to the desired dilution (*see* **Note 6**).
3. Fashion a small apparatus for holding mRNA capillaries from a small piece of cardboard, a glass slide, and a scotch tape (Fig. 4a–c) (*see* **Note 7**).
4. Place a dollop of grease in the center of the capillary holder fold (Fig. 4c) (*see* **Note 8**).
5. Prepare a special glass capillary tube by breaking it into half and placing it on the capillary holder grease dollop (Fig. 4e).
6. mRNA-containing capillaries need to be stored on ice. To prepare the capillary storage container, line the inside of a 15 cm petri dish with folded tissue paper (Fig. 4f).
7. Place a glass slide on the tissue paper and apply a dollop of grease on the slide to serve as a capillary holder (Fig. 4f).
8. Place the capillary storage petri dish on ice (*see* **Note 9**).

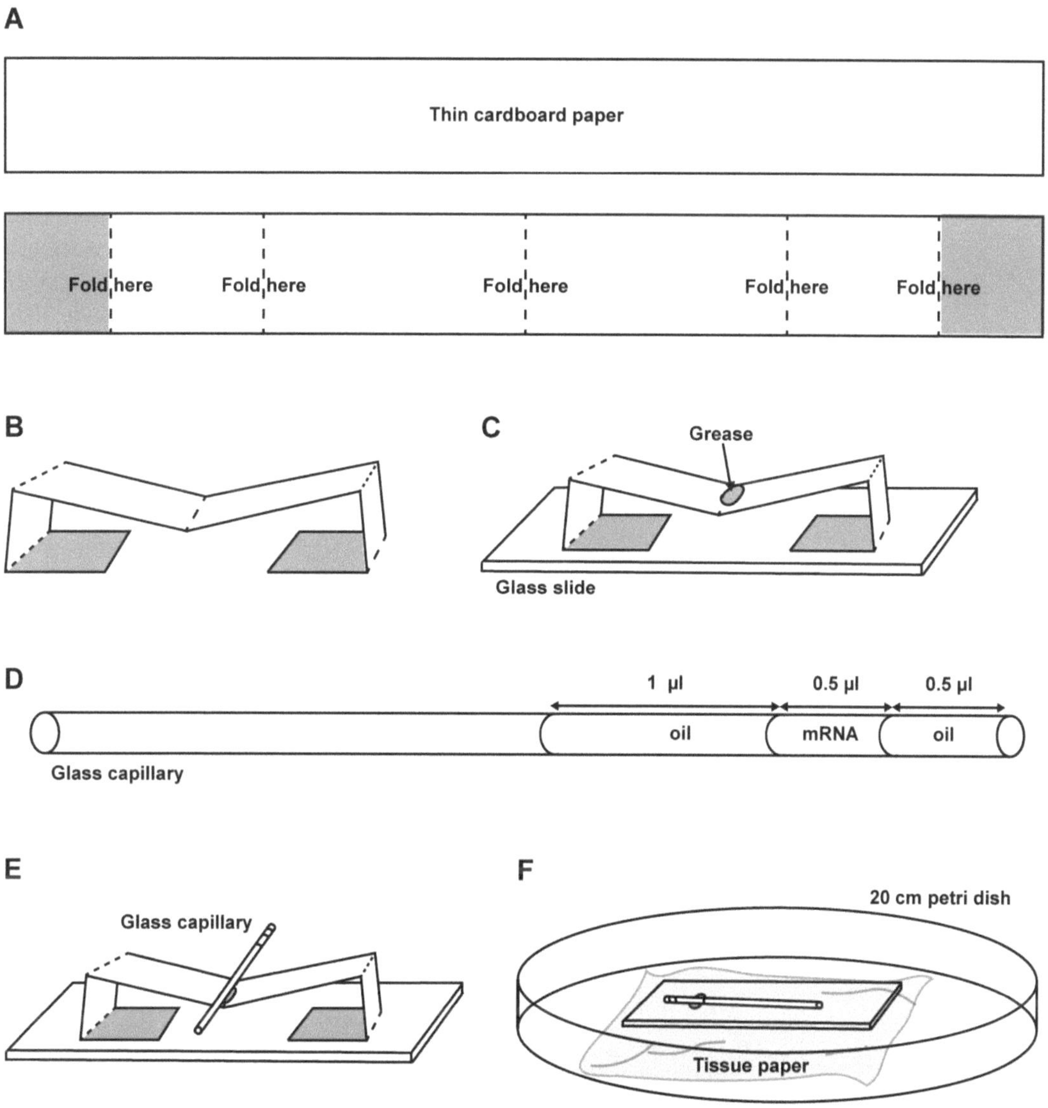

Fig. 4 Preparation of mRNA for microinjection. (**a**, **b**) Procedures to produce a small contraption to hold capillaries during mRNA loading. Fold a small piece of cardboard at the indicated perforations. (**c**) Stick folded capillary holder on a 25 × 75 mm glass slide using tape and place a dollop of grease to hold the capillary in place. (**d**, **e**) Load into the capillary the indicated volumes of oil, mRNA, and oil and place on capillary holder grease. (**f**) Place loaded capillary on 25 × 75 mm glass slide containing a dollop of grease. Transfer the capillary-containing slide into a 15 cm petri dish overlaid with tissue and place the petri dish on ice

9. Load 1 μL of capillary oil into the non-sharp end of the half-broken glass capillary from **step 5** and place it back on the holder.
10. Load 0.5 μL of mRNA into the capillary and place it back on the holder.

11. Load 0.5 μL of capillary oil into the capillary and place it on ice inside the capillary storage dish by gently laying it over the grease dollop (Fig. 4d).

3.2.2 Preparation of Microinjection Chamber

A U-shaped microinjection chamber can be constructed from clear plastic with little effort by most workshops according to the dimensions in Fig. 5a. In this chamber, oocytes are lined up for microinjection inside a small glass shelf created by spacing two specially cleaned coverslips with a double-sided tape. The cleanliness of coverslips used for microinjection strongly affects oocyte health. We use the following steps to thoroughly clean and store coverslips:

1. Fill a coverslip-washing plastic box (maximum volume 1 L) with 500 mL of Milli-Q water and bring to a boil in a microwave.
2. Dilute a dash of washing up liquid detergent in the heated water.
3. Using fine tweezers, drop 22 × 22 mm coverslips one by one into the detergent solution.
4. Swirl container box gently to mix and let it sit for 15–20 min.
5. Pour out the detergent solution taking care not to lose coverslips.
6. Wash three times with Milli-Q water while thoroughly but gently shaking at every wash.
7. Bring to a boil in microwave.
8. Wash three times with Milli-Q water while thoroughly but gently shaking at every wash.
9. Lay out paper towels on a clean bench and transfer coverslips from wash box to tissues using fine tweezers.
10. Prepare a 50% v/v HCl solution by adding 200 mL of 10 M HCl to 200 mL of Milli-Q water inside the wash box (*see* **Note 10**).
11. Using fine tweezers, transfer coverslips individually into the wash box and leave in 50% v/v HCl overnight.
12. Drain the acid solution into a glass bottle for reuse.
13. Wash five times with Milli-Q water while thoroughly but gently shaking at every wash and let it sit for 10 min.
14. Wash again five times with Milli-Q water while thoroughly but gently shaking at every wash.
15. Using fine tweezers, transfer coverslips individually into 50 mL Falcon tubes containing 80% v/v EtOH and 20% Milli-Q water for storage (*see* **Note 11**).

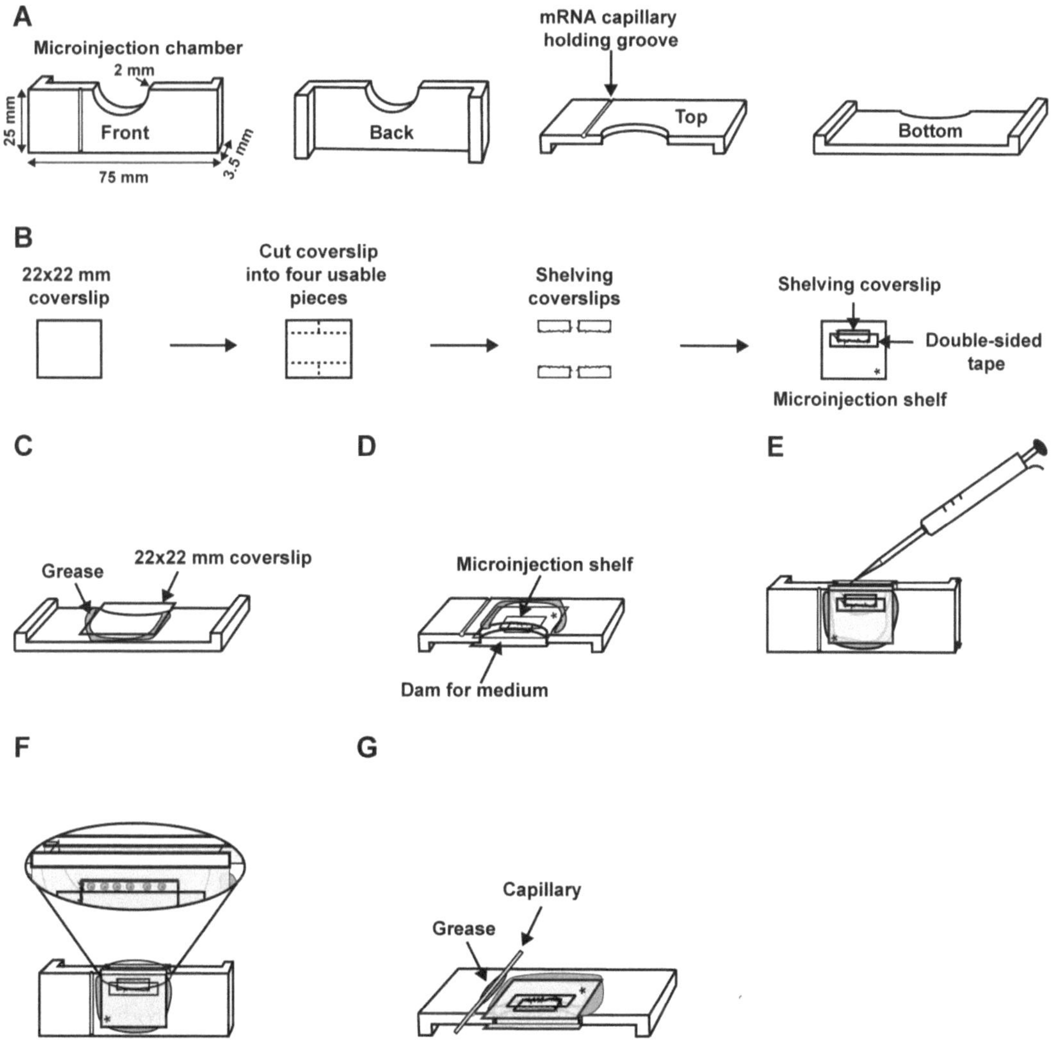

Fig. 5 Preparation of chamber and oocytes for microinjection. (**a**) Dimensions and sides of a microinjection chamber that can be constructed from plastic. (**b**) Cutting of coverslips into smaller pieces for construction of microinjection shelves. The non-cut (non-sharp) edge of shelving coverslips should be stuck on double-sided tape as an overhang to create space between the two coverslips. The side of the coverslip to which the shelf is affixed with double-sided tape is marked with asterisk symbol to aid with orientation. (**c**) Apply a thin layer of grease and mount a 22 × 22 mm coverslip on the back of the microinjection chamber. (**d**) Apply a thin layer of grease and mount a microinjection shelf on the front of the chamber. The shelving coverslip should face the medium dam. (**e**) Fill the dam space created on the chamber with M2 + dbcAMP. (**f**) Load oocytes into microinjection shelf by mouth pipetting. (**g**) Apply a thin layer of grease into the holding groove and mount mRNA capillary

We then prepare a microinjection shelf and chamber using the following steps:

(a) Using fine tweezers, place a double-sided tape roughly 5 mm away from either side of a 22 × 22 mm coverslip (Fig. 5b) and firmly stick it down by pressing on it using the non-sharp end of the tweezers.

(b) Using a ruler and a glass-cutting knife, cut another coverslip into smaller pieces as shown in Fig. 5b. Keep four of the small pieces for building shelves.

(c) Using fine tweezers, place a shelf piece coverslip from **step b** onto the double-sided tape affixed to the coverslip in **step a**. The smaller coverslip piece should have its cut end resting on the tape while its non-cut end should visibly hang over the tape (Fig. 5b), thus creating a space (the "shelf") between the two coverslips where oocytes will be loaded for microinjection.

(d) Press firmly on the small piece coverslip while taking care not to crack it (*see* **Notes 12** and **13**).

(e) Position the plastic microinjection chamber upside down and apply a thin layer of grease around its U-shape (Fig. 5c).

(f) Using fine tweezers, place a clean and uncut coverslip on the grease and press down on it using the non-sharp end of the tweezers to remove air bubbles from between the grease and coverslip (Fig. 5c).

(g) Flip the chamber over and apply a thin layer of grease around its U-shape (Fig. 5d).

(h) Using fine tweezers, place a coverslip containing the shelf, with the shelf facing inside (Fig. 5d), and press down on it using the non-sharp end of the tweezers to remove air bubbles. This concludes the assembly of a microinjection chamber consisting of a shelf and a medium dam.

3.2.3 Microinjection of Oocytes

We microinject oocytes with mercury-filled needles produced using a horizontal micropipette puller (*see* **Note 14**). Microinjection needles are connected to oil-filled hydraulic pumps through a micropipette holder. Due to high surface tension of mercury, pressure displacement of oil within the hydraulic pump translates into picoliter-scale displacement of mRNA samples within the micropipettes. When performed on calibrated microscopes fitted with binoculars bearing measurement rulers, this allows quantitative microinjection of mRNA into oocytes. We produce microinjection needles using a P-1000 micropipette puller (Sutter Instruments) equipped with a box filament (FB255B). The parameters provided in Table 1 yield microinjection needles that in our hands allow smooth microinjection and do not affect oocyte health and maturation. We find that adjustment of the parameters according to the manufacturer's instruction is necessary from time to time (*see* **Note 15**). For oocyte loading, capillary fitting, and oocyte microinjection, we use the following steps:

Table 1
Micropipette pulling settings for production of microinjection needles for box filament FB255B

Heat	Pull	Velocity	Time	Pressure	Ramp
525	30	40	250	200	515

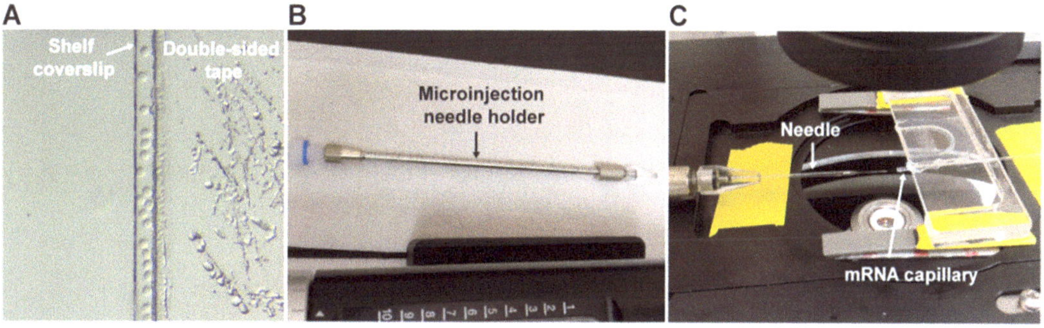

Fig. 6 Positioning of oocytes, mRNA capillary, and chamber for microinjection. (**a**) Oocytes loaded into microinjection shelf by mouth pipetting. (**b**) Microinjection needle holder connected to hydraulic manual injector. (**c**) Needle, chamber, and mRNA-loaded capillary positioned on microinjection microscope

1. Add 800 μL of M2 + dbcAMP into the medium dam of the microinjection chamber (Fig. 5e).
2. Load oocytes into microinjection shelf by mouth pipetting (Figs. 5f and 6a).
3. Place the microinjection chamber down with the shelf side away from the bench surface and apply two minimal amounts of grease onto the front and back of the capillary groove (Fig. 5g).
4. Using fine tweezers, place the mRNA capillary inside the capillary groove, supported by the grease (Fig. 5g).
5. Transfer the microinjection chamber-mRNA capillary assembly to a microinjection microscope and place it with the shelf side facing away from the stage (Fig. 6c).
6. Attach a mercury-filled needle to the micropipette holder of a hydraulic pump (Fig. 6b) and place on a micromanipulator.
7. Bring microinjection needle and mRNA capillary into focus (Fig. 7a).
8. Use the micromanipulator to slowly move the needle to one edge of the mRNA capillary (Fig. 7b) and break its tip by gently tapping on the capillary surface.
9. Push oil outward through the micropipette holder using the hydraulic pump until the mercury moves to the front of the needle (Fig. 7c). This should normally happen after a few rotations of the hydraulic pump dial. If not, it is an indication

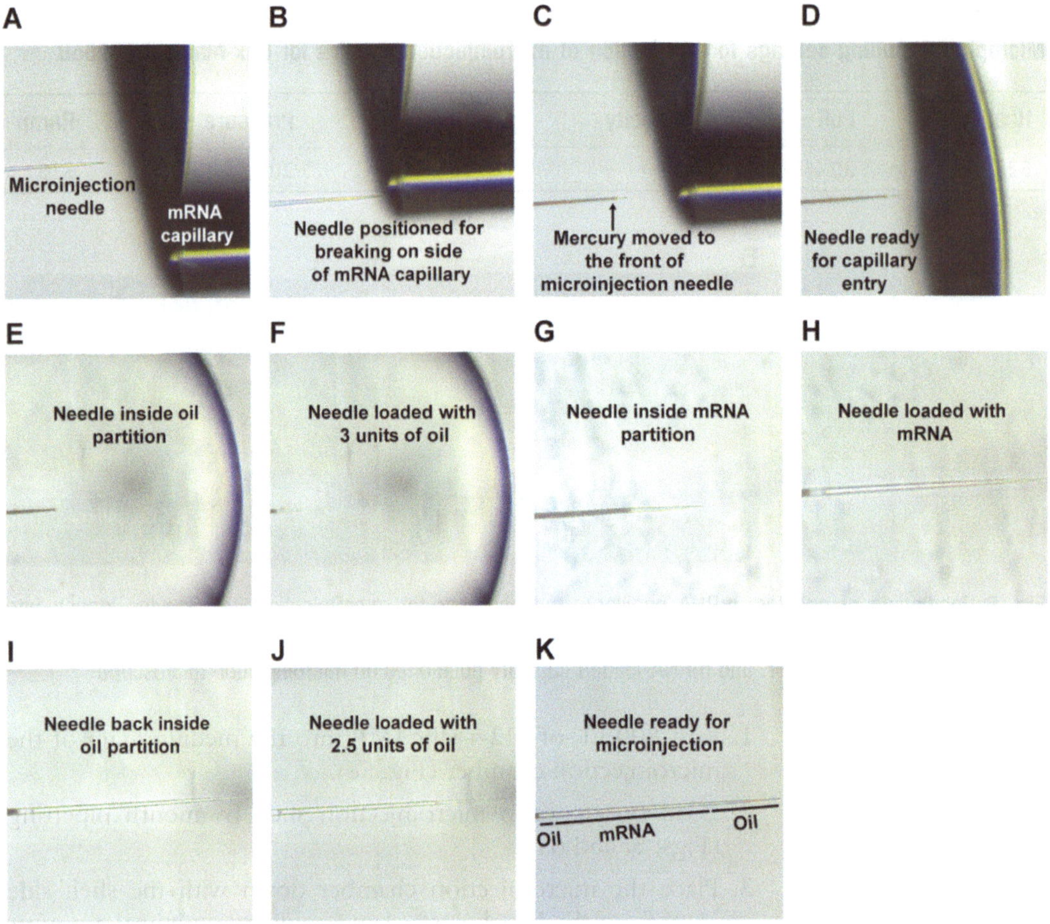

Fig. 7 Loading of microinjection needle with mRNA. (**a–c**) Steps of breaking the tip of the microinjection needle and moving mercury to the front of the needle. (**d–k**) Steps of needle loading with oil and mRNA

that the needle is not sufficiently broken. Repeat the tip breaking step and try to move the mercury to the front of the needle again.

10. Using the micromanipulator, move the needle into the front oil portion of the capillary and bring it into focus (Fig. 7d).
11. Move the needle into the capillary and take up three units of oil by applying negative pressure through the hydraulic pump (Fig. 7e, f).
12. Move the needle farther inside the capillary into the mRNA portion and bring it into focus (Fig. 7g).
13. Take up 6–7 pL unit equivalents of mRNA (Fig. 7h).
14. Move the needle back into the first oil portion of the capillary and bring it into focus (Fig. 7i).
15. Take up 2.5 units of oil (Fig. 7j).

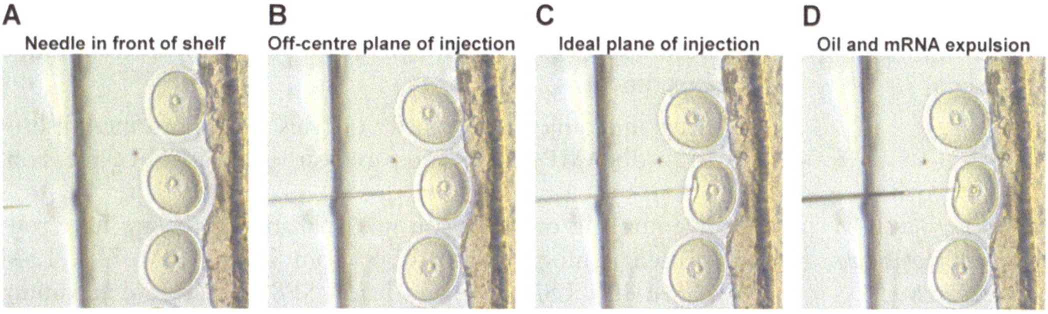

Fig. 8 Microinjection of mouse oocytes. (**a**) Positioning of microinjection needle in focus with oocytes. (**b**) Microinjection needle in contact with oocyte at a focal plane that is not ideal for injection (above the hemispheric region of the oocyte). (**c**) Microinjection needle in contact with oocyte at the ideal focal plane. (**d**) Piercing into oocyte with microinjection needle followed by expulsion of oil and mRNA into the oocyte cytoplasm. When the front oil portion and mRNA volumes are expelled, mercury can be seen returning to the needle's front

16. Move the needle out of the capillary and bring it into focus (Fig. 7h).
17. Using the stage and the micromanipulator, move the needle to the front of the microinjection shelf (Fig. 8a). Take extra care not to hit the needle against the chamber coverslip when entering into the chamber.
18. Move the needle into the shelf and slightly push the oocyte to be injected to determine if it is in the ideal microinjection plane (Fig. 8b, c) (*see* **Note 16**).
19. Use the micromanipulator to pierce through the zona pellucida and move the needle into the oocyte cytoplasm.
20. Apply positive pressure inside the needle through the hydraulic pump until the front oil portion of the needle is expelled into the oocyte cytoplasm. This is quickly followed by the mRNA portion, which may not be readily visible. Once the mercury has moved back to the front of the needle and an oil droplet is inside the oocyte (Fig. 8d), it can be assumed that the mRNA has been successfully introduced into the oocyte.
21. As soon as oil expulsion is seen, quickly maneuver the stage to withdraw it away from the needle holder until the needle comes out of the chamber (*see* **Note 17**).
22. Reposition the microinjection needle to the front of the mRNA capillary. If they are not both in focus, bring the capillary into focus by gently pressing down or lifting up its front end using fine tweezers.
23. Once needle and mRNA capillary are in focus, repeat microinjection steps for each oocyte. As long as the capillary is kept in

the adjusted position, it should not be necessary to move the needle in the Z-axis throughout the rest of the microinjection experiment.

24. After microinjection, transfer oocytes to a prewarmed dish of M2 + dbcAMP droplets prepared earlier for mRNA expression.

3.2.4 High-Resolution Live Imaging of Spindle Actin Assembly and Chromosome Segregation During Meiosis I and II

For live imaging of meiosis in mammalian oocytes, we have routinely used confocal microscopes from Zeiss (LSM 710, LSM 780, LSM 800, LSM 880) and Leica (SP8). To achieve maximum resolution during imaging, we use ×40 and ×63 water immersion objectives (1.1–1.2 NA). In addition, we are able to achieve high signal-to-noise ratio using the Airyscan super-resolution module on LSM 800 and LSM 880 microscopes. Combined, these setups allow us to perform four-dimensional (x,y,z,t) live-cell imaging of meiosis without compromising oocyte viability.

To perform live imaging of microtubules, chromosomes, and spindle actin, oocytes must first be released from prophase arrest by washing out dbcAMP. Depending on the mouse strain used, the release from prophase arrest can take 30 min to 1 h. We find that oocytes from 129-s mouse strains release from prophase arrest faster than oocytes from CD1, C57BL/6, and FVBN strains. Achieving a temperature of 37 °C inside the imaging dish, close to the objective lens where oocytes eventually settle, is absolutely critical for oocyte development (*see* **Note 18**). To visualize the entire process of meiosis starting from nuclear envelope breakdown, oocytes should be transferred to the microscope immediately after washing out dbcAMP.

1. Prepare a 35 mm culture dish containing nine droplets of M2 and a glass-bottom imaging dish containing a droplet of M2 covered with paraffin oil 30–45 min before starting a live imaging experiment and store at 37 °C.
2. If using SNAP-MAP4-MTBD to simultaneously image actin, microtubules, and chromosomes, prepare an additional dish with droplets of SNAP-Cell-647-SiR substrate diluted 1:200 in M2 + dbcAMP.
3. After 2–3 h of mRNA expression, remove dbcAMP by mouth pipetting oocytes through the nine droplets of M2 prepared in **step 1**. If labeling microtubules with SNAP-MAP4-MTBD, wash oocytes through droplets of M2 + dbcAMP+SNAP-Cell-647-SiR prepared in **step 2** (*see* **Note 19**).
4. Move washed oocytes into the imaging dish containing only M2 medium by mouth pipetting and transfer the dish to a confocal microscope (*see* **Note 20**).
5. Locate oocytes through the binoculars using only transmission light (*see* **Note 21**).

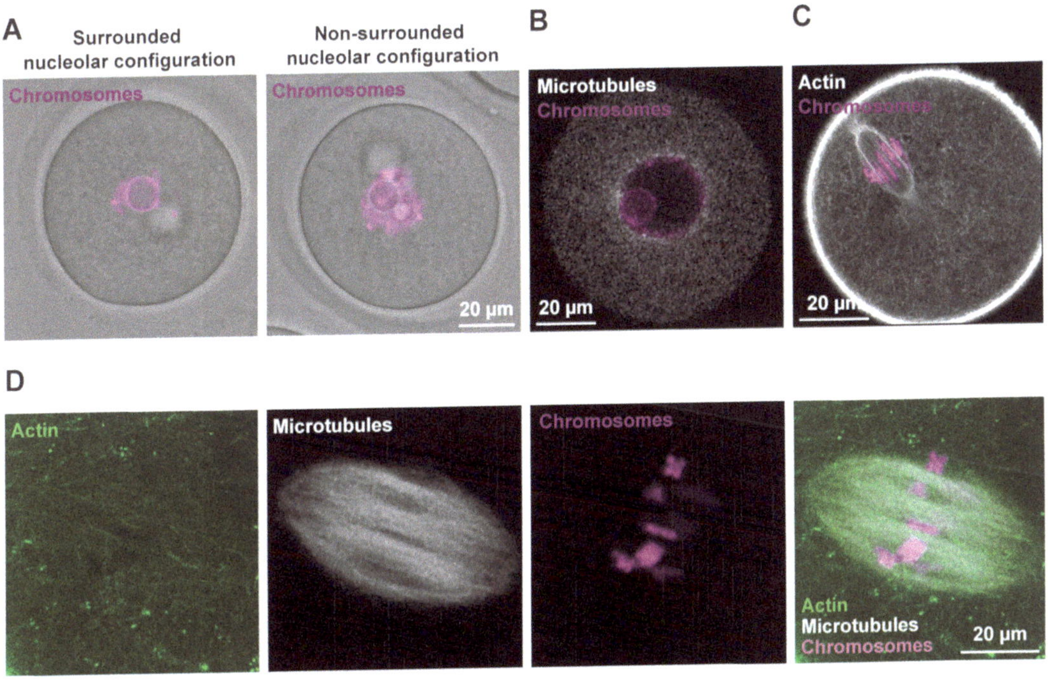

Fig. 9 Live imaging of spindle actin, microtubules, and chromosomes during mouse oocyte meiosis. (**a**) Single confocal section images of surrounded and non-surrounded nucleolar configurations of chromosomes (magenta) in prophase-arrested mouse oocytes. (**b**) Optimal signal-to-noise ratio of microtubules (gray) and chromosomes (magenta) in an oocyte-resuming meiosis after dbcAMP washout. (**c**) Spindle actin (white) and chromosomes (magenta) in a mouse oocyte progressing through meiosis I. The signal-to-noise ratio of spindle actin, cytoplasmic actin, and cortical actin indicates optimal mRNA expression level of EGFP-UtrCH that does not cause stabilization of actin filaments and perturb meiotic maturation. (**d**) Single-plane super-resolution (Airyscan) live imaging of actin (green), microtubules (gray), and chromosomes (magenta) in a mouse oocyte during spindle bipolarization. Faint labeling of actin filaments during spindle bipolarization indicates optimal expression of EGFP-UtrCH that will not interfere with spindle relocation and chromosome segregation due to actin stabilization

6. Switch to confocal mode and using coarse imaging settings (high scan speed, no line averaging, 512 × 512 pixels or less, low zoom) mark the positions of oocytes after centering them in the field of view.
7. Adjust the laser settings as necessary to determine the lowest laser intensities that give good signal-to-noise ratio for microtubules and chromosomes in prophase-arrested oocytes (Fig. 9b) or for spindle actin at the start of meiotic spindle bipolarization (Fig. 9d) (*see* **Note 22**).
8. For the actual imaging experiment, increase the image size to 512 × 512 pixels, set the pinhole size to 3 μm, and adjust zoom (to cover the entire oocyte), scan speed, and directionality (unidirectional or bidirectional scan) to achieve a time lapse of 3–6 min while acquiring a z-stack thickness of at least 20 μm at confocal sections of 1.5 μm. The identified imaging

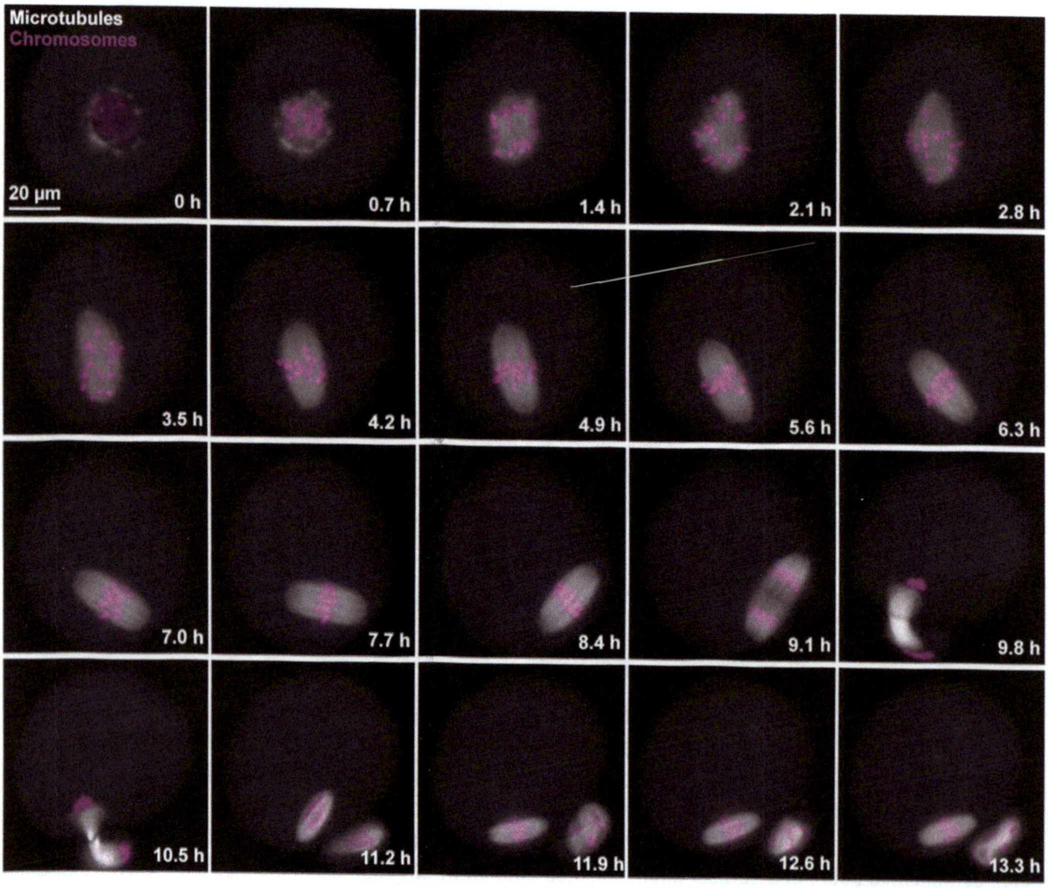

Fig. 10 High-resolution live imaging of meiotic spindle assembly and chromosome segregation in a mouse oocyte. Upon release from prophase arrest, a meiotic spindle is assembled from microtubules (gray). Homologous chromosomes (magenta) are captured by the spindle and transported to the oocyte surface where they are segregated. After a second meiotic spindle assembles and aligns the chromosomes, the egg is arrested in metaphase of meiosis II until fertilization

conditions should allow continuous live imaging of oocytes from nuclear envelope breakdown through meiosis I to anaphase I and metaphase II spindle assembly without compromising oocyte health (Fig. 10) (*see* **Note 23**).

3.2.5 Pharmacological Drug Addition Experiments

Cytoskeletal loss-of-function assays can be readily performed in oocytes by addition of widely used cytoskeletal inhibitors to the culture medium. Cytoskeletal disruption is reversible when oocytes are treated with drug concentrations we routinely use. We typically treat oocytes with 5 μg/mL cytochalasin D to disrupt the actin cytoskeleton, with 5 μM nocodazole to disrupt microtubules and 5 μM SiR-actin to stabilize actin. In addition, cytochalasin D and nocodazole can be combined to simultaneously disrupt actin filaments and microtubules in loss-of-function assays. Drugs can be added before releasing oocytes from prophase arrest. Alternatively,

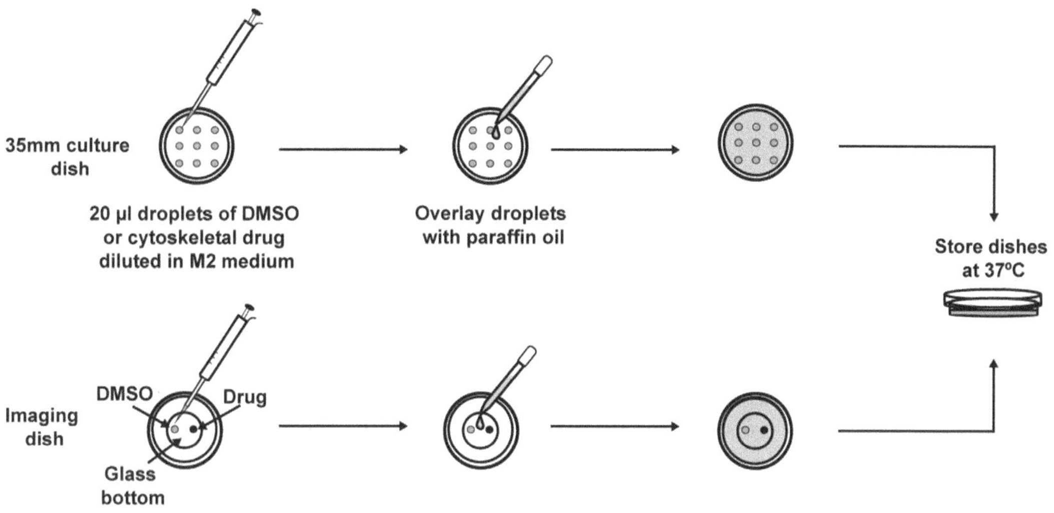

Fig. 11 Cytoskeletal disruption assays for studying spindle actin function in mouse oocytes. Oocytes should be washed through 9 prewarmed 20 μL droplets of DMSO (control), cytochalasin D (to disrupt spindle actin), or nocodazole (to disrupt microtubules) before being placed in a glass-bottom imaging dish containing a droplet of DMSO or the respective cytoskeletal drug. For simultaneous drug treatment experiments, as many as four droplets can be placed in the imaging dish without risking cross-contamination

acute cytoskeletal disruption or stabilization can be achieved by addition of drugs at any desired stage of meiosis (e.g., spindle relocation, immediately before anaphase I or II, or before and after metaphase II spindle assembly). For analysis of spindle actin function in meiosis II chromosome alignment and segregation, cytochalasin D or SiR-actin should be acutely added at least 4 h after metaphase II spindle assembly to ensure that all chromosomes are fully aligned first.

1. For drug addition experiments, dilute a stock of the drug of choice in M2 medium without dbcAMP. Make a similar dilution of DMSO for use as control (*see* **Note 24**).
2. Prepare control and experimental 35 mm culture dishes by placing nine droplets of diluted DMSO and drug, respectively, and overlay with paraffin oil (Fig. 11).
3. Prepare one glass-bottom imaging dish with one droplet of diluted DMSO and one droplet of diluted drug sufficiently spaced apart to avoid contamination of the control droplet by the drug. Overlay these carefully by gently applying paraffin oil between the two droplets so that it spreads out to the droplets (Fig. 11). This minimizes any chance of carryover of the drug to the control group by the spreading of paraffin oil.
4. Prewarm culture and imaging dishes at 37 °C.
5. Divide oocytes into two groups and wash them through the droplets of control and drug dishes.

6. Transfer oocytes to the imaging dish containing DMSO and drug droplets accordingly and proceed with microscopy.
7. If the experiment involves acute perturbation of spindle actin in meiosis II, perform the steps above with oocytes that have matured into eggs and have been arrested at metaphase II for at least 4 h (*see* **Note 25**).

3.2.6 Parthenogenetic Activation of Mouse Eggs to Visualize Spindle Actin and Chromosome Segregation During Anaphase II

After release from prophase arrest, mouse oocytes progress through meiosis I, segregate the homologous chromosomes in anaphase I, and become arrested in metaphase II until fertilized, when the sister chromatids are segregated in anaphase II. Importantly, fertilization can be chemically mimicked in vitro to induce anaphase II and observe sister chromatid separation. We follow these steps to efficiently achieve release from metaphase II arrest in mouse eggs:

1. Immediately before activation experiment, prepare a 1 M $SrCl_2$ stock solution in embryo transfer water.
2. Dilute $SrCl_2$ stock solution to 10 mM in calcium-free M2 medium (prepared by replacing stock D with equal volume of embryo transfer water; *see* Subheading 2).
3. If anaphase II is to be observed in control- versus drug-treated oocytes, $SrCl_2$ should be diluted to 10 mM in calcium-free M2 containing DMSO or the cytoskeletal drug of choice.
4. Prepare a 35 mm culture dish with nine droplets of 10 mM $SrCl_2$ alone or with DMSO and cytoskeletal drugs. Overlay droplets with paraffin oil and store at 37 °C.
5. Prepare a glass-bottom imaging dish with corresponding 10 mM $SrCl_2$ droplets, overlay with paraffin oil, and store at 37 °C.
6. Wash oocytes through $SrCl_2$ droplets and place in $SrCl_2$ droplet inside imaging dish.
7. Transfer immediately to confocal microscope and acquire time-lapse images as described earlier.

3.2.7 Quantification of Chromosome Alignment and Segregation Errors from Live Imaging Datasets

Drug-mediated disruption of spindle actin or its genetic disruption in formin-2 knockout oocytes leads to chromosome alignment and segregation errors that lead to oocyte aneuploidy [6]. High temporal resolution is critical for quantification of chromosome misalignment and segregation. 3–6-min time-lapse live imaging datasets allow us to capture oocyte chromosome alignment and segregation in detail and to reproducibly quantify errors. These chromosomal defects can be quantified in two ways—firstly through manual quantification using well-defined criteria and secondly through automatic detection of chromosomes in live imaging datasets using Imaris software (Bitplane).

The first method takes into account the meiotic spindle length in maximum intensity projection confocal images to define outliers of chromosome alignment. As such, it should be used for quantification only in oocytes where the meiotic spindle is positioned parallel to the imaging plane (Fig. 12a). Z-projection of meiotic spindles that are oriented at various angles relative to the imaging plane will lead to inaccurate measurement of spindle length and the perception of chromosome misalignment where there is none (Fig. 12a). Alignment should be quantified in the metaphase frame immediately before anaphase onset. Anaphase onset is defined as the first frame where homologous chromosomes (anaphase I) or sister chromatids (anaphase II) noticeably start to move apart (Fig. 12b).

1. Using ImageJ/Fiji or a suitable image analysis software such as Zen (Zeiss), perform maximum intensity projection of the acquired confocal sections.
2. Select time-lapse movies where the meiotic spindle is parallel to the imaging plane.
3. Measure the spindle length and determine the spindle equator position.
4. Draw a rectangle from the spindle equator to either spindle pole spanning one-third of half the spindle length.
5. Score as misaligned any chromosome that clearly lies outside this rectangle region in the last metaphase frame before anaphase onset (Fig. 12c).
6. Starting from the next frame, score as lagging or severely lagging those chromosomes that fail to clear the central spindle region within 12 or 18 min of anaphase I onset (Fig. 12c).

The second method uses the three-dimensional surface reconstruction module of Imaris software (Bitplane) to automate the detection of misaligned and lagging chromosomes from high-resolution live imaging datasets. This approach can be used to independently confirm results from manual analyses. We have confirmed that the steps described here can be performed using Imaris versions 7.0–9.2.

1. Navigate to the “Surpass” menu of Imaris and open a time-lapse imaging dataset (without maximum intensity projection) where the meiotic spindle is oriented parallel to the imaging plane at anaphase onset. Imaris can open a wide range of raw data file types including ZEN (Zeiss) and LAS X (Leica) files.
2. If acquired, switch off the transmission light channel in the “Display adjustment” pane for easier visualization of the dataset.

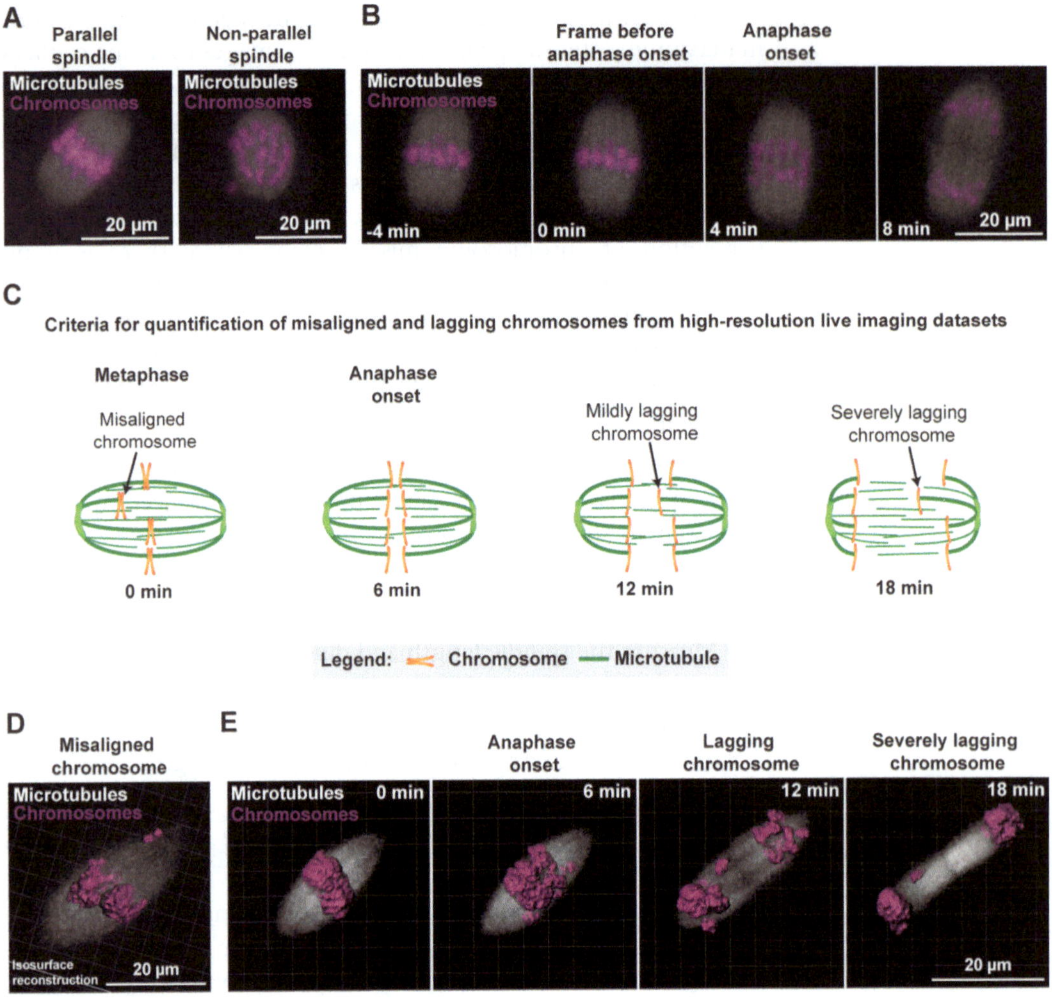

Fig. 12 Quantification of meiotic chromosome misalignment and segregation errors from high-resolution live imaging datasets. (**a**) Maximum intensity projections of microtubules (gray) and chromosomes (magenta) in a spindle that is parallel (left) and nonparallel (right) to the imaging plane. Because the spindle length appears shorter and the metaphase plate much wider, projection of nonparallel spindles gives the impression of chromosome misalignment where there is none, and thus should be avoided. (**b**) The frame of anaphase onset is determined as the time at which chromosomes (magenta) clearly begin to move toward opposite spindle poles. (**c**) Criteria for quantification of misaligned chromosomes and lagging chromosomes during meiosis are shown. Misaligned chromosomes are classified based on spindle length measurement as described in Subheading 3.2.7. Lagging chromosomes are defined as those chromosomes that fail to clear the central spindle region within 12 min of anaphase onset. Chromosomes that remain in the central spindle region 18 min after anaphase onset are classified as severely lagging. (**d**, **e**) Automated identification of misaligned (**d**) and lagging or severely lagging (**e**) chromosomes by three-dimensional iso-surface reconstruction of chromosomes (magenta) and using criteria described in (**c**)

3. Under the volume properties submenu, select "add new surface" to reconstruct the chromosomes.
4. In the automatic creation menu that shows up, untick "Track Surfaces (over time)" and press next (blue and white arrow button).

5. In the next window, select the chromosome channel as the source channel for reconstruction and press next without changing default settings (thresholding based on absolute intensity is the default setting for surface creation). If the signal-to-noise ratio for H2B-mRFP in the live imaging dataset is optimal, chromosomes will likely appear as a single large mass before anaphase onset (*see* **Note 26**).
6. In the thresholding adjustment window, make sure that the "Region Growing" box is unticked and press the finish button (green) to complete the creation process.
7. For ease of visualization, switch off the chromosome channel in the "Display adjustment" pane.
8. Navigate to the frame immediately before anaphase onset and score the number of misaligned chromosomes that appear separated from the main mass of chromosomes (Fig. 12d).
9. Navigate through the next frames and score lagging chromosomes that are disconnected from the main separating chromosome mass (Fig. 12e) as lagging and severely lagging using the same criteria applied in manual quantifications.

3.3 Studying Spindle Actin Function in Fixed Mouse Oocytes

For best results, mouse oocyte fixative solution should be prepared and prewarmed at 37 °C prior to fixation. In addition, the extraction and washing buffers should be prepared beforehand and stored at 4 °C when not being used.

3.3.1 Fixation and Preparation of Mouse Oocytes for Immunofluorescence

1. Roll a Kimwipe tissue and moisten with water. Squeeze out excess water and place moistened tissue on the side of a multiwell Terasaki plate to maintain humidity during fixation and immunostaining.
2. Add 18 μL of fixative solution to three wells of a Terasaki plate for each group of oocytes (e.g., DMSO-treated controls) that is to be fixed.
3. Collect oocytes in as little medium as possible and transfer to fixing well using a mouth pipette (*see* **Note 27**).
4. Fix oocytes at 37 °C in a designated immunofluorescence incubator for 30 min. If spindle actin is to be labeled, reduce this to 25 min.
5. Wash oocytes three times with 18 μL PBT extraction buffer to replace fixative solution.
6. Extract by incubation in PBT at 4 °C overnight. For urgent analyses, 1-h extraction may give sufficient results.

3.3.2 Labeling of Spindle Actin, Chromosomes, and Microtubules in Fixed Oocytes

1. Wash oocytes three times with PBT-BSA and block for 30–60 min in PBT-BSA at room temperature.
2. Dilute primary antibodies (e.g., anti-tubulin) in PBT-BSA.

3. Wash oocytes three times with 18 µL primary antibody solution and incubate for 1–3 h at room temperature or 4 °C overnight (*see* **Note 28**).
4. After incubation with primary antibodies, wash oocytes three times with 18 µL volumes of PBT-BSA.
5. Dilute secondary antibodies and other dyes (e.g., Hoechst to label chromosomes, Phalloidin-Alexa-488 or Rhodamine-Phalloidin to label actin) in PBT-BSA.
6. Wash oocytes three times with 18 µL volumes of secondary antibody or dye solution and incubate for 1–2 h at room temperature.
7. After incubation, wash oocytes three times with 18 µL volumes of PBT (*see* **Note 29**).
8. Place droplets of PBT-BSA on glass-bottom imaging dishes and cover with paraffin oil.
9. Transfer oocytes to droplets using mouth pipettes and proceed with imaging.

4 Notes

1. 6 mL of M2 + dbcAMP per mouse (two ovaries) is usually sufficient for isolation, preparation, and microinjection of oocytes.
2. Dissected ovaries should be kept in M2 + dbcAMP during transfer from animal facility to the lab.
3. Make sure to move ovaries around the entire culture dish surface during puncturing and avoid keeping them in a small area throughout to avoid excessive cell death.
4. During puncturing, needles often become bent and efficiency of oocyte expulsion from ovaries eventually drops. It is important to use fresh needles for each ovary to maximize the number of oocytes obtained.
5. Glass pipettes should always be wider than the diameter of mouse oocytes (about 80 µm) to prevent cell death due to squeezing.
6. For EGFP/SNAP-MAP4-MTBD, EGFP-UtrCH, and H2B-RFP, we find that the in vitro mRNA transcription kit from Thermo Fisher (AM1344) consistently produces similar mRNA yields from 10 µg DNA templates. We therefore routinely use these at final dilutions of 1:16 for EGFP/SNAP-MAP4-MTBD, 1:100 for H2B, and 1:8 for EGFP-UtrCH.

7. This apparatus can be reused for holding capillaries during mRNA loading and does not need to be fashioned for each experiment.
8. Due to reuse, the amount of grease on the capillary holder will eventually lower and will need to be topped up from time to time.
9. This capillary storage dish can be reused and should be kept while topping up the grease and cleaning or replacing the glass slide from time to time.
10. Concentrated HCl is highly corrosive and the appropriate personal protective equipment should be worn when handling it.
11. Before use, coverslips from the 80% v/v EtOH solution should be wiped off on tissue until dry and gently cleaned by scrubbing off with fine tissue such as Kimwipes.
12. It's essential to make sure that this coverslip is fully affixed on the double-sided tape. Improperly fixed coverslips will eventually float away into the M2 + dbcAMP medium during microinjection and lead to oocyte loss. In addition, oocytes will not be supported by the double-sided tape during microinjection if the shelf coverslip is not fully affixed on the tape. This will make microinjection a very cumbersome and frustrating task. Taking the time to assemble a very good shelf is key to a smooth microinjection experiment and maximizes postinjection oocyte survival rate.
13. The remaining small coverslip pieces can be used to construct shelves for future experiments. Preparing several shelves at once and storing them saves time during microinjection experiments.
14. Mercury is highly toxic and appropriate personal protective equipment should be worn when handling it. We typically backfill microinjection needles with 3–5 μL of mercury in a fume cupboard using Hamilton syringes.
15. It is important to calibrate the volume of needles produced to maintain the quantitative element of this microinjection setup. This is achieved by dispensing capillary oil into a microinjection chamber filled with water. By measuring the diameter of the oil droplet via the binocular ruler and assuming that it is a sphere, a ruler unit versus volume table can be constructed. Manufacturers provide the value of ruler units in the form of unit/μm or similar. This can be used to translate binocular ruler units into SI units for volume calculations. We routinely microinject 6–7 pL of mRNA to simultaneously label actin and chromosomes or microtubules and chromosomes. Mouse oocytes are about 200 pL in volume. The final concentration of the introduced mRNA inside an oocyte can therefore be calculated using this quantitative microinjection setup.

16. The ideal microinjection plane is the oocyte hemisphere as oocytes often stay stationary and do not roll when being pushed at this position. Postinjection, oocytes are also more likely to stay inside the shelf when injected at their hemisphere, which simplifies their retrieval after microinjection. Using the membrane curvature as a readout, move the needle (using the Z-axis component of the micromanipulator) to the oocyte's hemispheric region.
17. Oocytes may stick to the needle and come out of the shelf during withdrawal. However, they will detach at the medium-air interface and remain inside the chamber. It is important to periodically top up the M2 + dbcAMP medium during microinjection by mouth pipetting more medium into the chamber under a stereomicroscope.
18. Very often, temperatures displayed on microscope environmental chamber controllers vary considerably from the actual temperature inside the imaging dish. It is therefore important to first measure the temperature inside a glass-bottom imaging dish containing M2 medium covered with paraffin oil using a thermocouple thermometer. Once temperature controller settings that yield 37 °C inside the M2 medium are determined, the microscope should be equilibrated at this temperature for at least 1 h before each live imaging experiment.
19. Oocytes should be incubated in SNAP-tag substrates for 30 min at 37 °C and washed out and stored in M2 + dbcAMP without the substrate for an additional 30 min before imaging. The time needed for labeling should be taken into consideration when planning an experiment to avoid prolonged expression of microinjected mRNAs.
20. Mammalian oocytes are non-adherent cells. Extra care should be taken when transferring the imaging dish to a microscope to avoid them from moving randomly within the M2 droplet, which would make locating them all on the microscope very difficult. We bunch oocytes together by mouth pipetting them and they typically remain together as long as the imaging dish is moved gently during transfer to microscope. It is also critical to determine a microscope-stage speed and acceleration that does not cause oocytes to float away during multi-position imaging.
21. To easily locate oocytes, align the center of the objective lens with the edge of the M2 droplet through visual inspection. This greatly simplifies finding the oocytes in transmission light as they will be close to the focal plane where the edge of the droplet is in best focus.
22. For live imaging experiments, we perform simultaneous imaging of actin, microtubules, and chromosomes on a single track as sequential imaging of channels will produce time delays

between imaging each color. Before nuclear envelope breakdown, only some acentriolar microtubule organizing centers (aMTOC) can be seen with MAP4-MTBD. UtrCH will produce a clear labeling of cortical actin and fainter labeling of cytoplasmic actin filaments, which will allow for later detection of spindle actin without stabilizing actin filaments (Fig. 9c). In meiosis-competent oocytes, chromosomes labeled with H2B-mRFP generally adopt a surrounded nucleolar (SN) configuration (Fig. 9a). In some oocytes, a non-surrounded nucleolar (NSN) configuration will be seen (Fig. 9a). Oocytes that achieve the surrounded nucleolar configuration before nuclear envelope breakdown are most likely to progress through meiosis. Since computing storage space is often a limitation with long-term live imaging of meiosis, it might be beneficial to only select those oocytes that have SN configuration for the overnight live imaging experiment. Although the microinjection techniques described here are highly quantitative and reproducible, some oocytes will often express more fluorescently labeled proteins than others. It is important to avoid imaging oocytes with much brighter fluorescence of actin, microtubules, and chromosomes as overexpression of each marker by itself will significantly perturb meiosis. In particular, overexpression of EGFP-UtrCH will stabilize actin filaments and prevent spindle relocation while overexpressed EGFP-MAP4-MTBD and SNAP-MAP4-MTBD will stabilize microtubules and lead to chromosome segregation errors. Similarly, overexpression of H2B-mRFP will interfere with chromosome individualization and segregation.

23. It is critical to determine imaging conditions that do not cause phototoxicity to oocytes for each microscope since lasers can vary in power from one system to another, even between identical systems from a single manufacturer. It is advisable to measure the laser power using an external meter such as PM160/PM16 0 T/PM16 0 T-HP from Thorlabs.

24. We use stock concentrations of 1 mg/mL cytochalasin D, 5 mM nocodazole, and 1 mM SiR-actin. Stocks of drugs are prepared by dissolving powders in DMSO. High concentrations of DMSO alone can perturb meiosis and oocyte development. Therefore, control oocytes should be treated with DMSO diluted similarly to drug stocks in M2 medium.

25. During the wash, always clean out the glass pipette by repeatedly flushing it with water to avoid contamination of control groups with the drug.

26. In suboptimal resolution imaging datasets where the signal-to-noise ratio is low, surfaces will be created outside chromosome regions due to detection of noise. These have to be manually deleted (please refer to Imaris user manual on how to do this).

27. Fixative solution will exclude water-based medium so mix it by briefly stirring with the mouth pipette while oocytes are being pipetted into the fixative. After each fixation step, dispose of the glass pipette. Reusing formaldehyde-contaminated pipettes to transfer oocytes from culture medium into the fixative solution will lead to inconsistent fixation and is a health and safety hazard. Fixing oocytes in multiple wells will minimize the amount of M2 transferred into the fixative solution along with oocytes. Attempting to place too many oocytes in one well will inevitably dilute the fixative too much and lead to nonuniform and inefficient fixation.
28. Diluted primary antibody solutions can also be pipetted into new wells of a Terasaki plate using manual pipettes and oocytes can be transferred into antibody wells by mouth pipetting. This is particularly useful when multiple antibodies have to be used and fixed oocytes need to be separated into groups. Here, mouth pipetting has strong advantage over manual pipetting because the number of oocytes being pipetted per well can be precisely controlled.
29. Oocytes can be stored at 4 °C in PBT for several days. However, wells should be topped up with PBT regularly to prevent them from drying out.

Acknowledgments

I would like to thank Kathleen Scheffler and Sam Dunkley for their input and critical reading of this manuscript. This work was supported by a Wellcome Trust and Royal Society Sir Henry Dale Fellowship.

References

1. Hassold T, Hunt P (2001) To err (meiotically) is human: the genesis of human aneuploidy. Nat Rev Genet 2(4):280–291. https://doi.org/10.1038/35066065
2. Herbert M, Kalleas D, Cooney D, Lamb M, Lister L (2015) Meiosis and maternal aging: insights from aneuploid oocytes and trisomy births. Cold Spring Harb Perspect Biol 7(4): a017970. https://doi.org/10.1101/cshperspect.a017970
3. Mogessie B, Scheffler K, Schuh M (2018) Assembly and positioning of the oocyte meiotic spindle. Annu Rev Cell Dev Biol. https://doi.org/10.1146/annurev-cellbio-100616-060553
4. Lister LM, Kouznetsova A, Hyslop LA, Kalleas D, Pace SL, Barel JC, Nathan A, Floros V, Adelfalk C, Watanabe Y, Jessberger R, Kirkwood TB, Hoog C, Herbert M (2010) Age-related meiotic segregation errors in mammalian oocytes are preceded by depletion of cohesin and Sgo2. Curr Biol 20 (17):1511–1521. https://doi.org/10.1016/j.cub.2010.08.023
5. Nakagawa S, FitzHarris G (2017) Intrinsically defective microtubule dynamics contribute to age-related chromosome segregation errors in mouse oocyte meiosis-I. Curr Biol 27 (7):1040–1047. https://doi.org/10.1016/j.cub.2017.02.025
6. Mogessie B, Schuh M (2017) Actin protects mammalian eggs against chromosome segregation errors. Science 357(6353). https://doi.org/10.1126/science.aal1647

7. Schuh M, Ellenberg J (2008) A new model for asymmetric spindle positioning in mouse oocytes. Curr Biol 18(24):1986–1992. https://doi.org/10.1016/j.cub.2008.11.022
8. Azoury J, Lee KW, Georget V, Rassinier P, Leader B, Verlhac MH (2008) Spindle positioning in mouse oocytes relies on a dynamic meshwork of actin filaments. Curr Biol 18 (19):1514–1519. https://doi.org/10.1016/j.cub.2008.08.044
9. Jaffe LA, Terasaki M (2004) Quantitative microinjection of oocytes, eggs, and embryos. Methods Cell Biol 74:219–242

Chapter 18

Using the Four-Cell *C. elegans* Embryo to Study Contractile Ring Dynamics During Cytokinesis

Fung Yi Chan, Ana Marta Silva, and Ana Xavier Carvalho

Abstract

Cytokinesis is the process that completes cell division by partitioning the contents of the mother cell between the two daughter cells. It involves the highly regulated assembly and constriction of an actomyosin contractile ring, whose function is to pinch the mother cell in two. Research on the contractile ring has particularly focused on the signaling mechanisms that dictate when and where the ring is formed. In vivo studies of ring constriction are however scarce and its mechanistic understanding is therefore limited. Here we present several experimental approaches for monitoring ring constriction in vivo, using the four-cell *C. elegans* embryo as model. These approaches allow for the ring to be perturbed only after it forms and include the combination of live imaging with acute drug treatments, temperature-sensitive mutants and rapid temperature shifts, as well as laser microsurgery. In addition, we explain how to combine these with RNAi-mediated depletion of specific components of the cytokinetic machinery.

Key words *C. elegans*, Cytokinesis, Contractile ring, Embryonic cell divisions, Acute drug treatments, Temperature-sensitive mutations, RNAi-mediated depletion, Laser microsurgery

1 Introduction

Cytokinesis completes cell division by physically partitioning the contents of the mother cell into the two daughter cells, ensuring that each daughter cell retains one copy of the replicated genome. This process is accomplished via the formation and constriction of a contractile ring (CR) that assembles around the cell equator just beneath the plasma membrane, after anaphase onset. Constriction of the CR progressively draws the plasma membrane inwards, closing the gap between the two daughter cells (Fig. 1a). Failure of cytokinesis compromises cell division leading to polyploidy that is thought to contribute to aneuploidy and tumorigenesis [1]. Observations by electron microscopy show that the CR is a 0.1–0.2-μm-thick layer consisting of circumferentially oriented

Fung Yi Chan and Ana Marta Silva are first co-authors.

Helder Maiato (ed.), *Cytoskeleton Dynamics: Methods and Protocols*, Methods in Molecular Biology, vol. 2101, https://doi.org/10.1007/978-1-0716-0219-5_18,

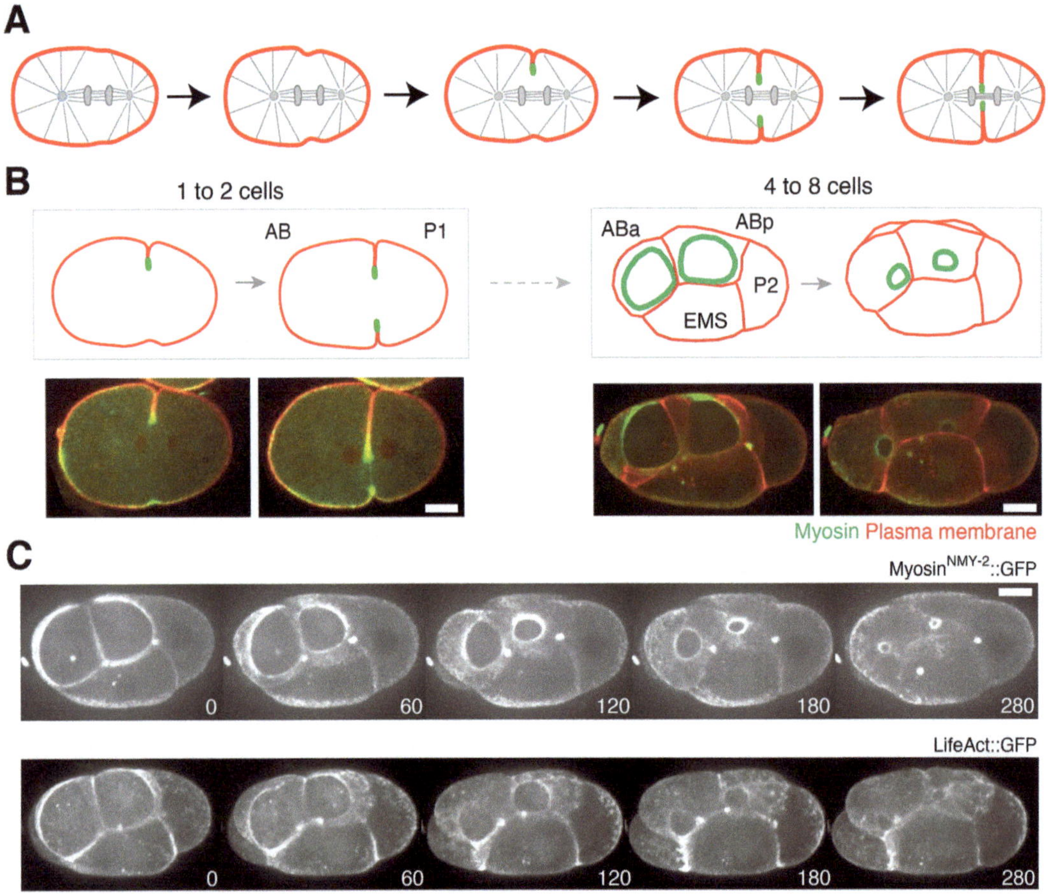

Fig. 1 End-on view of contractile ring circumference in ABa and ABp cells of the four-cell *C. elegans* embryo. (**a**) Schematics illustrating the process of cytokinesis. CR is in green, plasma membrane in red, and DNA, microtubules, and centrosomes in gray. (**b**) Schematics illustrating a one-cell (top, left) and a four-cell (top, right) *C. elegans* embryos with CRs in green and plasma membranes in red (top), and corresponding live images (bottom). Embryos at the bottom express a GFP-tagged marker of the CR, myosin^{NMY-2}::GFP (in green), and a mCherry-tagged marker of the plasma membrane (mCherry-tagged pleckstrin homology domain of PLC1∂1 in red). The one-cell embryo divides parallel to the imaging plane and two sides of the constricting ring are visualized when imaging a central plane of the embryo. ABa and ABp cells of the four-cell embryo divide almost perpendicularly to the imaging plane, and the entire ring circumferences can be visualized in a projection of a few central planes. (**c**) Stills of a time-lapse video of four-cell embryos expressing myosin^{NMY-2}::GFP (top row) or Lifeact::GFP, a marker for actin (bottom row). Both markers are enriched in the CR. Numbers are time in seconds and zero corresponds to beginning of ring constriction in ABa and ABp cells. Scale bars, 10 μm

actin filaments with interdigitating filaments of non-muscle myosin II, across a variety of systems [2–6]. Research on the CR has particularly focused on the signaling mechanisms that dictate when and where the ring is formed [7]. In contrast, studies on the structural and functional properties of the CR network itself and mechanisms of network contraction lag behind. The CR has been difficult to study largely because of its highly transient and dynamic nature. Moreover, many perturbations that affect cytokinesis preclude ring constriction, which hampers its study. Our research group has been focusing on the study of ring constriction using the early *Caenorhabditis elegans* embryo as experimental model. The early *C. elegans* embryo is uniquely suited to study the CR because of the high level of conservation of the machinery involved, highly stereotypical early embryonic divisions that allow for the detection of subtle defects when a component is perturbed, available fluorescently tagged CR components and mutants, effective depletion by RNAi, and established live imaging quantitative assays. Another important advantage is the nowadays routine use of genome editing to generate gene knockouts, specific insertions/deletions, and point mutations using CRISPR/Cas9.

1.1 Monitoring Cytokinesis in the Four-Cell *C. elegans* Embryo

From the four-cell stage onwards, some cells divide close to perpendicularly to the imaging plane. This means that an end-on view of the whole ring circumference can be obtained by imaging a few z-planes through the center of the embryo for the entire duration of constriction (Fig. 1b,c). Together with the stereotypical nature of early embryonic divisions, this greatly facilitates accurate measurements of ring size (circumference and width), and distribution, levels, and dynamics of fluorescently tagged CR components (Fig. 2). Ring constriction has been shown to be influenced by cell-cell contacts in epithelial cells [8–11] and filming cytokinesis in multicellular embryos also allows to take the impact of neighboring cells in consideration. Cytokinesis in one- and two-cell embryos is parallel to the imaging plane and consequently imaging through the center of the embryo only allows the visualization of two sides of the CR (Fig. 1b). In these cases, visualization of the whole ring circumference is possible when acquiring z-stacks through the width of the embryo that are then rotated by 90° and projected [12, 14]. Alternatively, special chambers can be used to hold the embryo in the vertical, or by placing the embryo into a hole in an agarose pad with the long axis of the embryo parallel to the objective body [12].

Ring constriction is a highly dynamic and transient process that takes ~5 min to occur in early *C. elegans* embryos, regardless of the initial size of the ring [13]. To specifically study ring constriction, we have been using a series of approaches that allow for CRs to form normally and only then be perturbed. To perturb the CR during constriction, we use rapid temperature shifts of temperature-sensitive mutant embryos, acute drug treatments, or laser microsurgery. All these approaches can be done in embryos depleted of a specific component by RNAi or in specific mutant backgrounds, where ring constriction still occurs. To follow the CR as it constricts, these experiments are performed in embryos expressing fluorescently tagged CR components.

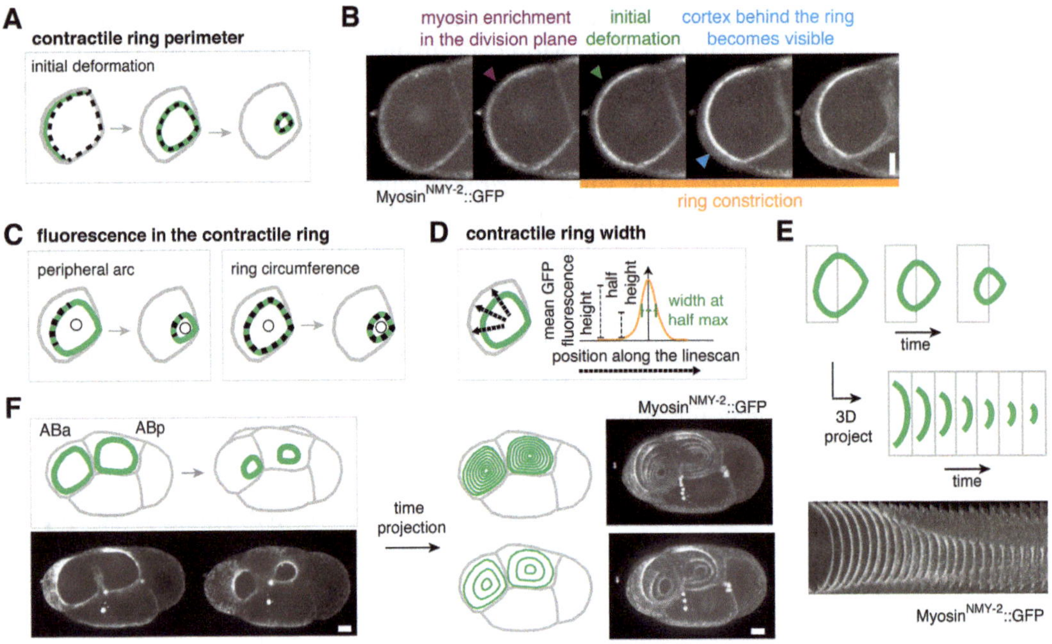

Fig. 2 Methods to characterize ring constriction in the four-cell embryo. (**a**) Schematic illustrating how to measure ABa or ABp CR perimeter over time, starting when the initial deformation of the cell is observed. Ring perimeter is determined by manually tracing its entire outline (dashed line) for each time point. (**b**) Stills of an ABa cell expressing myosin^{NMY-2}::GFP depicting myosin enrichment at the division plane (purple arrowhead), the initial deformation which corresponds to the onset of ring constriction (green arrowhead), and the cortex behind the ring that becomes clearly visible as constriction proceeds (blue arrowhead). (**c**) Fluorescence intensity in the ring during constriction is quantified by manually tracing a line with a fixed width over the peripheral arc or over the entire ring circumference (dashed lines) and subtracting the fluorescence in the cytoplasm (black circle) as represented in the scheme. (**d**) Ring width is determined by drawing lines with constant width from the center to the cell periphery across the ring peripheral arc for each time point (dashed arrows) and making an average line scan for GFP fluorescence (in orange). The width at half max corresponds to the ring width at a given time point. (**e**, **f**) Schematics and real images showing two different ways of displaying ring constriction in ABa and ABp cells over time: 3D projection of the peripheral arc of the ring (**e**) and time projection of maximum intensity Z-projections (**f**). In the schematics CR and plasma membrane are depicted in green and gray, respectively. Embryos in the images express myosin^{NMY-2}::GFP. Scale bars, 5 μm

1.2 Temperature-Sensitive Mutations

Temperature-sensitive (ts) mutants are variants of genes that encode for proteins with normal function at one temperature, or permissive temperature, and loss of function at a different temperature, or restrictive temperature. The use of ts mutants is invaluable to assess the requirement of a specific protein with high temporal resolution, which becomes particularly useful when dissecting the contribution of a protein to CR assembly versus ring constriction or late stages of cytokinesis. There are several ts mutants that can be used to learn more about cytokinesis in *C. elegans* (Table 1). Typically, these mutants are functional at 16 °C and nonfunctional at 25 °C. Intermediate temperatures lead to partial protein inactivation [15, 16]. When combined with live imaging, fast inactivation is

Table 1
Summary of cytokinesis temperature-sensitive mutants available in *C. elegans*

Gene	Allele	Substitution	References
Known to be fast-inactivating when temperature is rapidly upshifted from 16 to 26 °C			
Formin, cyk-1	*or596*	L1015H in FH2 domain	[50]
Myosin, nmy-2	*ne3409*	L981P in S2 neck, light-chain-binding domain	[51]
PRC-1, spd-1	*oj5*	R83W in CYK-4-binding domain	[52]
MKLP-1, zen-4	*or153t*	D520N in CYK-4-binding domain	[53, 54]
Aurora B, air-2	*or207t*	P265L in kinase domain	[53]
RacGAP, cyk-4	*or749t*	T546I in GAP domain	[15]
INCENP, icp-1	*or663t*	A551T in Aurora B interaction domain	[50]
Further testing necessary to determine how fast inactivation is when temperature is rapidly upshifted from 16 to 26 °C			
Actin, act-2	*or295*	G14R in the ATP-binding domain	[55]
Actin, act-2	*or621*	S15A in the ATP-binding domain	[55]
Rho kinase, let-502	*sb118*	Missense mutation in the catalytic domain	[56]
Myosin phosphatase, Mel-11	*it26*	No information found. Possible deletion of N-terminal region after Trp33	[57, 58]
PAR-1, par-1	*zu310*	I306N in the kinase domain (residue number as in isoform a)	[59]
PAR-2, par-2	*it5*	No information found	[57]
PAR-4, par-4	*it57*	P385S (residue number as in isoform c)	[60]
Atypical protein kinase C, pkc-3	*ne4246*	D386V	[61]

ideal to study ring constriction, as the protein of interest can be inactivated after ring assembly and its impact on ring constriction can be followed over time within the available time window [17]. The impact on ring constriction can be easily evaluated by following the decrease in ring perimeter over time in controls versus mutants after temperature upshift. As their effect is usually reversible, experiments with ts mutants can also include fast reactivation of the protein [18]. The combination of fast inactivation with live imaging is straightforward, especially if a device that allows for rapid temperature shifts directly on the sample is available.

1.3 Acute Drug Treatments

Drug treatments are particularly useful because a range of protein inhibitors are available that should impact cytokinesis (Table 2; [19]). The action of inhibitors has been mostly explored in tissue culture cells where delivery of the inhibitor is straightforward.

Table 2
Summary of compounds that interfere with the function of proteins involved in cytokinesis

Drug	Action, target	References[c]
Latrunculin A[a]	Prevents actin filament polymerization by sequestering actin monomers and increases filament depolymerization and severing	[20, 44, 45, 62]
Jasplakinolide[a]	Stabilizes actin filaments	[63–66]
Cytochalasin D[a]	Binds to actin filament barbed ends and prevents actin filament assembly and disassembly at that end	[67, 68]
Blebbistatin[b]	Inhibits non-muscle myosin II ATPase activity by blocking it in a low actin affinity state, sensitive to blue light and cytotoxic especially if imaging below 500 nm	[69, 70]
Azido-blebbistatin[b]	Photoreactive derivative of blebbistatin with enhanced activity under UV irradiation	[71]
Para-nitroblebbistatin[b]	Acts like blebbistatin but is noncytotoxic and is photostable	[72]
Swinholide-A	Severs actin filaments	[73, 74]
CK-666[a]	Inhibits actin polymerization mediated by ARP2/3 complex by stabilizing the inactive state of the complex	[17, 75]
SMIFH2[a]	Inhibits formin nucleation and barbed-end elongation	[76, 77]
BI-2536	Inhibits polo-like kinase 1	[78]
Y-27632[b]	Inhibits kinase activity of Rho kinases	[79]
ML-7	Inhibits the myosin light-chain kinase	[80]
MLN8054	Inhibits Aurora A by preventing autophosphorylation in T288 residue	[81, 82]
Hesperadin	Inhibits Aurora B	[83]
VX-680	Inhibits Aurora A and B	[84, 85]
C3 exoenzyme	Inhibits RhoA dissociation from GDI	[86–88]
Nocodazole[a]	Promotes tubulin depolymerization	[20, 89, 90]
Taxol[a]	Stabilizes microtubules by inhibiting tubulin disassembly	[91, 92]
Colchicine	Inhibits tubulin polymerization	[93]
Flavopiridol (L86-8275)	Inhibits CDK-1	[94, 95]
RO-3306	Inhibits CDK-1 through binding to the ATP pocket	[96]
CRT0103390[a]	Inhibits aPKC-3	[48, 97, 98]

[a]We or others have seen an effect when testing this compound in *C. elegans* embryos
[b]We have not seen an effect in cytokinesis when testing this compound in one-cell *C. elegans* embryos
[c]Mentioned references include papers where compounds were initially described, mechanism of action was reported, and experiments during cytokinesis and/or in *C. elegans* embryos were performed

However, experiments tend to consist of prolonged treatments that do not provide substantial insight because the impact on cytokinesis is usually evaluated by merely measuring cell multinucleation index at the end of the experiment. Acute treatments combined with live imaging are much more powerful in mechanistic studies because inhibitors are introduced in the system only when desired and their immediate or short-term effect can be thoroughly analyzed with a range of quantitative live imaging assays. The impermeable eggshell that surrounds *C. elegans* embryos prevented delivery of inhibitors in the past but a reliable method to reproducibly permeabilize the eggshell is nowadays routinely used for acute drug treatments in the one-cell embryo [20]. Here, we describe the protocol we have been using to permeabilize four-cell embryos without interfering with the timing of cell divisions.

1.4 Laser Microsurgery

Laser microsurgery allows the ablation of a specific structure inside the cell with high precision, without damaging the surrounding tissue. In addition, when combined with live imaging it also allows for precise temporal control. This spatiotemporal control is extremely useful to infer mechanical forces in vivo, by measuring the velocity of displacement of the severed ends of the cut structure in response to the laser cut. Laser microsurgery has been applied to study several cellular actomyosin structures such as rings in wound healing, stress fibers, cell cortex, and adhesions between epithelial cells during cell division [8, 21–25]. In *C. elegans*, laser ablation using pulsed nanosecond UV lasers has been used to study mechanisms of chromosome segregation and mechanical forces involved in asymmetric cell division through the ablation or cutting of structures such as the mitotic spindle, aster microtubules, and centrosomes [26–29]. Mechanical stress during embryo morphogenesis, gonad architecture, or neuronal function has also been studied using laser ablation [30–32]. Below, we describe the protocol we use to reproducibly cut constricting rings using a nanosecond Nd: YAG pulsed laser [33, 34]. Near-infrared femtosecond pulsed lasers have also been used to cut CRs in mammalian cells, fission yeast, and drosophila epithelial tissues [8, 25, 35].

1.5 RNAi-Mediated Protein Depletion

RNA interference (RNAi) is a powerful method of gene silencing mediated by double-stranded RNA, which reproducibly leads to depletion of the gene product of interest in *C. elegans* embryos [36]. In *C. elegans*, proteins present in the gonad are continually packaged into oocytes that become embryos after fertilization, so the protein pool that remains after RNAi-mediated degradation of the mRNA is gradually removed from the system irrespective of its intrinsic turnover properties. Depletion relies on the rate of embryo production and penetrant depletion of different targets is usually achieved within 40–48 h after administering dsRNA by injection of L4 worms at 20 °C [37]. This means that we can have precise control over depletion levels by choosing shorter incubation times. This control over the level of depletion is extremely convenient when partial depletions are intended. To study the contribution of specific proteins to ring constriction, achieving partial depletions may be particularly important because penetrant depletion of some components of the cytokinesis machinery is known to lead to (1) cytokinesis failure preventing the study of ring constriction, or (2) gonad malformation and sterility, preventing embryo production altogether [38]. In alternative to shorter RNAi treatments, slowing down the rate of egg laying by decreasing the temperature to 16 °C also delays penetrant depletion.

Here we describe methods for assessing the dynamics and behavior of the CR network during cytokinesis in *C. elegans* four-cell embryos, as well as approaches to acutely perturb the CR specifically during constriction in combination with live imaging.

2 Materials

2.1 Monitoring Cytokinesis in the Four-Cell *C. elegans* Embryo

2.1.1 Equipment

1. Microfabricated chambers from Muwells (MuChambers, Muwells Inc.).
2. Scalpel.
3. Worm pick: Picks can be made by inserting 0.5 cm of a 3–4 cm piece of platinum wire (0.5 mm thick) into the tip of an inoculation loop holder. The wire is then flattened at the end for picking worms.
4. Eyelash tool: Stick the end of an eyelash onto the narrowest end of a flat toothpick with a small drop of nail polish.
5. Stereomicroscope equipped with a transmitted illumination stand.
6. Coverslips (24 × 60 mm and 18 × 18 mm).
7. Microscopy slides.
8. Metal slide with a 30-mm-wide hole in the center.
9. Watch glass.
10. Forceps.
11. Mouth pipette: Pull a capillary tube (diameter 1 mm) by holding and pulling on the capillary ends while heating its center on a flame until two pulled capillary halves are obtained; fit the blunt end of one of the halves into an aspirator rubber tube (~35 cm long) with a connector for the capillary on one side and a mouth piece on the other.
12. Spinning-disk confocal microscope equipped with a ×60 and ×100, NA 1.4 objectives: For image acquisition, we use an electron multiplication back-thinned charge-coupled device camera.

2.1.2 Reagents

1. Nematode growth medium (NGM) agar plates (3 g/L NaCl, 17 g/L agar and 2.5 g/L peptone, 1 M $CaCl_2$, 5 mg/mL cholesterol in ethanol, 100 mM $MgSO_4$, and 100 mM KH_2PO_4).
2. Bacterium *Escherichia coli* OP50 [39] to feed worms.
3. Temperature-controlled incubators to maintain worms at 16, 20, or 25 °C.
4. Deionized water to prepare all media and buffers.
5. Autoclaved M9 buffer (86 mM NaCl, 42 mM Na_2HPO_4, 22 mM KH_2PO_4, and 1 mM $MgSO_4{\cdot}7H_2O$).
6. 2% Agarose.

2.2 Temperature-Sensitive Mutations

2.2.1 Equipment

1. As in Subheading 2.1.1.
2. Device for rapid temperature changes preferably mounted on the microscope. We use CherryTemp (CherryBiotech).
3. Thermometer.

2.2.2 Reagents

All mentioned in Subheading 2.1.2.

2.3 Acute Drug Treatments

2.3.1 Equipment

As in Subheading 2.1.1.

2.3.2 Reagents

1. All mentioned in Subheading 2.1.2.
2. Bacterium *Escherichia coli* HT115.
3. Autoclaved egg salt buffer (5 mM HEPES, pH 7.4, 118 mM NaCl, 40 mM KCl, 3.4 mM $CaCl_2$, 3.4 $MgCl_2$).
4. Luria-Bertani (LB) medium, resuspend 25 g of powder in 1 L of dH_2O and autoclave.
5. 1 M Isopropyl β-D-1-thiogalactopyranoside (IPTG) in dH_2O, sterilize with a 0.22 μm syringe filter, store aliquots at −20 °C.
6. 100 mg/mL Ampicillin in dH_2O, sterilize with a 0.22 μm syringe filter, store aliquots at −20 °C.
7. 10 mM Latrunculin A in DMSO, store aliquots at −80 °C.
8. Meiosis medium (25 mM HEPES, pH 7.4, 5 mg/mL insulin, 20% heat-inactivated FBS, and 60% Leibowitz-15 medium).
9. Fluorescent lipophilic dye.
10. Small-molecule inhibitors of interest.

2.4 Laser Microsurgery

2.4.1 Equipment

1. As in Subheading 2.1.1.
2. Pulsed Nd:YAG laser for laser microsurgery.

2.4.2 Reagents

All mentioned in Subheading 2.1.2.

2.5 RNAi-Mediated Protein Depletion

2.5.1 Equipment

1. As in Subheading 2.1.1.
2. Electronic microinjector.
3. Needle puller.

2.5.2 Reagents

1. All mentioned in Subheading 2.1.2.
2. Bacterium *Escherichia coli* HT115.
3. 3× Soaking buffer (32.7 mM Na_2HPO_4, 16.5 mM KH_2PO_4, 6.3 mM NaCl, 14.1 mM NH_4Cl).
4. Luria-Bertani (LB) medium, resuspend 25g of powder in 1 L of dH_2O and autoclave.
5. 1 M Isopropyl β-D-1-thiogalactopyranoside (IPTG) in dH_2O, sterilize with a 0.22 μm syringe filter, store aliquots at −20 °C.
6. 100 mg/mL Ampicillin in dH_2O, sterilize with a 0.22 μm syringe filter, store aliquots at −20 °C.
7. 5 mg/mL Tetracycline in 70% ethanol, store aliquots at −20 °C.
8. Halocarbon oil.
9. Genomic or cDNA from wild-type *C. elegans.*
10. Capillary tubes (1 mm diameter).

3 Methods

3.1 Monitoring Cytokinesis in the Four-Cell C. elegans Embryo

In this section, we provide detailed protocols to mount embryos with or without compression to then acquire time-lapse movies of cytokinesis in ABa and ABp cells of *C. elegans* four-cell embryos expressing at least one fluorescent contractile ring component. In addition, we describe methods to analyze ring behavior in the acquired movies. Stills of time-lapse movies of four-cell embryos expressing non-muscle myosin II tagged with GFP (myosin^{NMY-2}::GFP) [40] or Lifeact tagged with GFP, a probe for actin [34, 41], are shown in Fig. 1c.

3.1.1 Mounting Embryos in an Open Chamber (for Filming with No Embryo Compression)

1. Thoroughly rinse the chamber with water. Fill the chamber with M9 buffer by directly pipetting it over the microwells to remove air bubbles (*see* **Note 1**).

2. Use worm pick or forceps to pick adult worms from a NGM plate and place them on the dissection board inside the chamber under a stereomicroscope.
3. Use scalpel to dissect the adult hermaphrodites by cutting them through the middle for embryo release (*see* **Note 2**).
4. Use eyelash tool to drag the four-cell embryos onto the microwells (*see* **Note 3**). By gravity, embryos naturally drift down to settle at the bottom of the wells in only a few seconds. Multiple embryos can also be placed in the same well by using the eyelash tool to guide them to different positions within the same well.
5. Mount the chamber with embryos on the metal slide and place the whole setup on the microscope making sure that the sample and the objective are centered with the hole of the metal slide. The bottom of the microwells is made of glass, ideal for imaging through it (*see* **Note 4**).

3.1.2 Mounting Embryos in an Agarose Pad (for Filming with Compression)

1. Prepare assembly setup to make agarose pads of always the same thickness: use a flat platform and place three microscopy slides parallel to one another along their longest side; tape the outer slides to the platform (the thickness of the tape will dictate the thickness of the agarose pad); the agarose will be placed on the slide in the center.
2. Melt 1 mL of 2% agarose in a heating block at 95 °C (*see* **Note 5**).
3. Use a plastic pipette to transfer a drop of melted agarose to the center of the slide on the pad assembly platform and immediately place another slide on top, perpendicular to the bottom one, by sitting on the taped slides.
4. Wait for ~15 s until agarose solidifies; separate the two perpendicular slides by sliding them apart; the agarose pad will stick to one of the slides.
5. Place a worm inside a watch glass containing M9 buffer and use a scalpel to cut the worm into several pieces so that embryos are released.
6. Use a mouth pipette to aspirate the embryos of desired stage and transfer them to the agarose pad (*see* **Note 6**).
7. With the scalpel cut out a triangular shape on the agarose pad to indicate where the embryos are located.
8. Place a 18 × 18 mm coverslip over the embryos, transfer mounted sample onto the microscope, and image through the coverslip.

3.1.3 Image Acquisition

To set acquisition protocol start by optimizing laser power and exposure time for each fluorophore to be imaged in order to minimize light exposure and signal bleaching and avoid signal saturation. Several z-planes should be acquired to visualize the whole ring circumference over time (Fig. 1c). Camera, objectives, z-scan, and time lapse should be changed depending on the desired spatiotemporal resolution. An acquisition regime of 10 z-planes, 1 μm apart, every 10 s is routinely used in our laboratory. If filming more than one fluorophore, make sure that you alternate channels in each z-plane. To compare ring constriction rate in controls and differently perturbed embryos, the room or microscope has to be maintained at a stable and constant temperature, as small differences in temperature result in changed kinetics. We keep the microscope room at 20 °C.

3.1.4 Measurement of Contractile Ring Perimeter and Mean Constriction Velocity During Ring Constriction

1. Make a maximum intensity z-projection of the acquired movie using the tool Z-project in Fiji [42].
2. To determine the ring perimeter of dividing ABa or ABp cells expressing a fluorescent CR marker trace the entire CR in z-projections using the segmented line tool in Fiji and use the measure tool to determine the length of the total segmented line for each time point (*see* **Note 7** and Fig. 2a).
3. Align value series from different embryos relative to a specific ring perimeter or the beginning of furrowing (*see* **Note 8** and Fig. 2b).
4. Average corresponding perimeter values and plot against time.
5. Calculate overall ring constriction rate by determining the slope in the linear region of the curve.
6. Determine the rate of ring constriction over time, by dividing the difference in perimeter by the difference in time for pairs of consecutive time points. Pool individual measurements from all embryos analyzed and organize data according to descending ring perimeter. Calculate the mean rate of the values that fall in overlapping 5 μm intervals of ring perimeter and plot average instantaneous rate against the ring perimeter at the center of each interval.

3.1.5 Measurement of Fluorescence Intensity in the Contractile Ring Throughout Constriction

1. Make a maximum intensity z-projection of the acquired movie using the tool Z-project in Fiji [42].
2. Draw a segmented line of constant width over the peripheral arc of the ring or the entire ring circumference for each time point and use the measure tool to determine the mean fluorescence intensity (*see* **Note 9** and Fig. 2c).
3. Determine the cytoplasmic and camera signal background by drawing a 10-pixel diameter circle over the cytoplasm in the center of the ring or outside the embryo (Fig. 2c), respectively, and using the measure tool to determine mean fluorescence intensity inside the circles (*see* **Note 10**).

4. Subtract the background fluorescence from the mean fluorescence intensity in the CR for each time point measured (*see* **Note 11**).
5. Within each value series, normalize individual measurements to the mean fluorescence intensity of all measurements made in that embryo.
6. Pool individual measurements from all embryos analyzed and organize data according to descending ring perimeter.
7. Calculate the mean normalized fluorescence intensity of the values that fall in overlapping 5 μm intervals of ring perimeter and plot mean normalized fluorescence intensity against the ring perimeter at the center of each interval.

3.1.6 Measurement of Ring Width Throughout Constriction

1. Make a maximum intensity z-projection of the acquired movie using the tool Z-project in Fiji [42].
2. Draw two or three lines with constant width from the ring center to the cell periphery across the ring peripheral arc and generate a line scan of mean fluorescence intensity per pixel for each using the plot profile tool in Fiji (Fig. 2d).
3. Subtract the average cytoplasmic fluorescence background (measured as described above) from all values.
4. Align individual line scans from each time point to maximum intensity and average aligned values.
5. Normalize sequence of average line scans from each embryo by dividing all values by the average signal between the central five positions of each line scan collected in that embryo.
6. Average normalized line scans from different embryos that fall within a chosen perimeter range, for example within a 4 μm interval, to determine ring width for a given ring size.
7. Interpolate the width of each peak at half its height to measure the width at half maximum for each perimeter range (Fig. 2d).

3.1.7 Displaying Ring Constriction Over Time

1. Build kymographs from maximum intensity projections of constricting rings by cropping the region of the movie that contains the peripheral arc of a constricting ring and using the 3D project tool and the brightest point projection option in Fiji (Fig. 2e).
2. Build an overlay of different time points during ring constriction using the z-project tool on already z-projected movies in Fiji (Fig. 2f); time overlays can be color-coded by using the tool to make a composite in Fiji (*see* **Note 12**).

3.2 Temperature-Sensitive Mutations

In this section, we describe the protocol for how to use fast-inactivating ts mutants to study CR constriction. The embryos to be imaged need to co-express the ts mutant and a fluorescent marker of the CR. If such a strain does not exist start by generating one using genetic crosses. Mutant and control embryos expressing a CR fluorescent marker must be filmed in identical temperature regimes for comparison (*see* **Note 13**). In these experiments, temperature has to be carefully controlled. Commercially available or customized devices capable of rapidly changing the temperature on the sample can be used. In the protocol below, we describe the case of a device that is directly placed above the embryo to be imaged.

1. Keep worms and M9 buffer at permissive temperature (16 °C) by placing worm plates and tube with buffer on top of a layer of bubble wrap, placed above ice on an ice bucket (*see* **Note 14**).
2. Directly dissect the ts mutant or control adult hermaphrodites on a coverslip (24 × 60 mm) with a small amount (~10 μL) of cooled M9 buffer; alternatively, dissect the worms in a watch glass containing cooled M9 buffer and use a mouth pipette to transfer the embryo(s) of interest to the coverslip (24 × 60 mm) (*see* **Note 15**).
3. Immediately place the coverslip with embryo on the stage of the microscope, cover with device for rapid temperature changes, and start imaging at 16 °C.
4. When furrow ingression initiates at any time during ring constriction, rapidly shift to 25 °C and continue imaging.

3.3 Acute Drug Treatments

There are two ways of rapidly introducing small-molecule inhibitors into *C. elegans* embryos. One of them involves puncturing the impermeable eggshell that surrounds the embryo such that the drug can enter [43]. This method requires the availability of an ablation laser coupled to the microscope. The second method does not require any additional equipment and consists of genetically permeabilizing the eggshell by partially depleting PERM-1, a gene involved in setting the eggshell impermeability. This method was initially reported to permeabilize one-cell embryos [13]. We have been using the same method to reliably permeabilize four-cell embryos without affecting the timing of cell divisions. In Fig. 3, we show that the timing of ring constriction in ABa and ABp cells and the timing between divisions of ABa and EMS cells are not altered in PERM-1-depleted embryos. Here we describe the protocol to permeabilize four-cell embryos (Fig. 4a,b). As an example on how to use PERM-1 depletion for acute drug treatment in the study of ring constriction, we show that addition of Latrunculin A, a drug that prevents actin filament polymerization by sequestering actin monomers [44, 45] (Table 2), at anaphase onset precludes CR formation, whereas the impact of adding it during ring constriction depends on its concentration. High Latrunculin A doses result in ring rupture followed by disintegration (5 μM, Fig. 4c), and low doses result in constriction slowdown

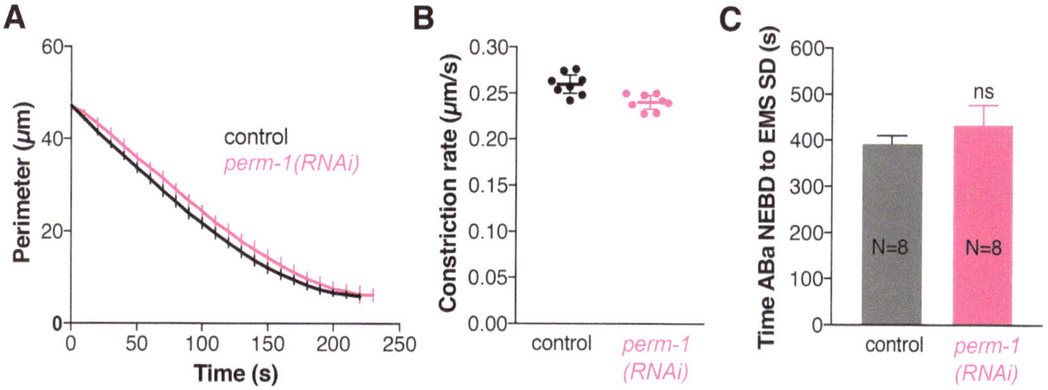

Fig. 3 Partial depletion of PERM-1 permeabilizes the eggshell without altering cell division timing of ABa and ABp cells in four-cell embryos. (**a**) Decrease in ring perimeter is plotted against time from the beginning of ring constriction, for impermeable wild-type (gray, $n = 8$) and permeable PERM-1-depleted ABa and ABp cells (*perm-1(RNAi)*; pink, N=8). (**b**) PERM-1 depletion does not affect ring constriction rate. Scatterplot shows the quantification of ring constriction rate in impermeable (gray, $n = 8$) and permeable (pink, N=8) embryos. (**c**) Time interval between NEBD of ABa cell and shallow deformation (SD) of EMS cell was not affected upon PERM-1 partial depletion. *N* is the number of embryos analyzed. Statistical significance was determined using *t*-test, ns = not significant ($P > 0.05$)

(0.5 μM, Fig. 4c). These results indicate that actin filament polymerization occurs during ring constriction and this is important for ring constriction pace and integrity of ring structure. Below, we also explain how to combine embryo permeabilization with depletion of your protein of interest.

3.3.1 Preparation of Permeabilized Embryos by Partial Depletion of PERM-1

Depletion of PERM-1 by RNAi can be achieved by soaking, feeding, or injection [20]. Our current method of choice is feeding RNAi. We have successfully used the conditions described below in a variety of strains expressing different fluorescent markers (*see* **Note 16**).

1. Transform the vector L4440 (RNAi feeding vector) containing part of the *perm-1* gene into HT115 *E. coli* cells (*see* **Note 17**).
2. Inoculate a single colony from a streaked plate containing HT115 *E. coli* expressing *perm-1* dsRNA in 50 mL of LB medium containing 50 μg/mL ampicillin and grow overnight at 37 °C.
3. Dilute the overnight culture 1:50 in 5 mL of LB medium containing 100 μg/mL ampicillin and incubate at 37 °C for 2.5–3 h until OD reaches 0.4.
4. In parallel, prepare NGM plates with 50 μg/mL ampicillin and 0.005 mM or 0.001 mM IPTG.
5. Spread 200 μL of bacterial culture on the NGM plates. Leave the plates with the lid opened to dry inside a cupboard for 1 h.

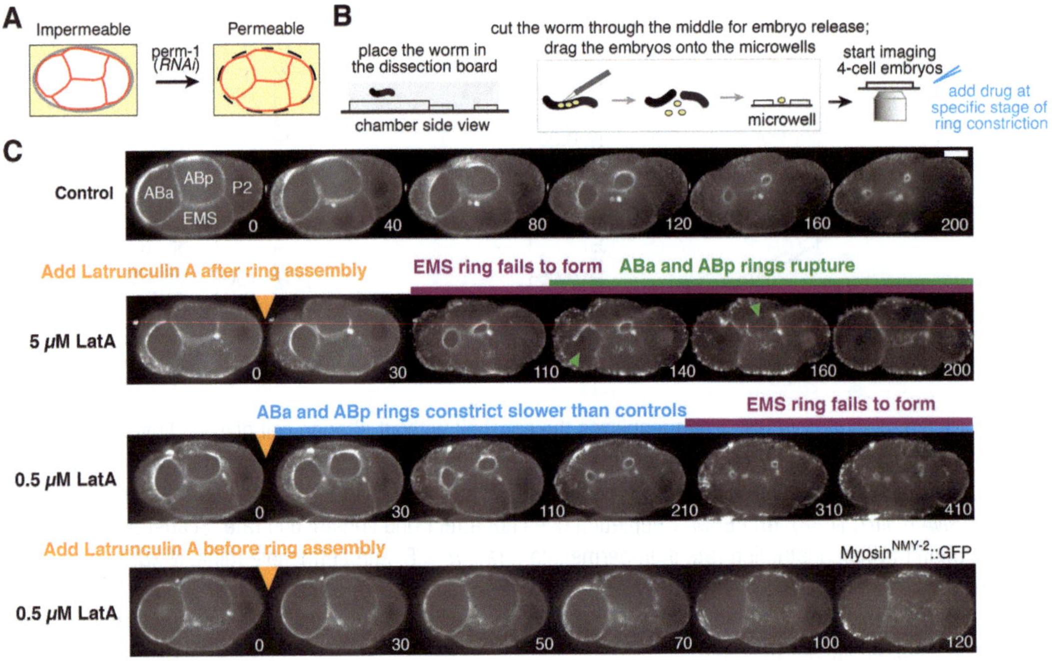

Fig. 4 Impact of adding different concentrations of Latrunculin A to constricting rings in ABa and ABp cells. (**a**) Schematic illustrating drug internalization in permeable embryos (in yellow). (**b**) Schematic illustrating the procedure for embryo dissection and mounting for imaging after permeabilization by PERM-1 partial depletion. (**c**) Stills of spinning disk confocal time-lapse movies of four-cell embryos expressing myosin^{NMY-2}::GFP and treated or not with different doses of Latrunculin A, added at different points during cytokinesis. Numbers on stills are time in seconds and zero corresponds to frame before Latrunculin A addition. Addition of 5 μM Latrunculin A after initiation of ring constriction in ABa and ABp cells leads to ring slowdown and ring rupture, whereas 0.5 μM Latrunculin A leads to ring slowdown but allows for completion of ring constriction without any signs of rupturing. Addition of 0.5 μM Latrunculin A before rings started constricting prevented CR formation. Note that the ring in the EMS cell, which normally forms later than the rings in ABa and ABp cells, does not form with 5 μM or 0.5 μM Latrunculin A. Scale bar, 10 μm

6. Leave plates at room temperature for 4 h to induce RNA expression.
7. Transfer 25–30 L4 hermaphrodites to the *perm-1*(*RNAi*) plates and incubate for 16 h at 20 °C if 0.005 mM IPTG plates are used or 21 h at 20 °C if 0.001 mM IPTG plates are used (*see* **Note 18**). This treatment can be easily combined with depletion of other protein of interest (*see* **Note 19**).

3.3.2 Acute Drug Treatment During Ring Constriction

1. Dilute drug of interest to the desired concentration in meiosis medium or egg salt buffer.
2. Dissect adult hermaphrodites treated with *perm-1* dsRNA in meiosis medium or egg salt buffer in open chambers as described in Subheading 3.1.1 (*see* **Note 20**).
3. Start imaging four-cell embryos. When ABa and ABp ABa or ABp CRs start constricting, remove all the medium from the

chamber, and gently replace with ~150 μL of diluted drug (*see* **Note 21**).

4. Continue imaging and at the end of the video add a fluorescent lipophilic dye to confirm that the imaged embryos were permeable and indeed divided in the presence of the drug (*see* **Note 22**).

3.4 Laser Microsurgery

An example of laser microsurgery in the CR using the protocol below is shown in Fig. 5. Using the setup described in [33, 34], we can perform laser microsurgery at different stages of ring constriction and two or more sequential cuts can be made in the same constricting ring (Fig. 5a). By cutting constricting rings labeled with myosin$^{NMY\text{-}2}$::GFP, we previously showed that rings respond to laser cutting by immediately snapping open, with severed ends pointing outwards, indicating that cortical tension resists ring constriction. Instead of furrow regression, we consistently observed that rings were able to rapidly recover the curvature they had before laser cutting, repair, and continuing constricting (Fig. 5b). By measuring the length of the gap and open ring just after laser cutting, as well as velocity of gap widening during snapping, it is possible to mechanically dissect ring tension and cortical relaxation throughout ring constriction in ABa and ABp cells (Fig. 5c–e). This biophysical dissection can also be done in perturbed backgrounds where a component of the cytokinesis machinery is depleted (Subheading 3.5) or in combination with acute drug treatments as described in Subheading 3.3.2 (Fig. 5f,g and [34]). The biophysical analysis, the capacity and time to recover, and the characterization of ring behavior after repair allow for the better understanding of the contribution of each component to the integrated mechanics of ring and surrounding cortex, as well as to the capacity of remodelling and integrity of ring structure.

1. Use a glass slide to precisely focus, align, and adjust the power of the laser necessary to reproducibly cut the glass (*see* **Note 23**).
2. Prepare the embryos for image acquisition as described in Subheading 3.1.1 with or in Subheading 3.1.2 without compression (*see* **Notes 24**, **25**, and **26**).
3. Test laser conditions in the CR of ABa or ABp cells and increase power until rings are reproducibly cut open (*see* **Notes 27** and **28**).
4. Set up the acquisition protocol to monitor CR behavior immediately after laser ablation: a single image can be acquired in the 488 nm channel with fastest acquisition mode to increase temporal resolution of the immediate response of the severed ends to the laser cutting; to follow constriction just before and after ring repair, the acquisition protocol can be changed to include more z-planes so that the visualization of the entire ring circumference is possible. Different acquisition regimes such as 5 × 0.5 μm, 7 × 0.5 μm, or 7 × 1.0 μm z-stacks collected in the

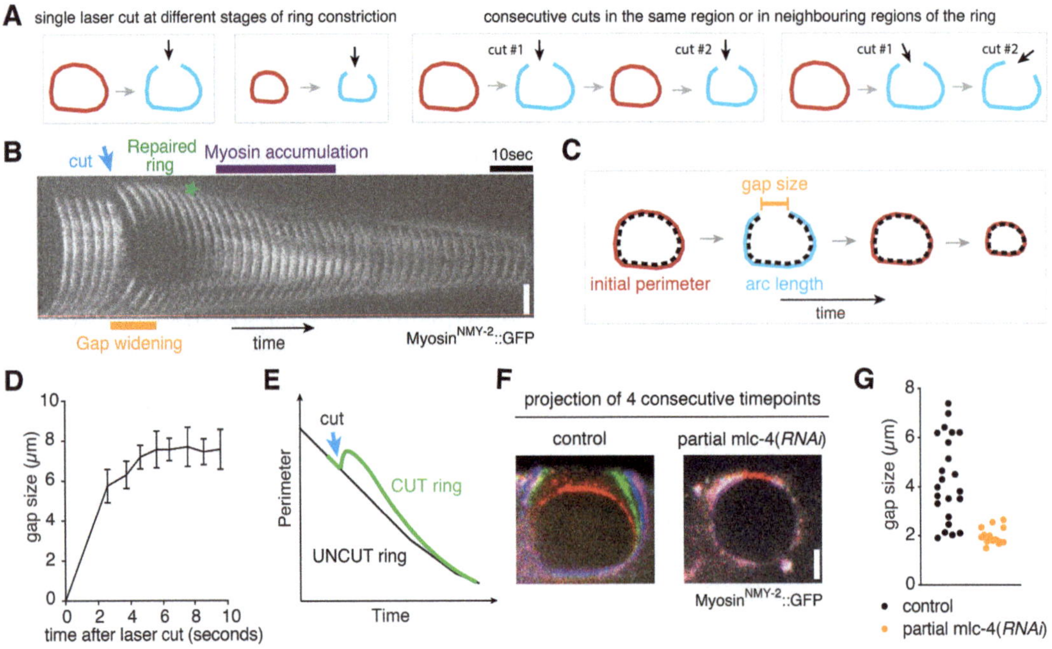

Fig. 5 Laser microsurgery as an approach to probe ring constriction in ABa and ABp cells. (**a**) Schematic illustrating laser cut at different stages of ring constriction (left) and consecutive cuts in the same or neighboring regions of the ring (right); red and cyan represent the ring before and after cut, respectively. (**b**) Kymograph of a constricting ring expressing myosin^{NMY-2}::GFP that has been cut once (blue arrow). After laser cutting, the ring snaps open and the gap between the two severed ends widens (orange line depicts gap widening time interval) before repair occurs and the ring becomes continuous again (green star); after repair the ring keeps constricting with local myosin hyperaccumulation (purple bar depicts time interval during which local myosin hyperaccumulation is visible). (**c**) Schematic illustrating that the immediate reaction of the CR to laser cutting can be characterized by measuring the initial ring perimeter before cut (in red), gap size between the two severed ends (orange line), and arc length after cut (blue). These values allow the calculation of the following parameters: (1) ring recoil after laser cut (initial perimeter minus ablated length minus arc length). The ablated length corresponds to the ring material removed by the laser spot; (2) the strain of the ring (ring recoil divided by the arc length); (3) cortical relaxation (gap size minus ring recoil). (**d**) Gap size, measured in cut rings with an average perimeter of 36 μm before laser cutting, plotted against time after laser cut, showing a fast phase of elastic recoil followed by a slow phase. After the first 10 s it is no longer possible to measure the gap size between the two severed ends due to initiation of ring repair. (**e**) Schematic illustrating the perimeter over time of uncut and cut rings, black and green curves, respectively. After laser cutting, the ring perimeter (includes arc length and gap) increases but rapidly starts decreasing at a higher rate than that of uncut rings, allowing ring constriction to complete within the same time in cut and uncut rings. (**f**) Time projections of four consecutive stills in control rings and rings depleted of MLC-4 (non-muscle myosin II regulatory light chain), before laser cutting (in red) and three consecutive time points after cutting (green, blue, and magenta). In the control ring, the severed ends are visible in the three projected time points, due to curvature change after laser cutting. In rings partially depleted of MLC-4, the ring does not snap open and, consequently, does not change curvature: three time points after laser cutting overlap. (**g**) Gap size of 23 cut control rings and 15 cut rings after partial depletion of MLC-4 measured in the first time point after laser cutting (3.9 s). The results in F and G reveal a decrease in cortical tension after partial depletion of MLC-4. Scale bars, 5 μm

488 nm channel in the fastest acquisition mode can be used; depending on the size of the z-stack acquired, the time lapse will have to be adjusted. Image acquisition of control uncut rings should be done in the same manner.

5. Initiate image acquisition before the ring reaches the desired stage of constriction (*see* **Note 29**). By that time, move the embryo to have the ring positioned in the laser target site, perform the laser cut, and continue following ring constriction (*see* **Note 30**).

3.5 RNAi-Mediated Protein Depletion

There are three common methods for delivery of double-stranded RNA (dsRNA) in *C. elegans*: injection, feeding, and soaking [46]. The RNAi method of choice depends on the budget and the availability of equipment. Microinjection of dsRNA requires a special microinjection system to guarantee consistent injection of small amounts of dsRNA. Feeding dsRNA is, in general, less effective and more variable than microinjection [37]. We describe protocols for RNAi by injection and feeding, the most commonly used delivery methods. To observe the CR, worms to be microinjected with dsRNA or fed with dsRNA-expressing bacteria need to express a fluorescent marker of the CR. A control RNAi experiment where the worm RNAi machinery is also engaged should be done in parallel. A control dsRNA against a target known not to affect embryogenesis or consisting of a random 500 bp sequence with no homology to the worm genome should be used. Uninjected worms or worms fed with bacteria carrying empty L4440 vector are also commonly used as controls in RNAi experiments.

3.5.1 RNAi by Microinjection

1. Generate double-stranded RNA by designing primers to amplify an exon region of 400–1000 base pairs of the gene of interest. One of the primers must contain the minimum promoter sequence for transcription with T7 RNA polymerase (TAATACGACTCACTATAGG) and the other the minimum promoter sequence for T3 RNA polymerase (AATTAACCCT-CACTAAAGG) on their 5′ end (*see* **Note 31**). N2 genomic DNA or cDNA can be used as a template for PCR amplification (*see* **Note 32**). Clean up the PCR product using a commercial kit (*see* **Note 33**) and transcribe both DNA strands using T3 or T7 RNA polymerases and ribonucleotides for 5 h at 37 °C. Add 1.0 μL of DNase to each reaction and incubate at 37 °C for 15 min to digest the PCR template. Clean up single-stranded RNA and anneal both single strands by mixing them at equal molar ratio with 1× soaking buffer for 10 min at 68 °C and then 37 °C for 30 min. Run an agarose gel to check the dsRNA and quantify the concentration by measuring the absorbance at 260 nm (*see* **Note 34**). Aliquot and store at −80 °C. Follow similar strategy to generate control dsRNA.

2. Before worm microinjection, dilute dsRNA to a final concentration of 0.5–1 mg/mL with 1× soaking buffer (*see* **Note 35**).
3. Spin down dsRNA in desktop centrifuge at maximum speed for 10 min at 4 °C and keep it on ice (*see* **Note 36**).
4. Load dsRNA onto a pulled needle and assemble it onto an electronic microinjector (*see* **Note 37**).
5. Have ~30 L4 hermaphrodite worms on a seeded plate ready to be injected.
6. Gently immobilize 1–3 L4 hermaphrodites on a small patch of halocarbon oil placed on a dry and thin 2% agarose pad using a worm pick and microinject dsRNA in the body near the head or tail (*see* **Notes 38** and **39**).
7. Add a small drop of M9 buffer over the injected worms to release them from the halocarbon oil and use the pick to gently scoop them out and transfer them to a drop of M9 on a new seeded NGM plate.
8. Incubate for 24–48 h at 20 °C or 36–72 h at 16 °C (*see* **Notes 40** and **41**).

3.5.2 RNAi by Feeding

1. Streak HT115 *E. coli* bacteria that express the dsRNA of interest from commercially available libraries on LB plate with 50 μg/mL ampicillin and 12.5 μg/mL tetracycline. Alternatively, the gene of interest can be cloned into the RNAi feeding vector L4440, which is then transformed into HT115 *E. coli* bacteria.
2. Inoculate single colony of HT115 expressing control or target dsRNA onto 50 mL of LB medium containing 50 μg/mL ampicillin and 12.5 μg/mL tetracycline. Grow until an OD of 1.6 at 37 °C overnight.
3. Centrifuge the overnight cultures for 10 min at 2500 × *g*. Remove the supernatant and resuspend the cell pellets in 2.5 mL LB medium containing 12.5 μg/mL tetracycline, 50 μg/mL ampicillin, and 1 mM IPTG.
4. Spread 100 μL of a 1:1:1 mix of 50 μg/mL ampicillin, 12.5 μg/mL tetracycline, and 1 mM IPTG onto dry NGM agar plates.
5. Add 75 μL of bacterial culture to the RNAi plates and leave plates at room temperature, in the dark, overnight for induction of dsRNA expression.
6. Transfer 25–30 L4 hermaphrodites to the RNAi plates and incubate for 24–72 h at 16 °C or 18–48 h at 20 °C (*see* **Note 42**).

4 Notes

1. Air bubbles will prevent embryos from settling at the bottom of the microwells.
2. During worm dissection, if early embryos remain inside of the cut worm, use the scalpel to gently squeeze the two halves of the worm body to release them.
3. Try to separate the embryos from each other when pushing them onto the wells with the eyelash tool to avoid that they land on top of each other, which will not be ideal for imaging.
4. If open chambers are not available, dissect the worms in a watch glass containing M9 buffer and use a mouth pipette to transfer the embryo of interest to a coverslip with a drop of M9 (24 × 60 mm). Make a circle of Vaseline around the drop of M9 buffer with embryos to serve as spacer and place another coverslip on top of the Vaseline to reduce buffer evaporation during imaging. Place the whole setup on a metal slide before transferring to the microscope.
5. Avoid keeping the 2% agarose at 95 °C for more than 2 h.
6. Minimize the amount of M9 buffer transferred to the agarose pad with the embryos; the embryos need to be covered in buffer but if too much buffer exists, use the mouth pipette to remove the excess or the eyelash tool to spread it over a larger area on the agarose pad.
7. Avoid measuring ring perimeters when smaller than 5 μm at the end of constriction; at this point CRs are small and difficult to measure in a precise manner.
8. The time point where furrowing initiates corresponds to the first point when a recognizable ingression is evident in z-projections. For instance in embryos expressing non-muscle myosin II::GFP it is straightforward to recognize furrow initiation because a front of myosin can be readily detected (Fig. 2b).
9. Measure fluorescence intensity in the entire ring circumference or only in the peripheral arc, depending on the fluorescent signal; after depletion of some actomyosin components the whole circumference may not be straightforward to distinguish either because of decreased signal, change in ring shape or direction of ring constriction.
10. To measure signal in the cytoplasm use a smaller circle for late constriction time points.
11. Camera or cytoplasmic background signal can be subtracted. In general, similar results are obtained with either.
12. This kind of representation allows the visualization of changes in ring shape, constriction direction, and ring connection to

cell-cell contacts when comparing control with perturbed situations.

13. Permissive and restrictive temperatures for each ts mutant can be slightly different and each ts mutant should be individually characterized.
14. Keep a thermometer next to the worm plates above the layer of bubble wrap to ensure that temperature is maintained at 16 °C.
15. If a device to rapidly change the temperature on the sample while on the microscope is not available, keep the four-cell embryo to be imaged in cooled buffer in an open chamber under a stereoscope until ABa or ABp cells start dividing; then quickly transfer sample to the heated microscope (and/or room) and start imaging.
16. We note that some strains may be more sensitive to perm-1 *(RNAi)* and may therefore require milder PERM-1 depletions. We recommend that permeabilized embryos are carefully inspected before performing drug treatments. IPTG concentration should be decreased and/or RNAi duration should be shortened when permeabilized embryos display an internalized polar body, present round shape, or reveal cytokinesis delays. A common defect indicating that PERM-1 depletion was more penetrant than desired is synchronous ring constriction in ABa, ABp, EMS, and P2 cells. This is in contrast to control impermeable embryos, where ABa and ABp cells divide first, then the EMS cell, and lastly the P2 cell.
17. We use the T01H3.4 clone from the Ahringer feeding RNAi library [47].
18. When optimizing/confirming ideal conditions for PERM-1 depletion in a specific strain, a lipophilic fluorescent dye can be used to check for embryo permeabilization. When the eggshell is permeable, the fluorescent lipophilic dye goes through and labels the plasma membrane. If eggshell is impermeable the dye will not access the plasma membrane. Alternatively, inspect embryo polar bodies in one-cell embryos: impermeable embryos present two polar bodies external to the embryo, one sitting by the eggshell (the first to be extruded during meiosis I) and the other by the plasma membrane, whereas permeabilized embryos present both polar bodies sitting by the plasma membrane. This is easily observed by differential interference contrast illumination.
19. When acute drug treatment is intended in perturbed backgrounds, perm-1 (*RNAi*) conditions should be optimized for each case. Conditions described here are likely to work well. Combining depletion of a protein of interest by RNAi with embryo permeabilization by PERM-1 depletion is also possible. We have been successfully combining depletion of PERM-

1 by feeding RNAi with depletion of other proteins by injection of the corresponding dsRNA. In this case we inject L4 hermaphrodites with the dsRNA of interest and place the injected worms in NGM plates with 50 μg/mL ampicillin and 0.001 mM IPTG and bacteria expressing perm-1 dsRNA for 21 h (if at 16 °C) or 16 h (if at 20 °C) before the RNAi of the protein of interest is completed. Feeding both dsRNAs simultaneously by mixing bacteria expressing them is also possible if perm-1 dsRNA is more diluted (1:9 ratios have been reported to work well [48]).

20. Permeable embryos are more fragile than embryos with an intact eggshell and cannot be transferred using a mouth pipette or compressed under a coverslip during imaging.
21. During solution exchange, the medium inside the microwells will not be disturbed and therefore embryos should not move or dry.
22. As chamber walls may absorb some inhibitor, the chamber needs to be properly washed after each use. To remove any residual drug in the chamber, clean it using a sonication bath for 20 min.
23. Experimental settings to perform laser ablation such as laser pulse energy and duration, speed of ablation, and coordination with live imaging need to be optimized (for more practical considerations check [49]). In the workstation we used for our studies, the second harmonic beam of a pulsed Nd:YAG laser (wavelength 532 nm), which was injected into the microscope stand through a secondary filter turret, was focused through a ×100 imaging objective (1.4 NA); pulse width was 10 ns, pulse energy was 1.5–2 μJ, and the sample was exposed to a 1 kHz pulse train for 800 ms. The laser spot at the sample was ~0.5 μm wide in the x–y plane and at least ~0.5 μm wide along the optical axis [33, 34]. We have not tested other ablation lasers and to our knowledge UV lasers have not been used to cut CRs.
24. The choice of the fluorescent marker(s) to be used for imaging needs to take into consideration the wavelength of the ablation laser: if the fluorescent probe used for imaging absorbs in the same region, its signal may be bleached. With the 532 nm laser ablation setup we used [33], we could not image mCherry-tagged CR markers because their signal would be completely bleached.
25. The fluorescent probe used to follow the CR in these experiments should show clear enrichment in the CR when compared to adjacent cytoplasm and cortex. This is crucial to perform the ablation in the correct spot. We preferentially use GFP-tagged non-muscle myosin II, myosin^{NMY-2}::GFP (Fig. 1c).

26. The outcome of CR laser ablation experiments in wild-type embryos is the same whether performed with or without compression, on agarose pads or in chambers, respectively.
27. To test if the laser is properly focused and aligned after setting the laser ablation conditions we first target the midbody, as this structure is not as dynamic as the CR and is easily distinguished from cortical and cytoplasmic signals.
28. With our setup, rings were successfully cut at 95% laser power, about 20% higher than that used to cut glass.
29. For laser microsurgery to be more precise and ring behavior easier to follow, we recommend that it is performed when ring perimeter of ABa or ABp cells is between 50 and 10 μm.
30. To verify that the laser ablation conditions being used are not damaging the plasma membrane where the CR is anchored, follow embryonic development until hatching; after laser ablation keep the embryos in the chamber wells, cover the chamber with a slide to avoid evaporation of the medium, and wait for approximately 12 h at 20 °C for the embryos to hatch. If detailed information is required follow development using differential interference contrast microscopy (DIC) microscopy.
31. Only one RNA polymerase can also be used: in that case perform two separate PCR reactions—one with forward primer containing the minimum promoter sequence for that RNA polymerase, and the other with reverse primer containing the minimum promoter sequence for that RNA polymerase. Transcribe both reactions in separate tubes, and proceed as described above.
32. Always use filter tips and wear gloves when working with RNAs.
33. Make sure that the concentration of PCR product is ~100 ng/μL; several PCR reactions can be pooled.
34. In the gel, both single-stranded RNAs should be well defined and reveal similar amounts of material. Annealed double-stranded RNA should be slightly shifted relative to single-stranded RNAs.
35. If dsRNA is too viscous, dilute it with 1× soaking buffer; for effective depletions use a minimum concentration of 0.5 mg/mL.
36. Centrifugation of the dsRNA is necessary to avoid any precipitate that could clog the injection needle.
37. Prepare microinjection needles by using a capillary tube (1 mm diameter) and a needle-pulling device. Needle-pulling conditions need to be optimized by changing the heating and pulling

strength in the needle puller in order to have adequate shape and length. Once defined, keep the settings constant. Keep pulled needles in a closed box, protected from dust. The end of the needle needs to be broken after loading the needle with dsRNA and assembling it onto the microinjector; gently push the needle against a pulled capillary tube to open it.

38. RNAi in *C. elegans* is systemic, spreads throughout the animal, and is transmitted to the following generation [36].
39. We recommend a beginner to watch free videos available online for better perception of the microinjection procedure. For additional details on microinjection of dsRNA, we direct the reader to [46].
40. Typically, penetrant protein depletion in embryos can be achieved ~40–48 h after injection of L4 hermaphrodites at 20 °C. However, penetrant depletion of some components of the cytokinesis machinery leads to gonad disorganization and worm sterility or prevents ring constriction during cytokinesis. In such cases, partially depleted embryos can be obtained after RNAi treatment of shorter duration and/or at lower temperature.
41. These RNAi experiments can be performed in embryos expressing a CR fluorescent marker in wild-type or mutant background. If microinjecting dsRNA into ts mutant worms, injected worms should be kept on top of a layer of bubble wrap placed over ice until all worms are microinjected and the plate should be placed in an incubator at 16 °C for the duration of the RNAi experiment.
42. Typically, penetrant protein depletion in embryos can be achieved ~72 h after feeding dsRNA to L4 hermaphrodites at 16 °C. As in **Note 40**, shorter treatments can be used when depleted protein prevents embryo production or prevents ring constriction during cytokinesis.

Acknowledgments

Research in our group receives funding from the European Research Council under the European Union's Horizon 2020 research and innovation programme to AXC (grant agreement 640553—ACTOMYO) and from Norte-01-0145-FEDER-000029—Advancing Cancer Research: from basic knowledge to application, supported by Norte Portugal Regional Operational Programme (NORTE 2020), under the PORTUGAL 2020 Partnership Agreement, through the European Regional Development Fund (FEDER). AXC has a Principal Investigator position from FCT (CEECIND/01967/2017). FYC and AMS hold FCT junior researcher positions DL 57/2016/CP1355/CT0013 and DL 57/2016/CP1355/CT0017, respectively. The authors declare no competing financial interests.

References

1. Lens SMA, Medema RH (2019) Cytokinesis defects and cancer. Nat Rev Cancer 19:32–45
2. Henson JH, Ditzler CE, Germain A et al (2017) The ultrastructural organization of actin and myosin II filaments in the contractile ring: new support for an old model of cytokinesis. Mol Biol Cell 28:613–623
3. Kamasaki T, Osumi M, Mabuchi I (2007) Three-dimensional arrangement of F-actin in the contractile ring of fission yeast. J Cell Biol 178:765–771
4. Maupin P, Pollard TD (1986) Arrangement of actin filaments and myosin-like filaments in the contractile ring and of actin-like filaments in the mitotic spindle of dividing HeLa cells. J Ultrastruct Mol Struct Res 94:92–103
5. Sanger JM, Sanger JW (1980) Banding and polarity of actin filaments in interphase and cleaving cells. J Cell Biol 86:568–575
6. Schroeder TE (1973) Actin in dividing cells: contractile ring filaments bind heavy meromyosin. Proc Natl Acad Sci U S A 70:1688–1692
7. Green R, Paluch E, Oegema K (2012) Cytokinesis in animal cells. Annu Rev Cell Dev Biol 28:29–58
8. Guillot C, Lecuit T (2013) Adhesion disengagement uncouples intrinsic and extrinsic forces to drive cytokinesis in epithelial tissues. Dev Cell 24:227–241
9. Pinheiro D, Hannezo E, Herszterg S et al (2017) Transmission of cytokinesis forces via E-cadherin dilution and actomyosin flows. Nature 545:103–107
10. Higashi T, Stephenson RE, Miller AL (2019) Comprehensive analysis of formin localization in Xenopus epithelial cells. Mol Biol Cell 30:82–95
11. Lázaro-Diéguez F, Müsch A (2017) Cell-cell adhesion accounts for the different orientation of columnar and hepatocytic cell divisions. J Cell Biol 216:3847–3859
12. Maddox AS, Lewellyn L, Desai A, Oegema K (2007) Anillin and the septins promote asymmetric ingression of the cytokinetic furrow. Dev Cell 12:827–835
13. Carvalho A, Desai A, Oegema K (2009) Structural memory in the contractile ring makes the duration of cytokinesis independent of cell size. Cell 137:926–937
14. Lewellyn L, Carvalho A, Desai A et al (2011) The chromosomal passenger complex and centralspindlin independently contribute to contractile ring assembly. J Cell Biol 193:155–169
15. Canman JC, Lewellyn L, Laband K et al (2008) Inhibition of Rac by the GAP activity of centralspindlin is essential for cytokinesis. Science 322:1543–1546
16. Jordan SN, Davies T, Zhuravlev Y et al (2016) Cortical PAR polarity proteins promote robust cytokinesis during asymmetric cell division. J Cell Biol 212:39–49
17. Chan FY, Silva AM, Saramago J et al (2019) The ARP2/3 complex prevents excessive formin activity during cytokinesis. Mol Biol Cell 30:96–107
18. Davies T, Sundaramoorthy S, Jordan SN et al (2017) Using fast-acting temperature-sensitive mutants to study cell division in Caenorhabditis elegans. Methods Cell Biol 137:283–306
19. Dhaliwal A (2013) Activators and inhibitors in cell biology research. Mater Methods 3:185–200
20. Carvalho A, Olson SK, Gutierrez E et al (2011) Acute drug treatment in the early C. elegans embryo. PLoS One 6:e24656
21. Mandato CA, Bement WM (2001) Contraction and polymerization cooperate to assemble and close actomyosin rings around Xenopus oocyte wounds. J Cell Biol 154:785–797
22. Kumar S, Maxwell IZ, Heisterkamp A et al (2006) Viscoelastic retraction of single living stress fibers and its impact on cell shape, cytoskeletal organization, and extracellular matrix mechanics. Biophys J 90:3762–3773
23. Colombelli J, Besser A, Kress H et al (2009) Mechanosensing in actin stress fibers revealed by a close correlation between force and protein localization. J Cell Sci 122:1665–1679
24. Tinevez J-Y, Schulze U, Salbreux G et al (2009) Role of cortical tension in bleb growth. Proc Natl Acad Sci U S A 106:18581–18586
25. Herszterg S, Leibfried A, Bosveld F et al (2013) Interplay between the dividing cell and its neighbors regulates adherens junction formation during cytokinesis in epithelial tissue. Dev Cell 24:256–270
26. Grill SW, Gönczy P, Stelzer EH, Hyman AA (2001) Polarity controls forces governing asymmetric spindle positioning in the Caenorhabditis elegans embryo. Nature 409:630–633
27. Bringmann H, Cowan CR, Kong J, Hyman AA (2007) LET-99, GOA-1/GPA-16, and GPR-1/2 are required for aster-positioned cytokinesis. Curr Biol 17:185–191
28. Cowan CR, Hyman AA (2004) Centrosomes direct cell polarity independently of microtubule assembly in C. elegans embryos. Nature 431:92–96

29. Grill SW (2003) The distribution of active force generators controls mitotic spindle position. Science 301:518–521
30. Kimble JE, White JG (1981) On the control of germ cell development in Caenorhabditis elegans. Dev Biol 81:208–219
31. Priti A, Ong HT, Toyama Y et al (2018) Syncytial germline architecture is actively maintained by contraction of an internal actomyosin corset. Nat Commun 9:4694
32. Vuong-Brender TTK, Amar, BM, Pontabry J, Labouesse M (2017) The interplay of stiffness and force anisotropies drives embryo elongation. elife 6:e23866
33. Pereira AJ, Matos I, Lince-Faria M, Maiato H (2009) Dissecting mitosis with laser microsurgery and RNAi in Drosophila cells. Methods Mol Biol 545:145–164
34. Silva AM, Osório DS, Pereira AJ et al (2016) Robust gap repair in the contractile ring ensures timely completion of cytokinesis. J Cell Biol 215:789–799
35. Wollrab V, Thiagarajan R, Wald A et al (2016) Still and rotating myosin clusters determine cytokinetic ring constriction. Nat Commun 7:11860
36. Fire A, Xu S, Montgomery MK et al (1998) Potent and specific genetic interference by double-stranded RNA in Caenorhabditis elegans. Nature 391:806–811
37. Oegema K, Hyman A (2006) Cell division. Wormbook 19:1–40
38. Green RA, Kao HL, Audhya A et al (2011) A high-resolution C. elegans essential gene network based on phenotypic profiling of a complex tissue. Cell 145:470–482
39. Brenner S (1974) The genetics of Caenorhabditis elegans. Genetics 77:71–94
40. Dickinson DJ, Ward JD, Reiner DJ, Goldstein B (2013) Engineering the Caenorhabditis elegans genome using Cas9-triggered homologous recombination. Nat Methods 10:1028–1034
41. Riedl J, Crevenna AH, Kessenbrock K et al (2008) Lifeact: a versatile marker to visualize F-actin. Nat Methods 5:605–607
42. Schindelin J, Arganda-Carreras I, Frise E et al (2012) Fiji: an open-source platform for biological-image analysis. Nat Methods 9:676–682
43. Hyman AA, White JG (1987) Determination of cell division axes in the early embryogenesis of Caenorhabditis elegans. J Cell Biol 105:2123–2135
44. Kashman Y, Groweiss A, Shmueli U (1980) Latrunculin, a new 2-thiazolidinone macrolide from the marine sponge latrunculia magnifica. Tetrahedron Lett 21:3629–3632
45. Morton WM, Ayscough KR, McLaughlin PJ (2000) Latrunculin alters the actin-monomer subunit interface to prevent polymerization. Nat Cell Biol 2:376–378
46. Conte D Jr, MacNeil LT, Walhout AJM, Mello CC (2015) RNA interference in Caenorhabditis elegans. Curr Protoc Mol Biol 109:26.3.1–26.330
47. Kamath RS, Fraser AG, Dong Y et al (2003) Systematic functional analysis of the Caenorhabditis elegans genome using RNAi. Nature 421:231–237
48. Rodriguez J, Peglion F, Martin J et al (2017) aPKC cycles between functionally distinct PAR protein assemblies to drive cell polarity. Dev Cell 42:400–415.e9
49. Zulueta-Coarasa T, Fernandez-Gonzalez R (2017) Tension (re)builds: biophysical mechanisms of embryonic wound repair. Mech Dev 144:43–52
50. Davies T, Jordan SN, Chand V et al (2014) High-resolution temporal analysis reveals a functional timeline for the molecular regulation of cytokinesis. Dev Cell 30:209–223
51. Liu J, Maduzia LL, Shirayama M, Mello CC (2010) NMY-2 maintains cellular asymmetry and cell boundaries, and promotes a SRC-dependent asymmetric cell division. Dev Biol 339:366–373
52. O'Connell KF, Leys CM, White JG (1998) A genetic screen for temperature-sensitive cell-division mutants of Caenorhabditis elegans. Genetics 149:1303–1321
53. Severson AF, Hamill DR, Carter JC et al (2000) The aurora-related kinase AIR-2 recruits ZEN-4/CeMKLP1 to the mitotic spindle at metaphase and is required for cytokinesis. Curr Biol 10:1162–1171
54. Pavicic-Kaltenbrunner V, Mishima M, Glotzer M (2007) Cooperative assembly of CYK-4/MgcRacGAP and ZEN-4/MKLP1 to form the centralspindlin complex. Mol Biol Cell 18:4992–5003
55. Willis JH, Munro E, Lyczak R, Bowerman B (2006) Conditional dominant mutations in the Caenorhabditis elegans gene act-2 identify cytoplasmic and muscle roles for a redundant actin isoform. Mol Biol Cell 17:1051–1064
56. Diogon M, Wissler F, Quintin S et al (2007) The RhoGAP RGA-2 and LET-502/ROCK achieve a balance of actomyosin-dependent forces in C. elegans epidermis to control morphogenesis. Development 134:2469–2479
57. Kemphues KJ, Kusch M, Wolf N (1988) Maternal-effect lethal mutations on linkage

group II of Caenorhabditis elegans. Genetics 120:977–986

58. Wissmann A, Ingles J, Mains PE (1999) The Caenorhabditis elegans mel-11 myosin phosphatase regulatory subunit affects tissue contraction in the somatic gonad and the embryonic epidermis and genetically interacts with the Rac signaling pathway. Dev Biol 209:111–127
59. Spilker AC, Rabilotta A, Zbinden C, Labbé J-C, Gotta M (2009) MAP kinase signaling antagonizes PAR-1 function during polarization of the early Caenorhabditis elegans embryo. Genetics 183:965–977
60. Morton DG, Roos JM, Kemphues KJ (1992) Par-4, a gene required for cytoplasmic localization and determination of specific cell types in Caenorhabditis elegans embryogenesis. Genetics 130:771–790
61. Fievet BT, Rodriguez J, Naganathan S et al (2013) Systematic genetic interaction screens uncover cell polarity regulators and functional redundancy. Nat Cell Biol 15:103–112
62. Fujiwara I, Zweifel ME, Courtemanche N, Pollard TD (2018) Latrunculin A accelerates actin filament depolymerization in addition to sequestering actin monomers. Curr Biol 28:3183–3192
63. Bubb MR, Spector I, Beyer BB, Fosen KM (2000) Effects of jasplakinolide on the kinetics of actin polymerization. J Biol Chem 275:5163–5170
64. Crews P, Manes LV, Boehler M (1986) Jasplakinolide, a cyclodepsipeptide from the marine sponge, Jaspis SP. Tetrahedron Lett 26:2797–2800
65. Zabriskie TM, Klocke JA, Ireland CM et al (1986) Jaspamide, a modified peptide from a Jaspis sponge, with insecticidal and antifungal activity. J Am Chem Soc 108:3123–3124
66. Scholze MJ, Barbieux KS, De Simone A et al (2018) PI(4,5)P2 forms dynamic cortical structures and directs actin distribution as well as polarity in Caenorhabditis elegans embryos. Development 145:164988
67. Cooper JA (1987) Effects of cytochalasin and phalloidin on actin. J Cell Biol 105:1473–1478
68. Berends CWH, Muñoz J, Portegijs V et al (2013) F-actin asymmetry and the endoplasmic reticulum-associated TCC-1 protein contribute to stereotypic spindle movements in the Caenorhabditis elegans embryo. Mol Biol Cell 24:2201–2215
69. Straight AF (2003) Dissecting temporal and spatial control of cytokinesis with a myosin II inhibitor. Science 299:1743–1747
70. Straight AF, Field CM, Mitchison TJ (2005) Anillin binds nonmuscle myosin II and regulates the contractile ring. Mol Biol Cell 16:193–201
71. Kepiro M, Varkuti BH, Bodor A et al (2012) Azidoblebbistatin, a photoreactive myosin inhibitor. Proc Natl Acad Sci U S A 109:9402–9407
72. Képiró M, Várkuti BH, Végner L et al (2014) Para-nitroblebbistatin, the non-cytotoxic and photostable myosin II inhibitor. Angew Chem 53:8350–8354
73. Bubb MR, Spector I, Bershadsky AD, Korn ED (1995) Swinholide A is a microfilament disrupting marine toxin that stabilizes actin dimers and severs actin filaments. J Biol Chem 270:3463–3466
74. Klenchin VA, King R, Tanaka J, Marriott G, Rayment I (2005) Structural basis of swinholide A binding to actin. Chem Biol 12:287–291
75. Nolen BJ, Tomasevic N, Russell A et al (2009) Characterization of two classes of small molecule inhibitors of Arp2/3 complex. Nature 460:1031–1034
76. Rizvi SA, Neidt EM, Cui J et al (2009) Identification and characterization of a small molecule inhibitor of formin-mediated actin assembly. Chem Biol 16:1158–1168
77. Panzica MT, Marin HC, Reymann A-C, McNally FJ (2017) F-actin prevents interaction between sperm DNA and the oocyte meiotic spindle in C. elegans. J Cell Biol 216:2273–2282
78. Steegmaier M, Hoffmann M, Baum A et al (2007) BI 2536, a potent and selective inhibitor of polo-like kinase 1, inhibits tumor growth in vivo. Curr Biol 17:316–322
79. Ishizaki T, Uehata M, Tamechika I et al (2000) Pharmacological properties of Y-27632, a specific inhibitor of rho-associated kinases. Mol Pharmacol 57:976–983
80. Saitoh M, Ishikawa T, Matsushima S et al (1987) Selective inhibition of catalytic activity of smooth muscle myosin light chain kinase. J Biol Chem 262:7796–7801
81. Hoar K, Chakravarty A, Rabino C et al (2007) MLN8054, a small-molecule inhibitor of Aurora a, causes spindle pole and chromosome Congression defects leading to aneuploidy. Mol Cell Biol 27:4513–4525
82. Manfredi MG, Ecsedy JA, Meetze KA et al (2007) Antitumor activity of MLN8054, an orally active small-molecule inhibitor of Aurora A kinase. Proc Natl Acad Sci U S A 104:4106–4111
83. Hauf S, Cole RW, LaTerra S et al (2003) The small molecule Hesperadin reveals a role for

Aurora B in correcting kinetochore–microtubule attachment and in maintaining the spindle assembly checkpoint. J Cell Biol 161:281–294

84. Harrington EA, Bebbington D, Moore J et al (2004) VX-680, a potent and selective small-molecule inhibitor of the Aurora kinases, suppresses tumor growth in vivo. Nat Med 10:262–267
85. Kotak S, Afshar K, Busso C, Gönczy P (2016) Aurora A kinase regulates proper spindle positioning in C. elegans and in human cells. J Cell Sci 129:3015–3025
86. Sekine A, Fujiwara M, Narumiya S (1989) Asparagine residue in the rho gene product is the modification site for botulinum ADP-ribosyltransferase. J Biol Chem 264:8602–8605
87. O'Connell CB, Wheatley SP, Ahmed S, Wang YL (1999) The small GTP-binding protein rho regulates cortical activities in cultured cells during division. J Cell Biol 144:305–313
88. Genth H, Gerhard R, Maeda A et al (2003) Entrapment of rho ADP-ribosylated by Clostridium botulinum C3 exoenzyme in the rho-guanine nucleotide dissociation inhibitor-1 complex. J Biol Chem 278:28523–28527
89. De Brabander M, Borgers M (1975) The formation of annulated lamellae induced by the disintegration of microtubules. J Cell Sci 19:331–340
90. Hoebeke J, Van Nijen G, De Brabander M (1976) Interaction of oncodazole (R 17934), a new antitumoral drug, with rat brain tubulin. Biochem Biophys Res Commun 69:319–324
91. Schiff PB, Fant J, Horwitz SB (1979) Promotion of microtubule assembly in vitro by taxol. Nature 277:665–667
92. Lee J-Y, Goldstein B (2003) Mechanisms of cell positioning during C. elegans gastrulation. Development 130:307–320
93. Sternlicht H, Ringel I, Szasz J (1980) The co-polymerization of tubulin and tubulin colchicine complex in the absence and presence of associated proteins. J Biol Chem 255:9138–9148
94. Losiewicz MD, Carlson BA, Kaur G et al (1994) Potent inhibition of CDC2 kinase activity by the flavonoid L86-8275. Biochem Biophys Res Commun 201:589–595
95. Potapova TA, Daum JR, Pittman BD et al (2006) The reversibility of mitotic exit in vertebrate cells. Nature 440:954–958
96. Vassilev LT, Tovar C, Chen S et al (2006) Selective small-molecule inhibitor reveals critical mitotic functions of human CDK1. Proc Natl Acad Sci U S A 103:10660–10665
97. Kjær S, Linch M, Purkiss A et al (2013) Adenosine-binding motif mimicry and cellular effects of a thieno[2,3-d]pyrimidine-based chemical inhibitor of atypical protein kinase C isoenzymes. Biochem J 451:329–342
98. Dorsey BD, Learn KS, Morris EL, et al (2017) Thienopyrimidine inhibitors of atypical protein kinase C. US Patent WO 2013078126 A1, 19 Nov, 2012

Chapter 19

Knocking Out Multiple Genes in Cultured Primary Neurons to Study Tubulin Posttranslational Modifications

Satish Bodakuntla, Carsten Janke, and Maria M. Magiera

Abstract

Microtubules, as integral part of the eukaryotic cytoskeleton, exert numerous essential functions in cells. A mechanism to control these diverse functions are the posttranslational modifications of tubulin. Despite being known for decades, relatively little insight into the cellular functions of these modifications has been gained so far. The discovery of tubulin-modifying enzymes and a growing number of available knockout mice now allow working with primary cells from those mouse models to address biological functions and molecular mechanisms behind those modifications. However, a number of those mouse models show either lethality or sterility, making it difficult to impossible to obtain a sufficient number of animals for a systematic study with primary cells. Moreover, many of those modifications are controlled by several redundant enzymes, and it is often necessary to knock out several enzymes in parallel to obtain a significant change in a given tubulin modification. Here we describe a method to generate primary cells with combinatorial knockout genotypes using conditional knockout mice. The conditional alleles are converted into knockout in the cultured primary cells by transduction with a lentivirus encoding cre-recombinase. This approach has allowed us to knock out the two main brain deglutamylases in mouse primary neurons, which leads to strongly increased polyglutamylation in these cells. Our method can be applied to measure different cellular processes, such as axonal transport, for which it can be combined with the expression of different fluorescent reporters to label intracellular proteins. Using a panel of conditional knockout mice, our method can further be applied to study the functions of a variety of tubulin modifications that require simultaneous knockout of multiple genes.

Key words Primary cells, Primary hippocampal neurons, Microtubules, Tubulin posttranslational modifications, Polyglutamylation, Mouse models

1 Introduction

Microtubules, together with actin and intermediate filaments, constitute the cytoskeleton of each eukaryotic cell. Microtubules assemble from evolutionarily conserved α- and β-tubulin heterodimers in a highly dynamic manner, and are involved in a plethora of cellular processes: they build meiotic and mitotic spindles during cell division, drive directed cell migration, determine cell shape, and serve as tracks for intracellular transport. Microtubules are

Helder Maiato (ed.), *Cytoskeleton Dynamics: Methods and Protocols*, Methods in Molecular Biology, vol. 2101, https://doi.org/10.1007/978-1-0716-0219-5_19, © Springer Science+Business Media, LLC, part of Springer Nature 2020

particularly crucial in neurons, where they play a pivotal role in the determination of neuronal polarity, dendritic arborization, and axon branching and in the transport of cellular cargoes from the cell body to synaptic terminals and vice versa [1, 2]. A key question is how microtubules are able to adapt their properties to participate in those highly diverse events. A mechanism to control microtubule properties that could potentially regulate these different functions are tubulin posttranslational modifications (PTMs), such as acetylation, detyrosination, or polyglutamylation, which are all highly enriched in neurons [3].

Though the presence of multiple PTMs on brain tubulin was identified decades ago, their biochemical and functional characterization became possible only after the enzymes catalyzing them had been discovered (reviewed in [4]). Overexpression or knockdown of those enzymes allowed performing experiments at the cellular level, thus providing first insights into the functions of tubulin PTMs (reviewed in [5, 6]). In parallel, the development of a variety of mouse models lacking modifying enzymes delivered novel insights into the physiological roles of tubulin PTMs [7–15]. Approaches allowing to combine these mouse models with cell-based assays can bolster the discovery of underlying mechanisms in physiologically relevant cell types. For instance, culturing primary neurons from mouse models with altered modification levels has shed light on the role of tubulin PTMs in several neuronal functions [7, 8, 11, 16–18].

Culturing primary cells with altered modification levels can be straightforward for some mouse models, and very difficult for others. Mice with no sterility phenotypes can be obtained from homozygous breeding pairs, which is the case for the two enzymes involved in tubulin acetylation, αTAT1/Mec-17 [19, 20] and HDAC6 [21]. Sterile mouse strains, or mice with perinatal lethality, such as the knockout for tubulin tyrosine ligase (TTL) [11] need to be bred from heterozygous progenitors, which subsequently requires single-embryo cell cultures followed by genotyping. More difficult still are the combinatorial knockout strains, in particular with one or more of the transgenes leading to sterility. These are almost impossible to use for generating sufficient amounts of cells, as the breeding schemes are very complex, and the relevant offspring is statistically too rare to perform systematic cell culture studies. Here we propose a method to overcome this problem, which we have developed to study the role of tubulin polyglutamylation in neurons.

Polyglutamylation is generated by multiple enzymes belonging to the family of tubulin tyrosine ligase-like proteins (TTLL) [22, 23] and removed by enzymes from the cytosolic carboxypeptidase (CCP) family [10, 24]. Several of these enzymes are involved in controlling polyglutamylation levels in neurons: TTLL1 and TTLL7 are the major glutamylases [8, 22], and CCP1 and CCP6 are the main deglutamylases [7, 10]. To strongly reduce polyglutamylation in the brain, both TTLL1 and TTLL7 need to be depleted concomitantly, while to increase polyglutamylation the inactivation of CCP1 and CCP6 is required in at least some parts of the brain, such as hippocampus [7]. In both cases, one of the two knockout mouse strains (CCP1 and TTLL1) is male sterile, and mice can thus only be bred heterozygous, which strongly reduces the probability of generating double-knockout mice.

To establish a reliable and sensitive cell biological approach for studying the impact of polyglutamylation on neuronal functions, we developed a cell-culture-based method to achieve increased polyglutamylation in primary hippocampal neurons, which can be easily extended to study the effects of increased or decreased PTMs other than polyglutamylation.

There are two possible strategies allowing to obtain primary neurons with perturbed tubulin PTM levels: overexpressing or silencing the corresponding enzyme(s). To obtain increased polyglutamylation levels one can overexpress glutamylating enzymes, or deplete the main deglutamylating enzymes, either by RNA interference (RNAi) or by gene knockout.

Overexpression of glutamylases has many drawbacks, the most important being that the amount of enzyme is far from being physiologically relevant, and could thus result in artifacts. For instance, all glutamylases overexpressed in Hela cells promiscuously glutamylated α- and β-tubulin, while normally these enzymes are specific to either α- or β-tubulin [23]. Moreover, in neurons, the most relevant enzyme is the glutamylase TTLL1, which could so far not be expressed as an active enzyme in cells due to the necessity to form a complex with at least four other proteins [22]. Finally, some tubulin-modifying enzymes have been observed to bind microtubules when overexpressed (our unpublished observations), which might alter microtubule dynamics and could impair the movement of organelles on the microtubules, independently of the PTM generated by the enzyme.

We thus strongly favor the strategy to remove the major deglutamylating enzymes, which in neurons are CCP1 and CCP6 [7, 10]. We decided against RNAi, as it rarely generates a complete loss of the target transcript, which in the case of the low-expressing tubulin-modifying enzymes would be very hard to control. Thus, our method of choice was to culture primary neurons from knockout mouse strains. The breeding scheme we developed to obtain *ccp1*$^{-/-}$*ccp6*$^{-/-}$ animals was very difficult for a number of reasons, such as the male infertility of *ccp1*$^{-/-}$ mice and a general lack of fitness of the double-knockout mice (which is why we used *Ccp6*$^{flox/flox}$ nestin-*cre* to breed the double-knockout animals [7]). The theoretical turnout of 1/16 of this breeding scheme was not appropriate for systematic studies using primary neuronal cultures; therefore we generated a mouse model in which both *CCP1* and

CCP6 can be conditionally knocked out (*Ccp1*$^{flox/flox}$*Ccp6*$^{flox/flox}$). These mice behave like wild type and thus can be easily bred to give homozygous embryos, which are the source of primary hippocampal neurons. Once neurons are cultured, they are transduced with a lentivirus expressing Cre recombinase to convert both flox alleles into knockout alleles, leading to the removal of deglutamylases, which results in tubulin hyperglutamylation. In parallel, the neurons from the same culture are transduced with a control virus, and can be used as an internal control with normal polyglutamylation levels. To verify the effect of the Cre recombinase alone, the equivalent experiment is performed using neurons from wild-type embryos transduced with the same GFP- and GFP-Cre viruses.

Thus, we describe a novel, physiologically relevant cell-based assay to attain excessive polyglutamylation and its possible applications. Our cell-culture-based method is not limited to studying microtubules or neurons, but has broader applications in the field of cell biology. As far as microtubule PTMs are concerned, our method can be expanded to study the impact of reduced polyglutamylation (using *Ttll1*$^{flox/flox}$*Ttll7*$^{flox/flox}$ neurons) or reduced detyrosination (with *Vash1*$^{flox/flox}$*Vash2*$^{flox/flox}$ neurons), all of which require the depletion of several genes.

2 Materials

2.1 Cloning of Cre Recombinase or Reporter Genes into Lentiviral Vector

1. pTRIP vector with CMV promoter, pTRIP-CMV-GFP-2A (modified from Addgene #102611).
2. Template for PCR reaction (brain cDNA or plasmid containing gene of interest).
3. KAPA HiFi HotStart 2× ReadyMix (Kapa Biosystems).
4. Specific primers:

   NLS-cre-FS-2A: 5′-tccactagtgtcgacATGCCCAAGAAGAAGAGGAAGGTG-3′

   NLS-cre-FS-2A: 5′-catgtttttctaggtTAATCGCCATCTTCCAGCAGGC-3′

   (*see* **Note 1**)
5. PCR machine.
6. DNase/RNase-free distilled water.
7. T4 DNA polymerase.
8. XhoI.
9. NheI.
10. Buffer 2.1 (New England BioLabs).
11. 1× Tris–borate–EDTA (TBE) buffer: Add 100 mL of 10× TBE to 900 mL of distilled water.

12. 0.8% Agarose gel: Add 4 g of agarose and 300 mL of 1× TBE in a 500-mL glass bottle. Heat the mix in the microwave oven until the agarose is dissolved and make up the volume to 500 mL with 1× TBE. Add 20 μL of 10 mg/mL ethidium bromide solution and store in incubator set at 55 °C for not more than 2 weeks. Prepare the gel tray and comb of desired size, pour the agarose, and let solidify at room temperature for at least 15 min.
13. 1 kb DNA ladder.
14. Gel extraction kit.
15. 1.5-mL Tubes
16. Chemically competent *E. coli* strain Dh5alpha.
17. Bacterial incubator at 37 °C.
18. LB-agar plates containing 100 μg/mL of ampicillin.
19. LB medium without antibiotic used for recovery of bacteria.
20. LB medium with 100 μg/mL of ampicillin.
21. 14-mL Round-bottom tubes.
22. Bacterial incubator with shaking set at 37 °C.

2.2 Lentivirus Production

1. X-Lenti 293T (Takara) (*see* **Note 2**).
2. Cell culture incubator at 37 °C with 5% CO_2.
3. Tissue culture-treated 75-cm^2 flasks.
4. Phosphate-buffered saline (PBS).
5. Trypsin.
6. Plain medium: Dulbecco's modified Eagle media (DMEM) without any supplements.
7. Complete culture medium: Supplement DMEM with a final concentration of 10% FBS (vol/vol), 200 mM L-glutamine, and 1× penicillin-streptomycin. Store at 4 °C.
8. Tissue culture-treated 6-well plates.
9. Tissue culture-treated 24-well plates.
10. Transfection reagent: TransIT-293 (Mirus Bio LLC) (*see* **Note 3**).
11. Expression and packaging plasmids: pTRIP lentiviral vector containing the gene of interest under the CMV or CMV-enhanced chicken beta-actin (CAG) promoter and packaging plasmids including psPAX2 (a gift from D. Trono, Addgene plasmid #12260) and pCMV-VSVG (a gift from B. Weinberg, Addgene plasmid #8454) (*see* **Note 4**).
12. Virus collection medium: Neurobasal medium-PS: supplement neurobasal medium with a final concentration of 1× penicillin-streptomycin. Store at 4 °C (*see* **Note 5**).

13. 15- and 50-mL screw-cap tubes.
14. 0.45-μm Filters and 20-mL syringes.
15. 1.5- and 2-mL tubes.

2.3 Mouse Primary Hippocampal Neuron Cultures and Lentiviral Transduction

1. DNase/RNase-free distilled water.
2. 1 mg/mL Poly-D-lysine: Add 50 mL of DNase/RNase-free distilled water to 50 mg of poly-D-lysine bottle. Aliquot by 5 mL or 10 mL in sterile 15-ml tubes and store at −20 °C.
3. Autoclaved distilled water.
4. Tissue culture-treated 75-cm^2 flasks.
5. Dissection medium: Add 10 mL of 10× HBSS and 1 mL of 1 M HEPES solution to 89 mL of DNase/RNase-free distilled water.
6. Dissection tools.
7. Tissue culture-treated 100-mm dishes.
8. Tissue culture-treated 60-mm dishes.
9. 15-mL Screw-cap tubes.
10. 10× Trypsin (2.5%), no phenol red: Aliquot by 500 μL in 1.5-mL tubes and store at −20 °C. Avoid repeated freeze-thaw cycles.
11. Plating medium: Add 5 mL of filter-sterilized heat-inactivated horse serum, 1.3 mL of glucose, and 1 mL of 100× penicillin-streptomycin to 92.7 mL of minimum essential medium. Prepare fresh before use.
12. 70-μm Cell strainers.
13. KOVA™ Glasstic™ Slide 10 with Grids (Fisher Scientific).
14. 35-mm Glass-bottom dishes.
15. Tissue culture-treated 6-well plates.
16. Tissue culture-treated 24-well plates.
17. 12-mm, No. 1.5H, round cover slips.
18. Tissue culture-treated 12-well plates.
19. B27 supplement (Thermo Fisher): Aliquot by 1 mL in 1.5-mL tubes and store at −20 °C. Avoid repeated cycles of freeze-thawing.
20. Neurobasal-PS: Supplement neurobasal medium with a final concentration of 1× penicillin-streptomycin. Store at 4 °C.
21. B27 medium: Add 2 mL of 50× B27 solution and 1 ml of Glutamax™ supplement to 97 mL of neurobasal-PS medium. Prepare fresh for each experiment.
22. 2× B27 medium: Add 2 mL of 50× B27 solution and 1 mL of Glutamax™ to 47 mL of neurobasal-PS medium. Prepare fresh for each experiment.

2.4 Genotyping

1. 2% Agarose: Add 10 g of agarose and 300 mL of 1× TBE in a 500-mL glass bottle. Heat the mix in the microwave oven until the agarose is dissolved and make up the volume to 500 mL with 1× TBE. Add 20 μL of 10 mg/mL ethidium bromide solution and store in incubator set at 55 °C for not more than 2 weeks. Prepare the gel tray and comb of desired size, pour the agarose, and let solidify at room temperature for at least 15 min.
2. 100 bp DNA ladder.
3. 10 mg/mL Proteinase K: Dissolve 100 mg of proteinase K in 10 mL of autoclaved distilled water. Aliquot and store at −20 °C indefinitely.
4. 1 M Tris–HCl pH 8: Dissolve 12.1 g of Tris in 100 mL autoclaved distilled water and adjust to pH 8 with HCl. Filter sterilize and store at room temperature.
5. 5 M NaCl: Dissolve 29.2 g of NaCl in 100 mL autoclaved distilled water. Filter sterilize and store at room temperature.
6. 0.5 M EDTA pH 8: Dissolve 14.6 g of EDTA in 100 mL autoclaved distilled water and adjust to pH 8 with KOH. Filter sterilize and store at room temperature.
7. Tail lysis buffer: 0.1 M Tris–HCl pH 8, 0.2 M NaCl, 5 mM EDTA, and 0.4% SDS. Mix 50 mL of 1 M Tris–HCl pH 8, 20 mL of 5 M NaCl, 5 mL of 0.5 M EDTA pH 8, and 10 mL of 20% SDS and complete the volume to 500 mL with distilled water.
8. Specific primers.
9. For CCP1, primer mix 1: Mix primers (5′-TTAAGCAGTGG CTGCCGGAGTGC-3′) and (5′-GTCTACAGCCACGTG CTCAGCAAAGG-3′) at 1:1 and primer mix 2: Mix primers (5′-GGTCATCTTGTAACAGTAGGGAAATCAAGCC-3′) and (5′-GGAAAGCATAGCATCATGTTAAAACTCCCTCC-3′) at 1:1.
10. For CCP6, primer mix 1: Mix primers (5′-GAATGGCAATG AGATCACCACTCTCCAGC -3′) and (5′-CTGTTGGGTG TCTGAGGCAAACACTTCC -3′) at 1:1 and primer mix 2: Mix primers (5′-AGGTGTTCAGTGTAAAGGAATCTGG TCCC-3′) and (5′-CTGTTGGGTGTCTGAGGCAAACAC TTCC-3′) at 1:1.
11. GoTaq® DNA polymerase (Promega): Use according to the manufacturer's instructions.
12. Electrophoresis cell: Mini-PROTEAN® cell with a PowerPac™ power supply (Bio-Rad).

2.5 SDS-Polyacrylamide Gel Electrophoresis and Immunoblotting

1. 5× Laemmli sample buffer (40 mL): 400 mM Tris–HCl pH 6.8, 450 mM DTT, 10% (wt/vol) SDS, and 50% (vol/vol) glycerol. Preheat 16 mL of 1 M Tris–HCl pH 6.8 to app. 40 °C, add 2.8 g of DTT, and 4 g of SDS. Stir the solution gently using a magnetic stirrer and add 20 mL of 100% glycerol. Continue stirring until the solution is homogenous and add a pinch of bromophenol blue to reach the desired intensity of color. Aliquot and store at −20 °C indefinitely.
2. Acrylamide/bis-acrylamide stock solution: 40% Acrylamide supplemented with 1.04% (wt/vol) bis-*N*,*N*′-methylene-bis-acrylamide. Store at 4 °C for up to 6 months.
3. 4× Resolving gel buffer: 1.5 M Tris–HCl pH 8.8 and 0.4% SDS. Dissolve 36.3 g of Tris in autoclaved distilled water and adjust to pH 8.8 with HCl for a final volume of 196 mL. Then, add 4 mL of 20% SDS stock solution. Filter-sterilize and store at room temperature indefinitely (*see* **Note 6**).
4. 4× Stacking gel buffer: 0.5 M Tris–HCl pH 6.8 and 0.4% SDS. Dissolve 12.1 g of Tris in autoclaved distilled water and adjust to pH 6.8 with HCl for a final volume of 196 mL. Then, add 4 mL of 20% SDS stock solution. Filter-sterilize and store at room temperature indefinitely.
5. *N*,*N*,*N*′,*N*′-Tetramethylethylenediamine (TEMED).
6. 10% Ammonium persulfate (APS): Dissolve 1 g of APS in 10 mL of autoclaved distilled water. Aliquot and store at −20 °C indefinitely.
7. SDS-PAGE electrophoresis equipment (Bio-Rad).
8. Electrophoresis cell: Mini-PROTEAN® cell with a PowerPac™ power supply.
9. Gel running buffer: 50 mM Tris–HCl, 384 mM glycine, and 0.1% SDS. Dissolve 6 g of Tris and 28.8 g of glycine in autoclaved distilled water for a final volume of 995 mL. Then, add 5 mL of 20% SDS stock solution.
10. Trans-Blot® Turbo™ Mini Nitrocellulose Transfer Packs (Biorad).
11. Trans-Blot® Turbo™ Transfer System (Biorad).
12. Tris-buffered saline–Tween 20 (TBST) solution: 20 mM Tris–HCl pH 7.4, 140 mM NaCl, and 0.1% Tween 20. Dissolve 2.42 g of Tris and 8.18 g of NaCl in autoclaved distilled water and adjust to pH 7.4 with HCl for a final volume of 999 mL. Add 1 mL of Tween 20 to the buffer and mix well (*see* **Note 7**).
13. TBST containing 5% (wt/vol) fat-free dry milk: Add 0.5 g of milk to 10 mL TBST.
14. Antibodies: 12G10, anti-α-tubulin antibody (used at 1/500 (vol/vol), developed by J. Frankel and M. Nelson, obtained

from the Developmental Studies Hybridoma Bank, developed under the auspices of the NICHD, and maintained by the University of Iowa). PolyE, anti-polyglutamylated tubulin antibody (used at 1/20,000 (vol/vol); AdipoGen). Anti-GFP antibody (used at 1/5000 (vol/vol); Acris Antibodies).

15. Clarity™ Western ECL Substrate (Biorad).
16. Western blot developer or imager.

3 Methods

Here, we describe a method to obtain primary neurons in which we can induce the knockout of tubulin-modifying enzymes, thus obtaining strong alterations in the levels of specific tubulin PTMs. As an example, we describe the protocol to generate hyperglutamylation in primary hippocampal neurons, which can be used to study the role of polyglutamylation in several neuronal functions, such as axonal outgrowth and branching, neuronal connectivity, or axonal transport.

Briefly, we culture hippocampal neurons from *Ccp1*$^{flox/flox}$*Ccp6*$^{flox/flox}$ E17.5 (embryonic day 17.5) mouse embryos and transduce them either with a GFP- or GFP-2A-*cre*-expressing lentivirus at DIV 0 (days in vitro). The use of lentivirus-mediated gene transfer allows a nearly complete transduction of all cells in the culture dish, and appears to not exert additional stress on the neurons, as we observe no neuronal death.

Cre-recombinase was cloned into the pTRIP vector under the CMV or the CMV-enhanced chicken beta-actin (CAG) promoter using one-step sequence and ligation-independent cloning (SLIC) method [25]. Considering the possibility that direct GFP-tagging of Cre protein might affect its recombinase activity, we cloned Cre behind a self-cleavable 2A peptide sequence [26] present in frame with the GFP protein sequence in the vector (Fig. 1a). In cells, the Cre recombinase is cleaved from GFP, which we confirmed by immunoblotting (Fig. 2b). Transduction of *Ccp1*$^{flox/flox}$*Ccp6*$^{flox/flox}$ neurons with GFP-2A-Cre resulted in "defloxing" of both alleles of the two genes (Fig. 4a), and a concomitant increase in tubulin polyglutamylation (Fig. 4b). In parallel, wild-type neurons were transduced with the same GFP- and GFP-Cre viruses to confirm that the Cre recombinase alone has no effect on polyglutamylation levels (Fig. 4c).

Thus, obtained neurons can be used to determine the impact of increased polyglutamylation on different neuronal processes including neuronal differentiation and intracellular transport, by staining organelles using commercially available dyes such as Mitotracker and Lysotracker. Furthermore, this protocol combined with lentivirus-based expression of reporter genes can extend the

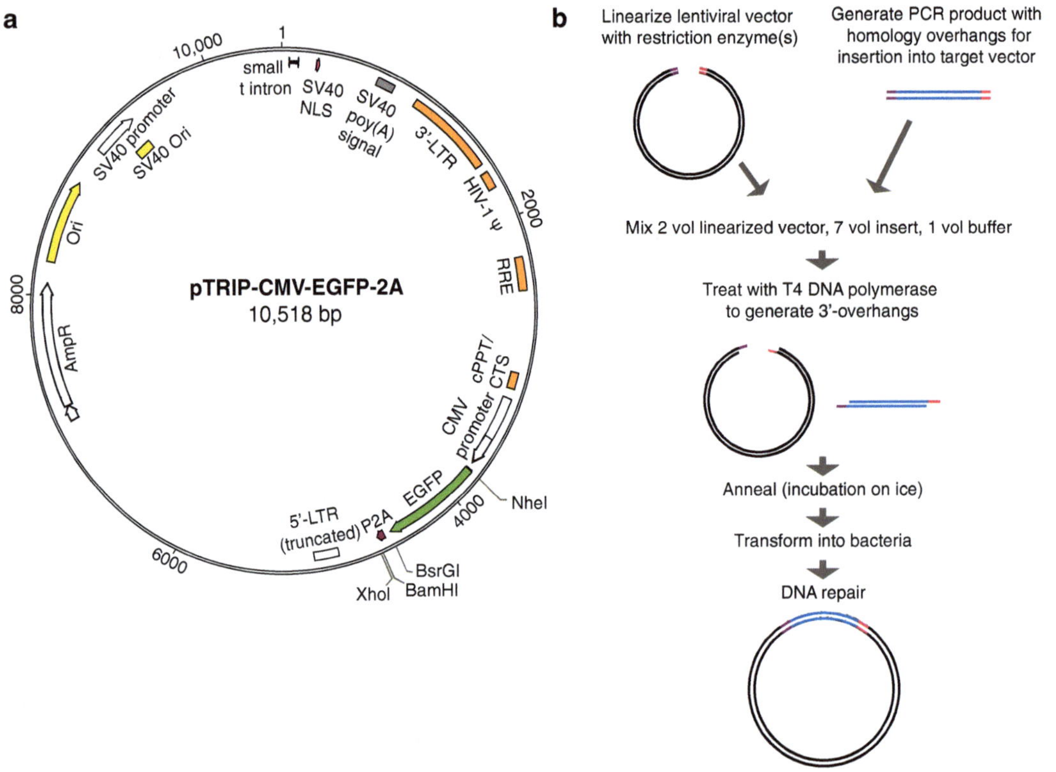

Fig. 1 Strategy used to clone the Cre recombinase and reporter genes into the pTRIP lentiviral vector. (**a**) Vector map of the pTRIP-CMV-EGFP-2A lentiviral plasmid. To allow the cleavage of Cre recombinase and GFP, Cre was inserted into the XhoI site. For further insertion of additional reporter genes, the backbone can be opened with NheI to generate a C-terminally GFP-tagged fusion protein, or with BsrGI to generate an N-terminally GFP-tagged fusion protein. (**b**) Flowchart of the sequence- and ligation-independent cloning (SLIC) method (adapted from [25])

applications of this method to study the trafficking of other vesicles such as endosomes (Rab5, Rab7, and LAMP1), BDNF vesicles (BDNF), autophagosomes (LC3), or synaptic vesicles (synaptotagmin, synaptophysin). Alternatively, the system can be used to study microtubule dynamics by the use of end-binding protein EB3 fused to GFP.

3.1 Cloning of Cre Recombinase and Reporter Genes into a Lentiviral Vector

3.1.1 Cloning of Cre Recombinase into the pTRIP Vector: Generation of the pTRIP-CMV-GFP-2A-Cre Plasmid (Fig. 1)

1. Linearize the pTRIP vector by restriction digestion with XhoI (Fig. 1a). Digest 3 μg of plasmid DNA for ~2.5 h with 0.5 μL of XhoI enzyme (20 U/μL) in a 50 μL reaction. Run the whole sample on a 0.8% agarose gel until the dye front runs out (*see* **Note 8**).
2. Purify the digested vector from the gel using the gel purification kit. Elute the vector DNA in 50 μL of elution buffer.
3. Amplify the Cre recombinase gene using the primers containing at least 15 bp of homology to the ends of the vector. In a

PCR tube, mix 25 μL of 2× KAPA ready mix, 0.2 μL of each primer (100 μM), and PCR substrate and make up the volume to 50 μL with DNase/RNase-free distilled water and perform the PCR reaction. Use the following PCR conditions: an initial step of denaturation at 95 °C for 3 min; 15 cycles of amplification (95 °C for 30 s, annealing of primers at 65 °C for 30 s, and extension at 72 °C for ~1.5 min) (*see* **Note 9**).

4. Run the whole sample on a 0.8% agarose gel and purify the PCR product from the gel using the gel purification kit. Elute the PCR product in 40 μL of elution buffer.
5. Mix 7 μL of insert or elution buffer (negative control), 2 μL of digested vector, and 1 μL of buffer 2.1 in to a 1.5-mL tube; centrifuge briefly; and place on ice (Fig. 1b).
6. Add 0.3 μL of T4 DNA polymerase (3 U/μL) to each tube and mix by firmly tapping the tube (Fig. 1b).
7. Centrifuge briefly and leave the reaction mix at room temperature for 2.5 min to generate 5′ overhangs (Fig. 1b) (*see* **Note 10**).
8. After the incubation, immediately transfer the tubes on ice to stop the activity of T4 DNA polymerase and leave it for ~5 min to facilitate the annealing of single-stranded overhangs (Fig. 1b).
9. Add 100 μL of chemically competent *E. coli* to the DNA reaction and mix by gently tapping at the bottom of the tube.
10. Leave the *E. coli* with the reaction mix on ice for 30 min, provide heat shock at 42 °C for 45 s, and bring back the tube on ice for 10 min. Add 1 mL of LB medium without antibiotics and let bacteria recover at 37 °C for at least 1 h.
11. After the incubation, pellet the *E. coli* cells at room temperature, 1000 RCF for 3 min; resuspend in ~100 μL of LB medium; plate the entire cell suspension on a LB-agar-ampicillin plate; and incubate at 37 °C for 16 h.
12. Next day, inoculate 5–10 colonies, each in a 14-mL round-bottom tube with ~ 5 mL of LB medium containing ampicillin (*see* **Note 11**).
13. Extract the plasmid DNA and test the clones by restriction digestion with appropriate enzymes. Finally, verify the positive clones by sequencing.

1. Linearize pTRIP-CMV-GFP-2A and pTRIP-CMV-GFP-2A-Cre lentivectors by restriction digestion with NheI (Fig. 1a). Digest 3 μg of plasmid DNA for ~2.5 h with 0.5 μL of NheI enzyme (20 U/μL) in a 50 μL reaction. Gel-purify the digested vector in 50 μL of elution buffer (*see* **Note 12**).

3.1.2 Cloning of Reporter Genes into pTRIP Lentivectors: Generation of the pTRIP-CMV-Reporter-GFP-2A and pTRIP-CMV-Reporter-GFP-2A-Cre Plasmids

2. Amplify reporter gene using the primers containing at least 15 bp of homology to the ends of the vector. In a PCR tube, mix 25 μL of 2× KAPA ready mix, 0.2 μL of each primer (100 μM), and PCR substrate and make up the volume to 50 μL with DNase/RNase-free distilled water.
3. Then follow the same procedure as in Subheading 3.1.1, from **step 4** onwards.

3.2 Lentivirus Production and Transduction (Fig. 2a) (See Note 13)

Grow X-lenti 293T cells in a 75 cm^2 flask to near confluence. It is critical not to let the cells become over-confluent as it will affect the efficiency of virus production. From a confluent 75-cm^2 flask, ~10 × 10^6 cells can be obtained. Pre-warm trypsin, PBS, and DMEM complete culture medium to 37 °C.

3.2.1 Day −1: Seeding Cells for Transfection

1. Remove the medium from the 75 cm^2 flask and gently wash the cells with 5 mL of pre-warmed PBS to remove any excess medium left in the flask.
2. Add 1 mL of trypsin to the flask and leave in the incubator for not more than 1 min. Collect all the trypsinized cells in culture medium.
3. Pellet down the cells by centrifugation at 150 RCF for 2 min.
4. Resuspend the cells gently in 1 mL of complete culture medium. It is important to dissociate the cells adequately, as leaving cell clumps will reduce the transfection efficiency in later steps.
5. Count and dilute the cells to 0.15 × 10^6 cells per ml in complete culture medium.
6. Add 3 mL of the cell suspension (0.45 × 10^6 cells) per well of a 6-well plate and leave the plate in the cell culture incubator (37 °C, 5% CO_2).

3.2.2 Day 0: Transfection of Plasmid DNA

Pre-warm the DMEM complete culture medium and plain DMEM to 37 °C. The following protocol gives volumes for transfection of one well of a 6-well plate for each lentiviral vector, GFP and GFP-2A-Cre.

1. Mix 1.6 μg of lentiviral plasmid of interest with 0.4 μg of pCMV-VSVG and 1 μg of psPAX2 in a 1.5-mL tube.
2. Mix the TransIT-293 transfection reagent by gently inverting the tube. Add 8 μL of TransIT-293 to 200 μL of plain DMEM medium in a 1.5-mL tube.
3. Mix the diluted TransIT-293 well by vortexing for ~10 s and spin shortly to collect all the liquid in the bottom of the tube.
4. Add 200 μL of the diluted TransIT-293 reagent to 3 μg of plasmid DNA mix (*see* **Note 14**).

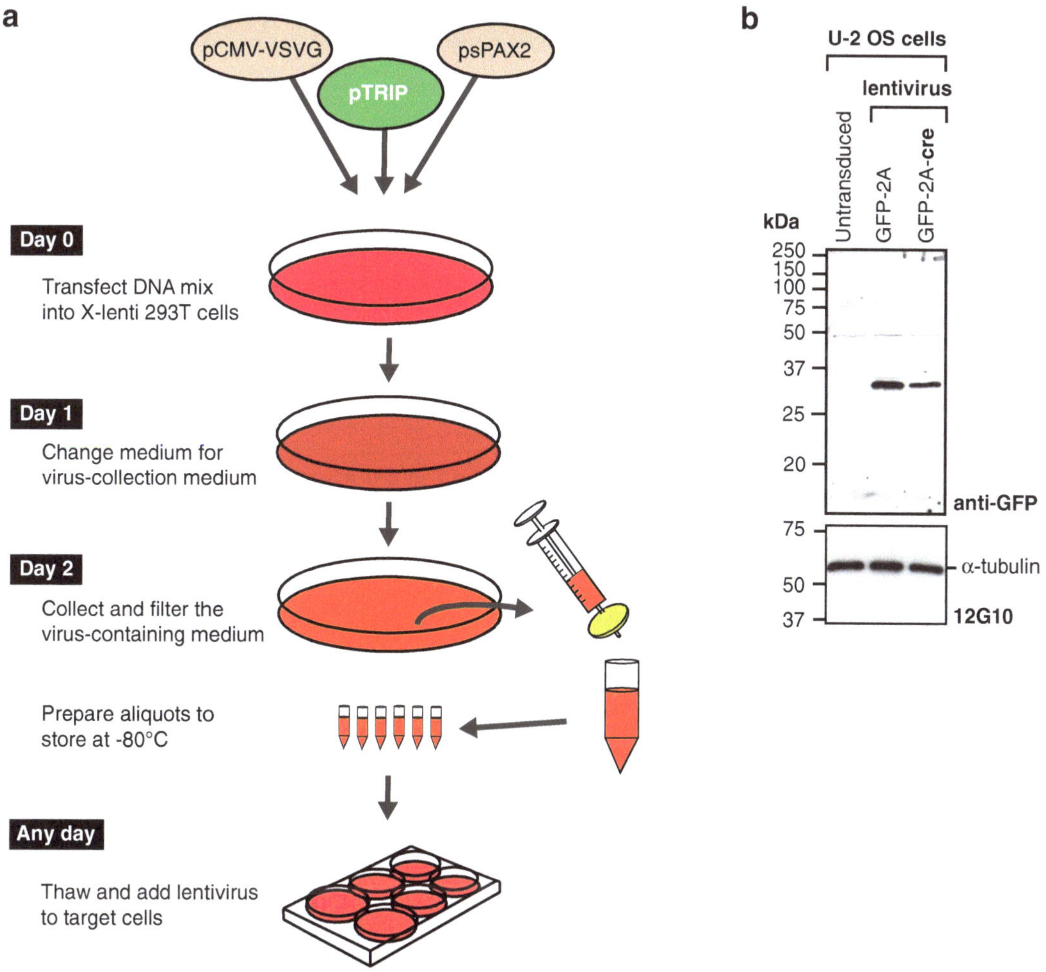

Fig. 2 Production and functional analysis of the lentiviral particles. (**a**) Flowchart of the production of lentiviral particles. X-lenti 293T cells are transfected with the envelope vector (pCMV-VSV-G), packaging vector (psPAX2), and backbone pTRIP vector containing the gene of interest. On the day following the transfection, the medium is replaced by the desired virus-collection medium, which is collected 24–36 h later, filtered, and aliquoted. The medium can either be directly used to transduce cells or frozen for later use. Note that the medium changes color to more orange during virus production. (**b**) Immunoblots showing the efficiency of the cleavage of the P2A site between the GFP and the Cre recombinase. Extracts from U-2OS cells transduced with lentivirus encoding either GFP or GFP-2A-Cre recombinase are immunoblotted with anti-GFP antibody. In both cases, only a single band of ~28 kDa, corresponding to GFP, is detected. The absence of a higher-molecular-weight band, corresponding to the fusion protein GFP-Cre recombinase, demonstrates that all GFP-Cre fusion protein was cut. Total α-tubulin (detected by the 12G10 antibody) is used as a loading control

5. Mix the tube gently by inverting and leave under the hood for 15–20 min to allow the formation of TransIT-293-DNA transfection complexes.
6. During the incubation time, replace the medium of the cells with 2 mL of fresh, pre-warmed complete culture medium.

7. After the incubation time, gently tap the tube containing the TransIT-293-DNA transfection mix and add 200 μL of the mix dropwise into each well.

3.2.3 Day 1: Replace Culture Medium with Virus Collection Medium

1. After 16 h of transfection, aspirate the complete culture medium and gently add 3 mL of prewarmed virus collection medium on the cells. To avoid keeping cells for too long outside the cell culture incubator, treat one plate at the time (*see* **Note 15**).

3.2.4 Day 2: Virus Collection

After 24–30 h of medium change, collect the medium containing the lentivirus (*see* **Note 16**).

1. Collect the medium containing the lentivirus in a 15 or 15 mL screw-cap tube.
2. Filter the medium through a 0.45-μm syringe filter to remove cell debris (*see* **Note 17**).
3. Use the virus fresh, or gently aliquot the virus-containing medium by desired volumes in 1.5-mL tubes and store at −80 °C.

3.2.5 Testing the Virus

The amount of virus to add on to the neurons has to be determined by testing the lentivirus on mammalian cell lines such as U-2 OS.

1. Prepare a cell suspension of 50,000 cells/mL, plate 500 μL of the cell suspension per well of a 24-well plate, and leave the plate in the incubator.
2. On the next day, thaw the lentivirus from −80 °C directly in the hand or under the hood (*see* **Note 18**).
3. Gently add different volumes (50, 100, 200, and 500 μL) of lentivirus, having removed the equivalent amount of media from the wells. Leave the cells with the virus in the incubator for the next 24–48 h.
4. Based on the GFP expression, visually estimate the number of cells transduced and the intensity of GFP fluorescence for the different volumes used. Accordingly, determine the desired amount of virus to be added to the neurons to achieve maximum transduction efficiency. Note that the same criteria of choosing the virus volume should be applied to both GFP and GFP-2A-Cre lentiviruses (*see* **Note 19**).

3.2.6 Testing the Cleavage of Cre Recombinase and GFP

Once the virus volume to be used is determined, test the posttranslational cleavage of Cre recombinase and GFP by immunoblotting.

1. Prepare a cell suspension of 50,000 cells/mL, and add 2 mL of this suspension per well of a 6-well plate and leave it in the incubator.

2. Next day, thaw the required amount of lentivirus and transduce the cells.
3. After 48 h of transduction, wash the cells at least two times with fresh medium or PBS to remove any leftover lentivirus.
4. Collect cells from each well in 500 μL of 2× Laemmli buffer and boil for 5 min.
5. Perform an immunoblot with an anti-GFP antibody. If the GFP and Cre are cleaved, GFP band will appear at ~27 kDa; else the fusion protein will migrate at ~65 kDa. In case you test the construct containing another reporter gene (vesicular marker or EB3 protein), take into account the size of this protein.

3.3 Mouse Primary Hippocampal Neuron Cultures and Lentiviral Transduction

Primary mouse hippocampal neurons are cultured from embryos at 17.5 days of gestation. In our hands, a single hippocampus gives app. 0.3×10^6 neurons. For each experiment in which untransduced, GFP- and GFP-Cre-transduced neurons will be analyzed, app. 2.2×10^6 cells (4 embryos) will be necessary: 6 wells of a 12-well plate (40,000 cells per well) for genotyping, 6 wells of a 6-well plate (80,000 cells per well) for immunoblotting, 6 glass-bottom dishes (80,000 cells per well) for imaging, and two 6-well plates (80,000 cells per well) for conditioned medium. If immunostaining analysis of the cells is also required, the necessary number of neurons should be seeded (20,000 cells per 12-mm-diameter coverslip). Accordingly, coat the required number of culture plates in forehand (*see* **Note 20**).

3.3.1 Poly-D-lysine Coating (See **Note 21**)

1. Thaw frozen aliquots of poly-D-lysine directly in the hands.
2. Add 500 μL of 1 mg/mL poly-D-lysine solution to the 35-mm glass-bottom dishes ensuring that it covers only the glass surface of the dish. For coating each well of a 12-well and 6-well dishes, 500 and 800 μL of poly-D-lysine is used, respectively. For coating glass coverslips, put them one by one in 24-well plate wells and add 80 μL of poly-D-lysine to cover the surface of coverslips.
3. Leave the dishes in the cell culture incubator for at least 4 h or overnight.
4. Remove the poly-D-lysine, wash the dishes by adding ~2 mL of autoclaved distilled water to the dishes, and plate wells. Add 1 mL of distilled water to coverslips in 24-well plates. Repeat the wash three times.
5. Finally, remove any remaining droplets of water by using the vacuum pump.

6. Leave the dishes open under the cell culture hood until they are completely dry. Use the poly-D-lysine-coated dishes immediately or store at 4 °C, sealed with parafilm until further use.

3.3.2 Neuronal Cell Culture

Before proceeding, prepare freshly anticipated amounts of dissection medium, plating medium, 1× B27 medium, and 2× B27 medium in tissue culture flasks. Keep the plating medium and B27 medium in the cell culture incubator for 2 h to allow it to equilibrate its temperature and CO_2 content. Thaw adequate amount of 10× trypsin aliquots and filter-sterilized horse serum. For up to five embryos, we typically prepare 100 mL of both dissection and plating media, 30 mL of 1× B27 media, 20 mL of 2× B27 medium, 500 μL of trypsin, and 5 mL of horse serum.

1. Sacrifice a 17.5-day-pregnant dam by cervical dislocation and collect the embryos in a 100-mm dish containing ~5 mL of dissection medium.
2. Rapidly decapitate the embryos, dissect the brains, and separate the two hemispheres (Fig. 3a). Collect all the hemispheres in a 60-mm dish containing ~3 mL of dissection medium.
3. With the help of a stereomicroscope, carefully remove the meninges off the hemisphere and dissect out the hippocampus (Fig. 3a). Collect the hippocampi in a 15-mL tube containing 4.5 mL of dissection medium (*see* **Note 22**).
4. Add 0.5 mL of 2.5% trypsin (10×) to the tube containing hippocampi, mix gently, and place it in the incubator or water bath at 37 °C for 20–25 min. During the incubation time, gently mix the tube regularly by inverting.
5. Meanwhile, fire-polish the edges of two glass Pasteur pipettes using Bunsen burner to make their sizes successively smaller (Fig. 3b). Label the treated pipettes according to their size and fix sterile rubber bulbs at their ends.
6. Coat the inner walls of the glass pipettes with horse serum by pipetting it up and down several times, to avoid the sticking of cells to the pipette walls while triturating.
7. After the trypsinization, use the largest triturating pipette to carefully remove the trypsin from the 15-mL tubes (*see* **Note 23**).
8. Wash off the excess trypsin from the tissue by incubating with ~5 mL of plating medium for 5 min. Repeat the wash once more and resuspend the hippocampi in ~1 mL of plating medium.
9. Gently aspirate and expulse hippocampi 6–10 times using the largest triturating pipette, avoiding formation of air bubbles. Repeat the procedure using the smaller pipette (*see* **Note 24**).

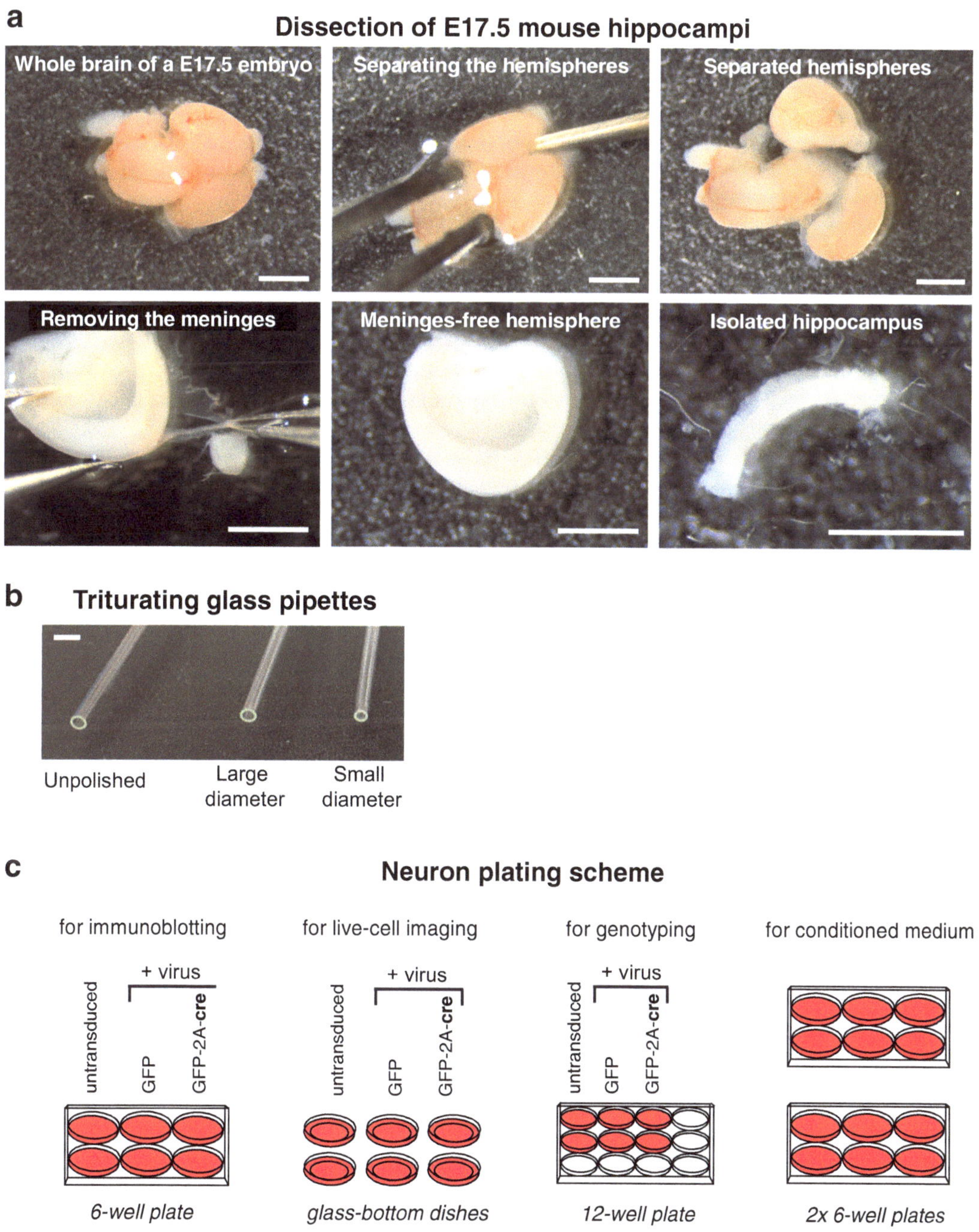

Fig. 3 Mouse primary hippocampal neuron culture and the experimental design. (**a**) Representative pictures of successive steps of hippocampus dissection from an E17.5 embryonic brain. Scale bar: 2 mm. (**b**) Representative images of unpolished, fire-polished, and reduced-diameter Pasteur pipettes used to dissociate hippocampi. Scale bar: 2 mm. (**c**) Schematic representation of the culture dishes needed for one experiment including non-transduced, GFP-, and GFP-2A-Cre-transduced neurons

10. Pass the cell suspension through a 70-μm cell strainer to remove any tissue clumps (*see* **Note 25**).
11. Count cells and prepare a cell suspension at 40,000 cells/mL. Add 500 μL of the cell suspension on each glass coverslip (for immunostaining), 1 mL in each 12-well plate well (for genotyping), and 2 mL in each 6-well plate well (for immunoblotting and for conditioned medium) and in each 35-mm glass-bottom dish (for live-cell imaging). The day of seeding is considered as DIV 0 (*days* in vitro *0*) (*see* **Note 26**).
12. Two to three hours after plating, change the medium for the neurobasal-B27. For the neurons for conditioned medium, simply replace the plating medium by 1× B27 medium. For neurons to be transduced with the lentiviral particles, see next chapter.

3.3.3 Transduction of Neurons

In this protocol we assume that the volume of the virus added to the neurons constitutes 50% of the total volume of the culture medium. As the virus-containing-medium (neurobasal) does not contain the supplements necessary for neurons to grow, it is essential to add twice the final amount of these supplements to the medium (2× B27 medium) used to complement the virus-containing medium volume.

1. Prepare fresh (*see* Subheading 3.2), or thaw the required amount of the desired lentiviruses (for example GFP- and GFP-2A-Cre-encoding lentiviruses) under the hood.
2. In the meantime, remove the plating medium and add 0.5 volume of 2× B27 medium to the neurons. For example, for a 35-mm glass-bottom dish add 1 mL of 2× B27 medium.
3. Once the virus is completely thawed, add 0.5 volume of the lentivirus-containing medium (for a 35-mm glass-bottom dish, add 1 mL of lentivirus). For untransduced samples, add neurobasal-PS medium. Rapidly return the cells to the cell culture incubator.

3.4 Genotyping

Prior to the analysis of the transduced neurons, confirm the genotype of neurons in order to verify the activity of the Cre recombinase.

1. Add 100 μL of the lysis buffer containing proteinase K to each well of 12-well dish and lyse the neurons using a p200 micropipette.
2. Incubate the lysate at 55 °C for at least 4 h or overnight.
3. Boil the samples at 95 °C for 10 min.
4. In order to determine the genotype of CCP1 and CCP6, perform PCR reactions using primer mix 1 and primer mix 2 for both genes.

5. Set up separate PCR reactions for each primer mix (25 μL): Add 18 μL of water, 5 μL of 5× buffer, 0.5 μL of 10 mM dNTPs, 0.5 μL of GoTaq® DNA polymerase (1 U/μL), 0.1 μL of primer mix, and 1 μL of DNA sample.

   Use the following PCR conditions: an initial step of denaturation at 96 °C for 5 min; 35 cycles of amplification (96 °C for 30 s, annealing of primers at 63.5 °C for 30 s, and extension at 72 °C for ~30 s).

6. Run 10 μL of the PCR product on the 2% agarose gel (*see* **Note 27**).

7. For CCP1, primer mix 1 amplifies ~400 bp fragment in the flox allele (and no product in the knockout allele) and the primer mix 2, ~600 bp fragment in the knockout allele (and no product in the flox allele). For CCP6, primer mix 1 amplifies ~440 bp fragment in the flox allele (and no product in the knockout allele) and the primer mix 2, ~800 bp fragment in the knockout allele (and no product in the flox allele) (Fig. 4a). Note that the primer mix 1 still amplifies a faint band in both CCP1 and CCP6, suggesting a residual presence of flox alleles, which could come from untransduced cells.

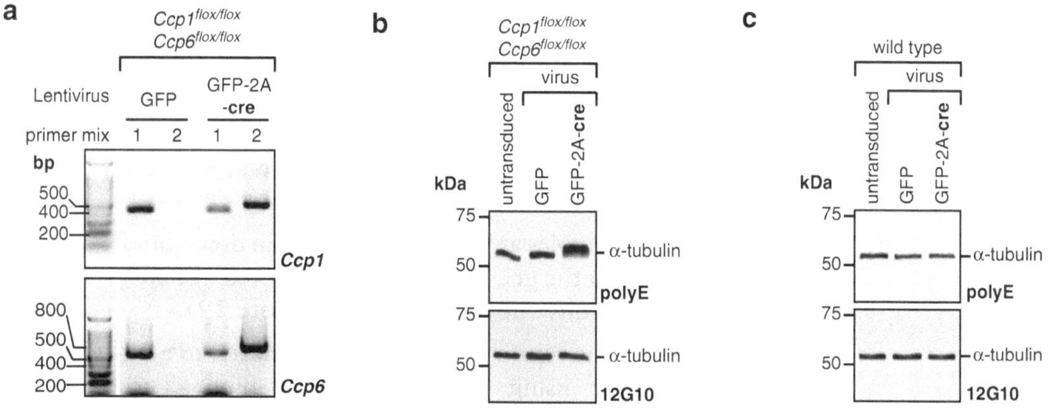

Fig. 4 Analysis of efficiency of the Cre-mediated recombination in primary hippocampal neurons. (**a**) An agarose gel showing the genotypes of *Ccp1* $^{flox/flox}$ *Ccp6* $^{flox/flox}$ neurons transduced with GFP- and GFP-2A-Cre viruses. Note that in the presence of GFP alone, only flox alleles are detected, while with GFP and Cre recombinase the major band comes from the KO alleles. Note the presence of a faint band (corresponding to the flox alleles) in the presence of the Cre recombinase, which could come from the incomplete transduction of neuronal population. (**b**) Immunoblot from *Ccp1* $^{flox/flox}$ *Ccp6* $^{flox/flox}$ neurons after transduction with GFP- or GFP-2A-Cre viruses. In the cells transduced with GFP-2A-Cre virus, knockout of *Ccp1* and *Ccp6* results in a strong increase of tubulin polyglutamylation levels (polyE antibody). Equal loading is verified using the total α-tubulin antibody (12G10). (**c**) Immunoblot showing that the Cre recombinase expressed in wild-type neurons has no effect on tubulin polyglutamylation levels (polyE antibody). Equal loading is verified using the total α-tubulin antibody (12G10)

3.5 SDS-Polyacrylamide Gel Electrophoresis and Immunoblotting

Prior to the analysis of the transduced neurons, collect the samples to later verify the effect of the depletion of CCP1 and CCP6 on the levels of tubulin polyglutamylation in transduced neurons by immunoblot (Fig. 4b).

1. Add 200 μL of the Laemmli sample buffer to each well of 6-well dish and lyse the neurons using a p200 micropipette (*see* **Note 28**).
2. Boil the samples at 95 °C for 5 min and spin down briefly.
3. Set up the gel-casting apparatus according to the manufacturer's description. For one 1-mm-thick 10% acrylamide gel, mix 1.25 mL of acrylamide/bis solution, 1.25 mL of 4× resolving buffer, and 2.5 mL of distilled water in a 15-mL tube. Add 40 μL of 10% APS and 5 μL of TEMED, mix gently by inverting the tube, and immediately pipette the solution into the preassembled casting apparatus. Add a small layer of distilled water (~300 μL) on the gel solution and leave it undisturbed for at least 30 min.
4. Remove the overlaid water layer by inverting the gel and drain off the remaining water droplets with a tissue paper.
5. Prepare the stacking gel: For one 1-mm-thick resolving gel, mix 175 μL of acrylamide/bis solution, 406 μL of 4× stacking buffer, and 1044 μL of distilled water. Add 10 μL of 10% APS and 4 μL of TEMED, mix gently by inverting the tube, and immediately pipette the solution onto the polymerized resolving gel. Carefully insert a comb into the stacking gel and allow it to polymerize for 30 min.
6. Assemble the electrophoresis tank and fill the chamber with the gel-running buffer. Remove the comb and wash the wells with the buffer.
7. Boil the samples at 95 °C for 5 min, spin down briefly, load the samples on the gel, and separate the proteins at 200 V until the dye front runs out from the gel (*see* **Note 29**).
8. Transfer the separated proteins from the gel to a nitrocellulose membrane using the Biorad Trans-Blot® Turbo system, according to the manufacturer's instructions.
9. Block the membrane in TBST containing 5% skimmed milk for 1 h, with gentle shaking, at room temperature.
10. Incubate the membrane with primary antibodies (12G10, PolyE or anti-GFP) diluted in TBST containing 2.5% skimmed milk for 1 h, with gentle shaking, at room temperature.
11. Wash the membrane 3 × 5 min in TBST with vigorous shaking.
12. Incubate the membrane with secondary, HRP-conjugated antibodies (anti-mouse for 12G10; anti-rabbit for poly E and

anti-GFP) diluted in TBST buffer for 1 h, with gentle shaking, at room temperature.

13. Wash the membrane 3 × 5 min in TBST with vigorous shaking.
14. Reveal the chemiluminescence signal on the membrane by incubating it with the ECL solution according to the manufacturer's instructions and using the available equipment (either chemical or digital imager).

4 Notes

1. The primer sequence written in lower case shares homology with the ends of the target vector, pTRIP-CMV-GFP-2A.
2. Always use cells which have been in culture for no more than 5–8 passages, as using cells after multiple passages may reduce the viral titer. It is very important to ensure that X-lenti 293 T cells are mycoplasma free, as mycoplasma contamination can affect the virus production efficiency.
3. We have used other transfection reagents such as jetPEI for virus production and found that the virus titer produced is lower than when using TransIT-293.
4. We used pTRIP vector with CAG and CMV promoters for the experiments. In our experience, there is no difference in their expression levels in neurons at young age. However, if transduction is to be performed at a differentiated stage, we suggest to use vectors with CAG promoter, as it expressed better than CMV promoter in mature neurons.
5. The choice of medium used for virus production depends on the medium preferred by the cells you intend to transduce. In this case, we used neurobasal medium as our target cells were neurons.
6. If there is a need to separate alpha- and beta-tubulins on the SDS-PAGE, we recommend using a different gel composition [27].
7. As Tween 20 is highly viscous, add it to the TBS solution using a cut p1000 pipette tip and rinsing the tip by repeated pipetting.
8. It is crucial to run the gel for a sufficiently long time, as it will aid in separating the cut from the uncut vector, reducing the contamination with uncut vector during purification of digested vector.
9. Extension time needs to be adjusted based on the length of the amplicon and speed of the polymerase. In our case, we consider that polymerase requires ~1 min to amplify ~1 kb of DNA template.

10. The timing of the reaction mix is to be strictly followed. Leaving the reaction mix for less or more time will reduce the efficiency of SLIC.
11. Usually, when the SLIC reaction is efficient, we get ~80 colonies in the positive plate and ~5 colonies in negative control plate. If the SLIC efficiency is poor, you will get less colonies in the positive plate. If the vector is not properly digested, you will get many colonies in the negative control plate.
12. NheI is selected when the reporter gene is to be C-terminally GFP tagged. Use BsrGI enzyme when the reporter gene is to be N-terminally GFP tagged.
13. Rules and regulations regarding working with lentiviral particles are specific to each country. Please enquire about and follow the rules that apply in your country.
14. Always add the diluted TransIT-293 into DNA mix, and not the other way around.
15. Usually we add 3 mL of virus collection medium per well of a 6-well plate. It is possible to obtain a more concentrated virus by adding a lower volume of medium. The minimal volume possible is 1.5 mL.
16. Avoid exceeding this time, as it might reduce the competence of produced lentivirus.
17. Do not use 0.2-μm filter, as it might damage the lentivirus.
18. As viruses are very sensitive to mechanical stress, do not accelerate thawing by pipetting or shaking.
19. As a general rule, transduction of primary cell cultures is difficult compared to cell lines. Therefore, the intensity of GFP fluorescence in U-2 OS cells and neurons might be different. If the GFP intensities are very different, then we suggest to use wild-type neurons for testing and determining the amount of virus to be added.
20. For cleaning coverslips, place a maximum of 200 coverslips in a 100-mL glass bottle. Wash them by shaking for at least 30 min in 3 M HCl, 2 × 10 min in distilled water, 3 × 10 min in 95% EtOH, and 2 × 10 min in distilled water. Let the coverslips dry under the hood or in an oven.
21. Proper coating of the surface with poly-D-lysine is crucial for attachment and development of the neurons. If the dishes are not coated properly, it will cause the neurons to clump or not adhere well.
22. Success of primary neuron culture depends on the time of dissection (from sacrificing the dam to collecting all the hippocampi) and should take no longer than 90 min.

23. Be careful as the hippocampi tend to be aspirated by the suction created by pipetting. If it is difficult to remove the trypsin without the hippocampi, proceed immediately to washes. The tendency of hippocampi to be aspirated with the pipette suction reduces after washes with plating medium.
24. Do not exceed the number of trituration passages through the Pasteur pipettes, even if pieces of tissue are still visible at the end of the procedure.
25. Some labs use 40 μm cell strainers to pass the cell suspension. In our hands, 70-μm cell strainers worked best without affecting the survival of neurons.
26. For live-cell imaging and immunostaining, it is important to plate neurons at a low density. Neurons seeded at high density will have overlapping axons and dendrites, making it difficult to use them for analysis.
27. Usually, 10 μL of PCR product is sufficient to visualize the amplification. If there is an unexpected problem with the gel, make a fresh gel and run another 10 μL from the remaining PCR product.
28. Depending on the day (DIV) of lysate collection, the sample might need to be further diluted before immunoblotting in order to not overload the gel with protein.
29. When analyzing the differences in levels of tubulin modifications, we strongly recommend to load low amounts of protein on the gel [27].

Acknowledgments

This work was supported by the ANR-10-IDEX-0001-02, the LabEx CelTisPhyBio ANR-11-LBX-0038. CJ is supported by the Institut Curie, the French National Research Agency (ANR) awards ANR-12-BSV2-0007 and ANR-17-CE13-0021, the Institut National du Cancer (INCA) grant 2014-PL BIO-11-ICR-1, and the Fondation pour la Recherche Medicale (FRM) grant DEQ20170336756. MMM is supported by the EMBO short-term fellowship ASTF 148-2015 and by the Fondation Vaincre Alzheimer grant FR-16055p, and SB by the FRM grant FDT201805005465. We thank C. Alberti, E. Belloir, F. Bertrand, V. Dangles-Marie, I. Grandjean, C. Caspersen, H. Hermange, A. Thadal, G. Buhagiar, C. Serieyssol, S. Gadadhar, and M. Sittewelle (Institut Curie) for technical assistance. We are grateful to M.-N. Soler, C. Lovo, and L. Besse from the PICT-IBiSA@Orsay Imaging Facility of the Institut Curie supported by the ANR through the "Investment for the future" program (France-BioImaging, ANR-10-INSB-04), and to N. Manel

(Institut Curie, Paris) for material and advice for the lentivirus production. We would like to thank F. Del Bene, V. Marthiens (Institut Curie), and C. González-Billault (University of Chile, Santiago, Chile) for instructive discussions and advice.

References

1. Mandell JW, Banker GA (1995) The microtubule cytoskeleton and the development of neuronal polarity. Neurobiol Aging 16 (3):229–237; discussion 238
2. Brady ST, Morfini GA (2017) Regulation of motor proteins, axonal transport deficits and adult-onset neurodegenerative diseases. Neurobiol Dis 105:273–282. https://doi.org/10.1016/j.nbd.2017.04.010
3. Janke C (2014) The tubulin code: Molecular components, readout mechanisms, and functions. J Cell Biol 206(4):461–472. https://doi.org/10.1083/jcb.201406055
4. Janke C, Bulinski JC (2011) Post-translational regulation of the microtubule cytoskeleton: mechanisms and functions. Nat Rev Mol Cell Biol 12(12):773–786. https://doi.org/10.1038/nrm3227
5. Magiera MM, Singh P, Janke C (2018) SnapShot: functions of tubulin posttranslational modifications. Cell 173(6):1552–1552 e1551. https://doi.org/10.1016/j.cell.2018.05.032
6. Gadadhar S, Bodakuntla S, Natarajan K, Janke C (2017) The tubulin code at a glance. J Cell Sci 130(8):1347–1353. https://doi.org/10.1242/jcs.199471
7. Magiera MM, Bodakuntla S, Ziak J, Lacomme S, Marques Sousa P, Leboucher S, Hausrat TJ, Bosc C, Andrieux A, Kneussel M, Landry M, Calas A, Balastik M, Janke C (2018) Excessive tubulin polyglutamylation causes neurodegeneration and perturbs neuronal transport. EMBO J 37(23):e100440. https://doi.org/10.15252/embj.2018100440
8. Ikegami K, Mukai M, Tsuchida J-i, Heier RL, Macgregor GR, Setou M (2006) TTLL7 is a mammalian beta-tubulin polyglutamylase required for growth of MAP2-positive neurites. J Biol Chem 281(41):30707–30716
9. Kalebic N, Sorrentino S, Perlas E, Bolasco G, Martinez C, Heppenstall PA (2013) AlphaTAT1 is the major alpha-tubulin acetyltransferase in mice. Nat Commun 4:1962. https://doi.org/10.1038/ncomms2962
10. Rogowski K, van Dijk J, Magiera MM, Bosc C, Deloulme J-C, Bosson A, Peris L, Gold ND, Lacroix B, Bosch Grau M, Bec N, Larroque C, Desagher S, Holzer M, Andrieux A, Moutin M-J, Janke C (2010) A family of protein-deglutamylating enzymes associated with neurodegeneration. Cell 143(4):564–578. https://doi.org/10.1016/j.cell.2010.10.014
11. Erck C, Peris L, Andrieux A, Meissirel C, Gruber AD, Vernet M, Schweitzer A, Saoudi Y, Pointu H, Bosc C, Salin PA, Job D, Wehland J (2005) A vital role of tubulin-tyrosine-ligase for neuronal organization. Proc Natl Acad Sci U S A 102(22):7853–7858
12. Rocha C, Papon L, Cacheux W, Marques Sousa P, Lascano V, Tort O, Giordano T, Vacher S, Lemmers B, Mariani P, Meseure D, Medema JP, Bièche I, Hahne M, Janke C (2014) Tubulin glycylases are required for primary cilia, control of cell proliferation and tumor development in colon. EMBO J 33 (19):2247–2260. https://doi.org/10.15252/embj.201488466
13. Bosch Grau M, Masson C, Gadadhar S, Rocha C, Tort O, Marques Sousa P, Vacher S, Bieche I, Janke C (2017) Alterations in the balance of tubulin glycylation and glutamylation in photoreceptors leads to retinal degeneration. J Cell Sci 130:938–949. https://doi.org/10.1242/jcs.199091
14. Gadadhar S, Dadi H, Bodakuntla S, Schnitzler A, Bieche I, Rusconi F, Janke C (2017) Tubulin glycylation controls primary cilia length. J Cell Biol 216(9):2701–2713. https://doi.org/10.1083/jcb.201612050
15. Giordano T, Gadadhar S, Bodakuntla S, Straub J, Leboucher S, Martinez G, Chemlali W, Bosc C, Andrieux A, Bieche I, Arnoult C, Geimer S, Janke C (2019) Loss of the deglutamylase CCP5 perturbs multiple steps of spermatogenesis and leads to male infertility. J Cell Sci 132(3). https://doi.org/10.1242/jcs.226951
16. Silva CG, Peyre E, Adhikari MH, Tielens S, Tanco S, Van Damme P, Magno L, Krusy N, Agirman G, Magiera MM, Kessaris N, Malgrange B, Andrieux A, Janke C, Nguyen L (2018) Cell-intrinsic control of interneuron migration drives cortical morphogenesis. Cell 172(5):1063–1078. https://doi.org/10.1016/j.cell.2018.01.031
17. Gilmore-Hall S, Kuo J, Ward JM, Zahra R, Morrison RS, Perkins G, La Spada AR (2019)

CCP1 promotes mitochondrial fusion and motility to prevent Purkinje cell neuron loss in pcd mice. J Cell Biol 218(1):206–219. https://doi.org/10.1083/jcb.201709028

18. Marcos S, Moreau J, Backer S, Job D, Andrieux A, Bloch-Gallego E (2009) Tubulin tyrosination is required for the proper organization and pathfinding of the growth cone. PLoS One 4(4):e5405
19. Akella JS, Wloga D, Kim J, Starostina NG, Lyons-Abbott S, Morrissette NS, Dougan ST, Kipreos ET, Gaertig J (2010) MEC-17 is an alpha-tubulin acetyltransferase. Nature 467 (7312):218–222. https://doi.org/10.1038/nature09324
20. Shida T, Cueva JG, Xu Z, Goodman MB, Nachury MV (2010) The major alpha-tubulin K40 acetyltransferase alphaTAT1 promotes rapid ciliogenesis and efficient mechanosensation. Proc Natl Acad Sci U S A 107 (50):21517–21522. https://doi.org/10.1073/pnas.1013728107
21. Hubbert C, Guardiola A, Shao R, Kawaguchi Y, Ito A, Nixon A, Yoshida M, Wang X-F, Yao T-P (2002) HDAC6 is a microtubule-associated deacetylase. Nature 417(6887):455–458
22. Janke C, Rogowski K, Wloga D, Regnard C, Kajava AV, Strub J-M, Temurak N, van Dijk J, Boucher D, van Dorsselaer A, Suryavanshi S, Gaertig J, Eddé B (2005) Tubulin polyglutamylase enzymes are members of the TTL domain protein family. Science 308 (5729):1758–1762. https://doi.org/10.1126/science.1113010
23. van Dijk J, Rogowski K, Miro J, Lacroix B, Eddé B, Janke C (2007) A targeted multienzyme mechanism for selective microtubule polyglutamylation. Mol Cell 26(3):437–448. https://doi.org/10.1016/j.molcel.2007.04.012
24. Tort O, Tanco S, Rocha C, Bieche I, Seixas C, Bosc C, Andrieux A, Moutin M-J, Xavier Aviles F, Lorenzo J, Janke C (2014) The cytosolic carboxypeptidases CCP2 and CCP3 catalyze posttranslational removal of acidic amino acids. Mol Biol Cell 25(19):3017–3027. https://doi.org/10.1091/mbc.E14-06-1072
25. Jeong J-Y, Yim H-S, Ryu J-Y, Lee HS, Lee J-H, Seen D-S, Kang SG (2012) One-step sequence- and ligation-independent cloning as a rapid and versatile cloning method for functional genomics studies. Appl Environ Microbiol 78(15):5440–5443. https://doi.org/10.1128/AEM.00844-12
26. Kim JH, Lee S-R, Li L-H, Park H-J, Park J-H, Lee KY, Kim M-K, Shin BA, Choi S-Y (2011) High cleavage efficiency of a 2A peptide derived from porcine teschovirus-1 in human cell lines, zebrafish and mice. PLoS One 6(4): e18556. https://doi.org/10.1371/journal.pone.0018556PONE-D-11-01024
27. Magiera MM, Janke C (2013) Investigating tubulin posttranslational modifications with specific antibodies. In: Correia JJ, Wilson L (eds) Methods cell biol, Microtubules, in vitro, vol 115. Academic Press, Burlington, pp 247–267. https://doi.org/10.1016/B978-0-12-407757-7.00016-5

Chapter 20

Measuring the Impact of Tubulin Posttranslational Modifications on Axonal Transport

Satish Bodakuntla, Maria M. Magiera, and Carsten Janke

Abstract

Axonal transport is a process essential for neuronal function and survival that takes place on the cellular highways—the microtubules. It requires three major components: the microtubules that serve as tracks for the transport, the motor proteins that drive the movement, and the transported cargoes with their adaptor proteins. Axonal transport could be controlled by tubulin posttranslational modifications, which by decorating specific microtubule tracks could determine the specificity of cargo delivery inside neurons. However, it appears that the effects of tubulin modifications on transport can be rather subtle, and might thus be easily overlooked depending on which parameter of the transport process is analyzed. Here we propose an analysis paradigm that allows detecting rather subtle alterations in neuronal transport, as induced for instance by accumulation of posttranslational polyglutamylation. Analyzing mitochondria movements in axons, we found that neither the average speed nor the distance traveled were affected by hyperglutamylation, but we detected an about 50% reduction of the overall motility, suggesting that polyglutamylation controls the efficiency of mitochondria transport in axons. Our protocol can readily be expanded to the analysis of the impact of other tubulin modifications on the transport of a range of different neuronal cargoes.

Key words Axonal transport, Mitochondria, Cargo, Microtubules, Posttranslational modifications of tubulin, Polyglutamylation

1 Introduction

Intracellular transport is an essential process for all eukaryotic cells, as it allows to deliver organelles, mRNA, and proteins to their predestined locations in different cellular compartments [1]. The complex morphology of neurons (very long and ramified axons and many dendrites) is particularly challenging for the intracellular transport machinery, which has to work efficiently and adapt to varying physiological requirements throughout the entire life span of an organism. In particular axonal transport, which can cover long distances, needs to be stringently controlled [2]. Defects in axonal transport have been linked to many neurodegenerative disorders, and are suspected to be one of the early molecular defects in many

Helder Maiato (ed.), *Cytoskeleton Dynamics: Methods and Protocols*, Methods in Molecular Biology, vol. 2101, https://doi.org/10.1007/978-1-0716-0219-5_20,

of them [3–5]. One of the key mechanisms to control microtubule properties, and their interactions with molecular transport machineries, are posttranslational modifications (PTMs) of tubulins—the building blocks of microtubules [6].

Neuronal microtubules are highly enriched with a variety of tubulin PTMs such as acetylation, detyrosination, and polyglutamylation [7]. Polyglutamylation is a particularly complex PTM, as it can generate branch chains of variable numbers of glutamate residues selectively on α- or β-tubulin. Thus, polyglutamylation can potentially encode a large variety of information into one and the same microtubule [6]. The signals generated by polyglutamylation are determined by the specificity of the modifying [8, 9] and demodifying enzymes [10, 11], which in turn can be regulated by cell-type-specific expression, subcellular localization, as well as adapting the cellular response to external cues [12]. While little is so far known on how polyglutamylation is regulated by signaling pathways, it has been shown that the major polyglutamylases in neurons are TTLL1 and TTLL7 [8, 13], and the key deglutamylases are CCP1 and CCP6 [10, 14].

Deregulation of tubulin PTMs in cells can result in microtubule dysfunctions and eventually in disease conditions [15]. Indeed, analyses of several mouse models have demonstrated that aberrations in the levels of tubulin polyglutamylation are linked to cilia-related disorders, such as respiratory problems [16], defects in spermatogenesis [17–19] and sperm motility [20], retinal degeneration [21], or neurodegeneration [10, 14, 22, 23].

Mutation or loss of CCP1, one of the major neuronal deglutamylases, leads to unopposed, and thus accumulating polyglutamylation, which results in degeneration of several neuronal populations in the brain, the most emblematic being Purkinje cells [10, 14, 18]. Strikingly in these mice, some parts of the brain do not degenerate, nor show signs of increased polyglutamylation. We showed that this was due to the compensation by a second deglutamylase, Ccp6, which was selectively expressed in cerebral cortex and hippocampus, thus protecting these brain regions [14]. These findings were the first to indicate that cell-specific expression of enzymes involved in polyglutamylation can result in cell-specific functions, or defects.

Despite first insights in a widespread implication of tubulin polyglutamylation in pathological conditions [15], underlying molecular mechanisms remained largely elusive, primarily because of a number of technical limitations. These limitations are exemplified when studying sensitive biological processes such as axonal transport. Although there are reports linking transport changes to polyglutamylation [24–26], the direct role of polyglutamylation in controlling transport in neurons has only been recently addressed on primary neurons of the now available knockout mouse models. One study measured axonal transport in cerebellar granule cells from *ccp1*$^{-/-}$ mice, as the cerebellum is the most affected brain region in this mouse model [22]. In order to determine the impact of increased polyglutamylation in hippocampal neurons, we developed a specific approach to convert *Ccp1*$^{flox/flox}$*Ccp6*$^{flox/flox}$ neurons into *ccp1*$^{-/-}$*ccp6*$^{-/-}$ in the culture dish (*see* Chap. 19). Measuring axonal transport of mitochondria, we found clear aberrations; however, only one parameter—the overall motility of these organelles—was affected. This could have easily been overlooked when analyzing only the motile mitochondria, which is why we used a paradigm that included all mitochondria, even the nonmotile ones [14].

Here we describe our procedure for imaging axonal transport in neurons with altered tubulin PTMs, as well as our analysis paradigm to apprehend a range of different, potentially subtle changes in axonal transport. As an example, we show the impact of increased polyglutamylation on the mitochondrial transport. This method can be readily extended to study the transport of other cellular cargoes, and different tubulin PTMs.

One way to study axonal transport is to follow selected, single particles throughout the imaging period, and determine their complex movement behavior (e.g., overall displacement, frequency of changes in direction, mean anterograde and retrograde speed). However, some cargoes, such as mitochondria or endosomes, are very dense in axons, making it difficult to reliably track the entire path of every moving particle throughout the imaging period. Analyzing only the "trackable" particles, however, can introduce an unwanted bias in the study.

To avoid this, we based our analyses paradigm on previously published work [27, 28], analyzing individual "runs," rather than tracing the movement of single particles throughout the imaging period. A "run" is defined as an uninterrupted, unidirectional motility event at constant speed. Thus, a change in the speed of a moving particle marks the beginning of a new run (Fig. 1). While this analysis paradigm does not inform us on the complex movements of single particles (changes between anterograde and retrograde movements, final destination of a particle), it provides unbiased, statistically exploitable data for parameters that might be altered by tubulin PTMs for both anterograde and retrograde transport: the speed at which a transport cargo moves along differentially modified microtubules, or its persistence (how long a particle moves before stopping). By taking into account the total number of particles in the field of observation, we can further determine the overall motility within the particle population, thus also taking into account all the nonmotile particles in the imaged axon.

Using this analysis paradigm, we investigated the impact of abnormally increased polyglutamylation on axonal transport in hippocampal neurons at DIV4 (4 days in vitro) (*see* Chap. 19). In neurons with hyperglutamylated microtubules, both average speed

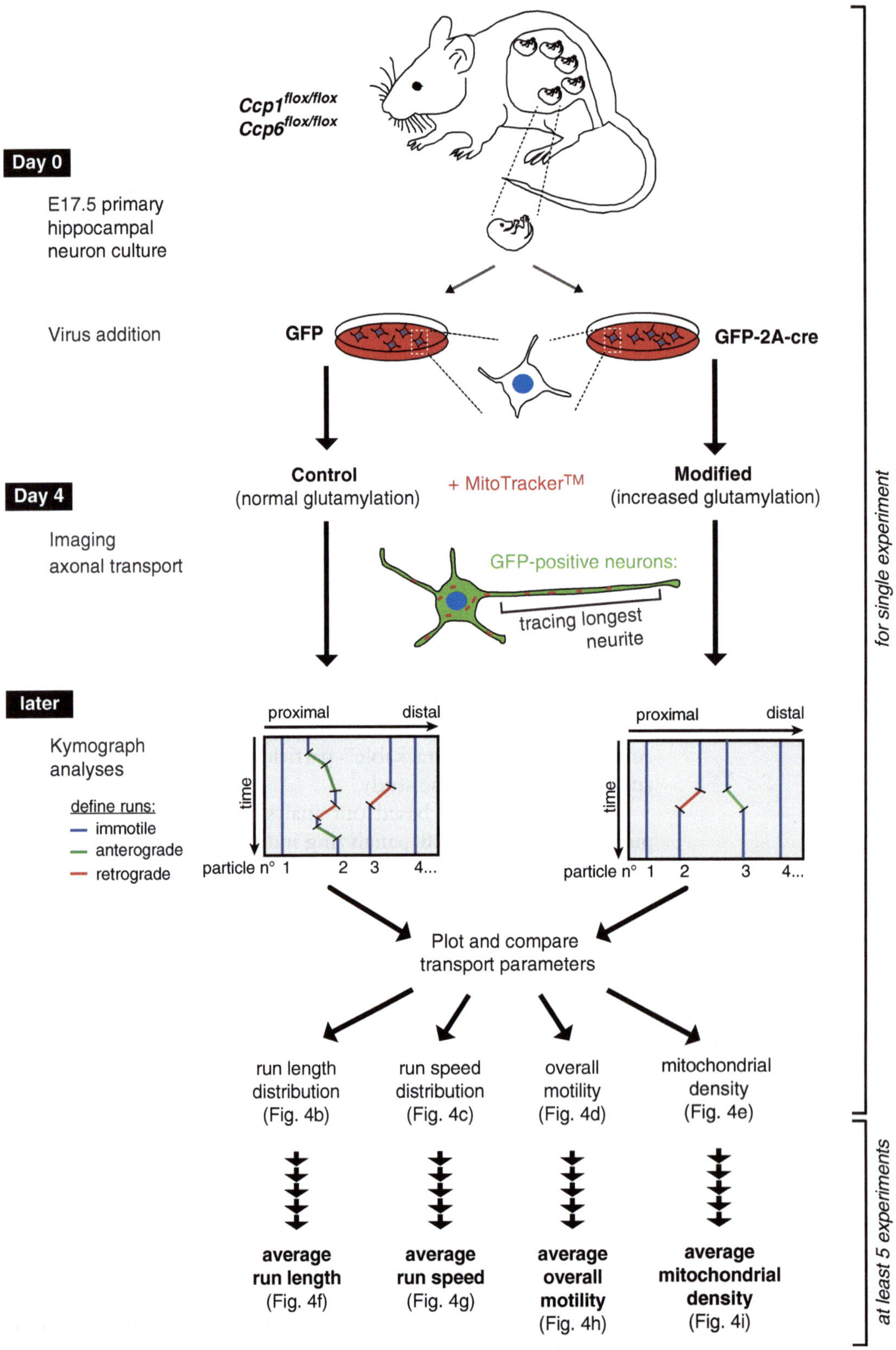

Fig. 1 Schematic representation of the successive steps of a single transport experiment. Primary *Ccp1*$^{flox/flox}$*Ccp6*$^{flox/flox}$ neurons were transduced with either GFP or GFP-2A-Cre lentivirus on DIV0 (0 days in vitro), and

and time of single runs were unaltered in anterograde and retrograde direction, while the overall motility of mitochondria was reduced by 50% relative to controls [14] (*see* Fig. 4). This effect could not have been discovered with an approach focusing on tracing only motile mitochondria. We therefore believe that our paradigm for axonal transport analysis can be used in a broader range of applications, especially when studying subtle changes such as those induced by tubulin PTMs [14].

2 Materials

2.1 Imaging of Axonal Transport

1. CO_2-testing equipment (Thermo Scientific).
2. Thermometer.
3. Primary mouse hippocampal neurons with normal and modified microtubules cultured on 35-mm glass-bottom dishes (e.g., Ibidi #81158).
4. B27 medium: Add 2 mL of 50× B27 solution (Thermo Fisher), 1 mL of 100× Glutamax, and 1 ml of 100× penicillin-streptomycin to 96 mL of plain neurobasal medium (Thermo Fisher).
5. Conditioned medium: B27 medium incubated with neurons from DIV0, which is not only equilibrated with the cell culture incubator's atmosphere but importantly with any signaling/growth molecules released by the neurons into the medium. *See* Chapter XXX for preparing conditioned medium.
6. Nikon Ti-E spinning disk inverted confocal laser microscope equipped with a 60× oil immersion objective (N.A. 1.40) and an ORCA-Flash 4.0 camera (Hamamatsu) controlled through MetaMorph software.
7. 1 mM MitoTracker™ Red CMXRos (Thermo Fisher): Dissolve 50 μg of MitoTracker™ in 94 μL of dimethyl sulfoxide (DMSO) (Sigma #D8418). Aliquot by 10 μL and store at −20 °C. Avoid repeated cycles of freeze-thaw.

Fig. 1 (continued) analyzed at DIV4. Mitochondria (stained by the MitoTracker® dye) movements were recorded for 1 min and their transport parameters were analyzed using the KymoToolBox plug-in. Run speed, run time, run length, and overall motility were extracted and compared between control (GFP) and increased polyglutamylation (GFP-2A-Cre) conditions. At least five experiments are necessary to calculate the average values of transport parameters obtained from individual experiments

2.2 Analysis of Axonal Transport

1. ImageJ 1.47v equipped with "nd Stack Builder" and "KymoToolBox" plug-ins.
2. Microsoft Excel and GraphPad Prism or equivalent software to plot and statistically test the obtained data.

3 Methods

In the current protocol, we describe general criteria to follow during the imaging of axonal transport and the analysis paradigm we developed to characterize the transport in neurons in which tubulin PTMs are modulated (Figs. 1 and 2). Using this method, we have analyzed the mitochondria transport in neurons with normal and abnormally increased polyglutamylation [14]. To obtain primary neurons with increased polyglutamylation, we developed a cell-culture-based system, in which primary neurons are isolated from *Ccp1*$^{flox/flox}$*Ccp6*$^{flox/flox}$ mice, and subsequently transduced with either a GFP-encoding (control) or a GFP-2A-*Cre* recombinase-encoding lentivirus (to induce the knockout of the two deglutamylases and induce hyperglutamylation) (*see* Chap. 19).

We analyzed mitochondria transport using the commercially available dye, MitoTracker™. On DIV4, mitochondrial movements were recorded in the GFP-expressing neurons for 1 min, and subsequently converted to kymographs and analyzed using the "KymoToolBox" plug-in [29] for ImageJ [30]. When redrawing the particle movements in the kymographs, we separately assigned each uninterrupted movement (defined by a constant speed = slope in the kymograph) as one "run" (Figs. 1 and 3). Runs with speed below 0.10 μm/s were considered as immotile.

For each single experiment (Figs. 1 and 4a–e), we analyzed the distribution of run speed and run time for both anterograde and retrograde transport. Based on the total amount of particles in the field of observation, we then determined the overall motility of the particle population. Of all the parameters measured, only the overall motility of the mitochondria was strongly decreased in neurons with hyperglutamylation, as compared to control cells (Fig. 4) [14].

To exclude that the expression of Cre recombinase in the neurons affects the transport, we performed an equivalent set of experiments in wild-type neurons. We demonstrated that the expression of GFP-2A-*Cre* did not affect any of the previously measured parameters of mitochondrial transport, thus demonstrating that the decrease of motility in *Ccp1*$^{flox/flox}$*Ccp6*$^{flox/flox}$ neurons transduced with GFP-2A-*Cre* was indeed a result of the knockout of the two deglutamylases, and thus of the microtubule hyperglutamylation it induced [14].

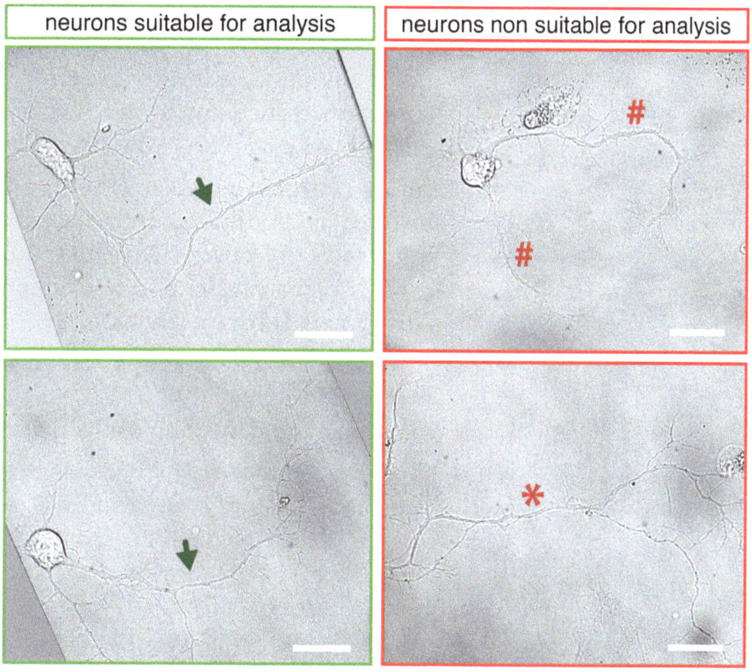

Fig. 2 Examples of neuron morphologies and their suitability for analysis. Only neurons with a distinct single axon (longest neurite emerging from the cell body) should be selected for analysis (green arrows). Neurons with ambiguous morphologies such as two axons (red #) or overlapping axons (red $*$) should not be considered for analyses

3.1 Imaging of Axonal Transport

1. Before starting the experiment use a thermometer and CO_2-testing device to manually verify the temperature and CO_2 levels of the microscope imaging chamber. If needed, calibrate the device such that the imaging stage is at 37 °C and 5% CO_2 (*see* **Note 1**).
2. On the day of imaging, remove the medium from the glass-bottom dishes and gently wash the transduced neurons with conditioned medium to remove any remains of the lentivirus from the dish. Pipette the medium gently, always against the edge of the dish and never directly on the neurons (*see* **Note 2**).
3. Thaw an aliquot of 1 mM MitoTracker™ solution (main stock). Prepare an intermediate 2 μM stock solution by adding 1 μL of the main stock to 500 μL of PBS.
4. Dilute the intermediate stock solution to a final concentration of 2 nM MitoTracker™ by adding 1 μL of 2 μM intermediate stock solution to 1 mL of conditioned medium.
5. Remove the neurobasal culture medium from the dish and quickly, yet gently, add 1 mL of conditioned medium containing 2 nM of MitoTracker™. Leave the dish in the incubator for 1 min (*see* **Note 3**).

6. Remove the medium containing MitoTracker™ and gently wash the neurons with 500 μL of conditioned medium to get rid of any unbound MitoTracker™ dye. Finally, add 1 mL of conditioned medium to the neurons and bring them to the microscope chamber for live imaging.
7. Image neurons immediately using a Nikon Ti-E spinning disk inverted confocal laser microscope equipped with a 60× oil immersion objective (N.A 1.40) and an ORCA-Flash4.0 camera (Hamamatsu), set at 2 × 2 binning operated by Leica MM AF imaging software (*see* **Note 4**).
8. After the imaging stage is set, focus on the neurons using bright-field light. DO NOT USE the laser to prevent photobleaching and damaging of the neurons.
9. Once the neurons are in focus, use bright field to find a suitable neuron and identify the axon by its typical characteristics—the longest neurite extending from the cell body. Dendrites at DIV4 are much shorter and thicker compared to axons (Fig. 2) (*see* **Note 5**).
10. If it is not possible to capture the cell body and axon in the image, write down the position of the cell body, and its distance from the actual imaging area.
11. Switch to laser fluorescence and verify that the chosen neuron is GFP positive. Change the excitation wavelength (in the case of MitoTracker™ Red CMXRos) and bring the transport particles in focus (*see* **Note 6**).
12. Set the gain or intensity of laser to a minimal value at which the transport particles are still clearly visible. Keep the laser settings constant for imaging all neurons in a single experiment (*see* **Note** 7).
13. Record the particle movements in the axon in streaming mode every 300 ms for a total of 1 min (*see* **Note 8**).
14. Acquire differential interference contrast (DIC) and GFP fluorescence image of the axon used for imaging transport to assess the general appearance of the neuron, which can eventually give some indications on the health of the neuron.
15. For the settings we use, acquisition of each neuron will generate one DIC image of the neuron, one image of the GFP expression in the cell, 201 time-lapse images corresponding to 201 time frames of the movie, and one ".nd" file.
16. For each dish, image a minimum of five neurons; however we prefer to image around ten neurons if possible. This is important as the dynamics of axonal transport are inherently heterogenous among individual neurons in culture, and can furthermore be affected by small variations in the experimental conditions.

17. Image each dish for no more than 30 min, as constant laser exposure is deleterious to the neurons.

3.2 Analysis of Axonal Transport

We recommend creating a separate folder for each analyzed neuron and saving there all the files generated during the different steps of analysis. In this example, we assume that the name of the cell and the folder is "Name."

1. Open the "Name.nd" MetaMorph file in the ImageJ software by drag-dropping the file on the ImageJ icon or toolbar to visualize the tiff images of the corresponding movie (Fig. 3a). In the "Bio-formats Import Options" window, select view stack with "hyperstack" and uncheck all other options (*see* **Note 9**).
2. Calibrate the hyperstack by using "Image > Properties" menu in ImageJ. Unit of length for defining pixel size is μm and for the frame interval time is seconds. If ImageJ automatically calibrates the hyperstack, confirm that the values are correct (*see* **Note 10**).
3. Save the file as "Name_stack.tiff" in its dedicated folder (Fig. 3a). Once the stack file is generated and saved, the individual files corresponding to all the time points of the movie (Name-t0.tiff–Name-t201.tiff) and the corresponding "Name.nd" file can be deleted.
4. Create a maximum intensity projection of the tiff stack file by using "Image > Stacks > Z Project." Subsequently, select "projection type" as maximum, and start slice and stop slice as 1 and 201, respectively (Fig. 3b).
5. Select the "segmented line" or "freehand" line tool and manually identify the axon, in which the cargo movement will be quantified (Fig. 3c). If needed, adjust the intensity of the maximum projection image beforehand (by using "Image > Adjust > Brightness/Contrast") to better visualize the axon. As a general rule, always trace the axon from the cell body toward the axonal terminal (proximal–distal) (*see* **Note 11**).
6. Add the traced path of the axon to "ROI manager" by pressing "t." Save the ROI as "Name_axon.roi" by using "More » save" option.
7. Close the maximum intensity projection image and apply the ROI to the "Name_stack.tiff" file.
8. Use the "KymoToolBox" by selecting "Plugins > KymoToolBox > Draw Kymo" menu to generate the kymograph and the kymostack of the cargo movements in the selected axon. Define the "Width" as 15 pixels (for the width of identified axon in **step 5**) and select "Get kymo" and "Get kymoStack" (Fig. 3d).

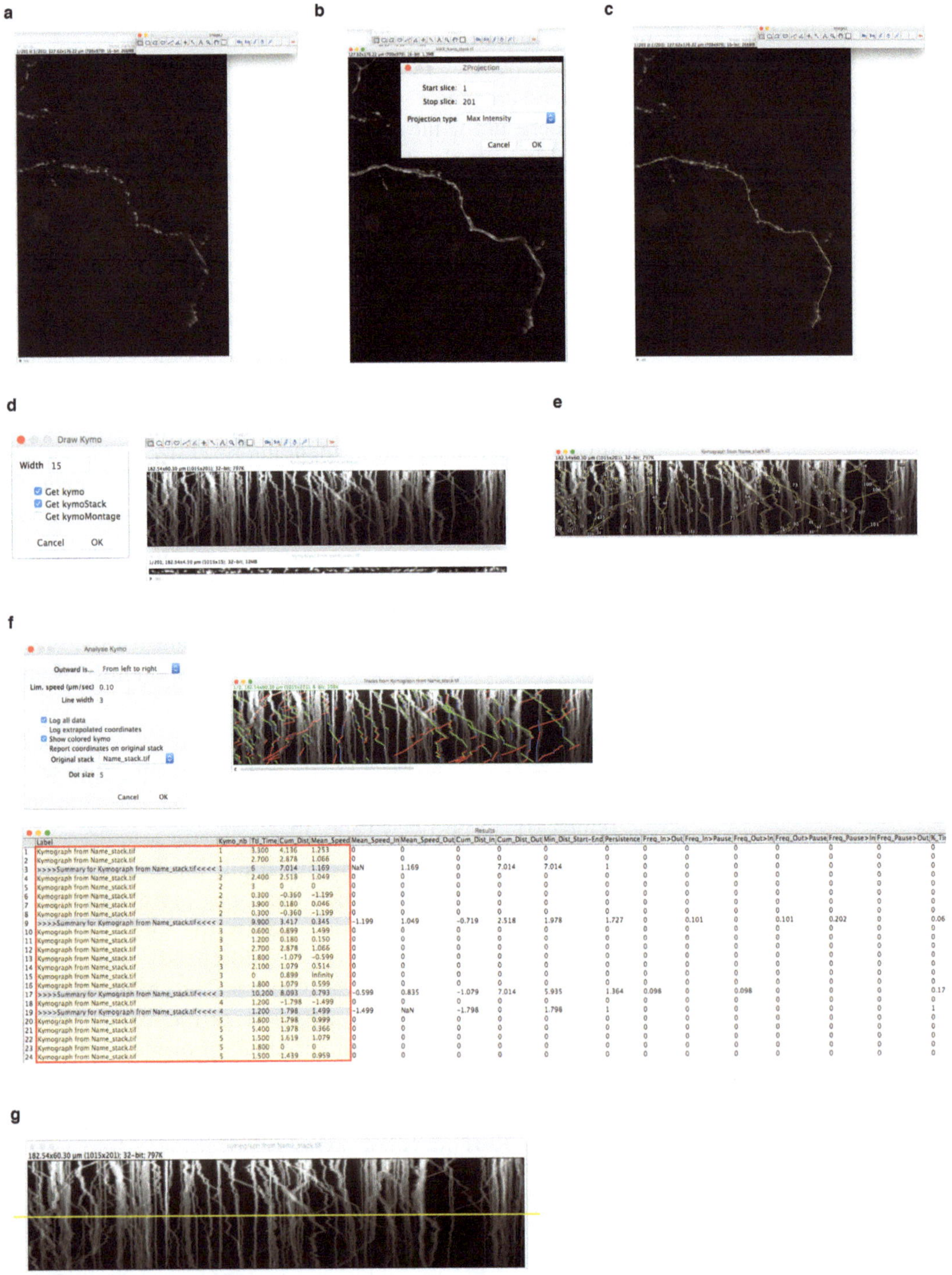

Fig. 3 Flowchart of the analysis paradigm to investigate the impact of differential polyglutamylation on axonal transport. (**a**) Hyperstack created from individual frames of the movie. (**b**) The maximum intensity projection of the mitochondria movements throughout the imaging period. (**c**) Hand-drawn line identifying the longest neurite (the axon) in which mitochondria transport will be analyzed. (**d**) Kymograph and kymostack of imaged

The kymograph represents particle movement in time vs. space, while the kymostack generates a straightened movie of the identified axon, which can be later helpful in the identification of particle's movement. Refer to "https://github.com/fabricecordelieres/IJ_KymoToolBox" for detailed description of the plug-in (*see* **Note 12**).

9. Save the generated kymograph and kymostack as "Kymograph from Name_stack.tif" and "Kymostack from Name_stack.tif," respectively, in the dedicated folder.
10. On the kymograph, trace all possible individual runs (Fig. 3e) of the particle trajectories. If necessary use the kymostacks to help identifying the trajectories. Add each tracing to the "ROI manager" by pressing "t." To keep track of the traced trajectories, select "Show All" on "ROI manager." Save all the ROIs as "Name_stack_kymo.zip" file in the dedicated folder (*see* **Note 13**).
11. Select "Plugins > KymoToolBox > Analyse Kymo" to retrieve the parameters of each run. Select "Outward is" as "From left to right" and "Line width" as 3 (line width for representation of color-coded runs). Set "Lim. speed" as 0.10 μm/s—runs with speed lower than 0.10 μm/s are considered immotile. Select "Log all data" and "Show colored kymo" to generate a table with the values of movement parameters of trajectories traced in **step 10** and a color-coded kymograph (Fig. 3f). As this analysis paradigm focuses on individual runs, use the data highlighted in yellow, which correspond to individual runs with constant speed and direction, and discard the data highlighted in gray, which correspond to summary of several runs. For the same reason, we only use run length (Cum_Dist), run speed (Mean_Speed), and run duration (Ttl_Time) parameters for further analysis (Fig. 3f, **red box**). Immotile trajectories are shown as vertical lines in blue while motile trajectories (runs) were shown as diagonal lines (green for anterograde and red for retrograde).

Fig. 3 (continued) mitochondria created by the KymoToolBox plug-in. (**e**) Hand-drawn trajectories of mitochondria runs in the kymographs. (**f**) Color-coded trajectories identified from (**e**) (blue: immotile, green: anterograde, red: retrograde). 0.10 μm/s threshold is applied for defining immotile particles. File with values of transport parameters of all identified trajectories (table format—can be saved as Excel file). For further analysis, the "Summary" lines highlighted in gray should be discarded, as they refer to trajectories containing several runs. Lines highlighted in yellow give the parameters of single runs and should be used for further analyses. The red box highlights the parameters used for further analysis. (**g**) A horizontal line (yellow) is drawn in the middle of the kymograph and the movement trajectories crossing this line were counted as the number of mitochondria in the analyzed axon

12. Save the color-coded kymograph as "Tracks from kymograph from Name_stack.tiff" and the results table as "Results_Name.xls" in the dedicated folder.
13. Use the "Kymograph from Name_stack.tif" to determine the number of particles and the length of the axon (in μm) used for analysis (Fig. 3g) (*see* **Note 14**).
14. To calculate one of the parameters to be analyzed, the particle density (mitochondria, vesicles) in the axons, use the following equation:

$$\text{Density of particles } (/100\ \mu\text{m of axon}) = \frac{\text{Number of particles} \times 100}{\text{Length of analyzed axon } (\mu\text{m})}$$

15. Repeat **steps 1–14** for all imaged neurons of a single experiment (one culture of hippocampal neurons from one $Ccp1^{flox/flox}Ccp6^{flox/flox}$ mouse). Pool data from all neurons of a single experiment (separately for control and hyperglutamylated neurons) (Figs. 1 and 3).
16. Use Prism or equivalent software to plot scatter dot plots for obtained parameters (run speed, run length, and run duration) of all the runs and compare the medians of the respective distributions between control and hyperglutamylated neurons for anterograde and retrograde movements (Figs. 2 and 4b–e). Exclude displacements of less than 0.9 μm (average size of a mitochondrion, corresponding to 5 pixels in our system) (*see* **Note 15**).
17. To measure the overall motility of the particles, determine the time all imaged particles in one axon spend in movement within the imaging period using the following equation:

$$\text{Overall motility } (\%) = \frac{\text{Sum of all run durations in the kymograph (s)} \times 100}{\text{Total number of particles} \times \text{Imaging time (s)}}$$

18. Repeat the experiments (starting with cell cultures from independent mice) and analyses at least five times (Fig. 1).
19. Finally, calculate the "average run length" and "average run speed" by calculating the mean of the medians (obtained in **step 15**) of all experiments (Figs. 1 and 4b–e). Calculate the *p*-values using an unpaired *t*-test, or another appropriate statistical method, to determine to which extend the differences between control and knockout conditions are statistically significant.
20. Calculate the "average overall motility" and "average vesicle density" from all the experiments by calculating the mean of

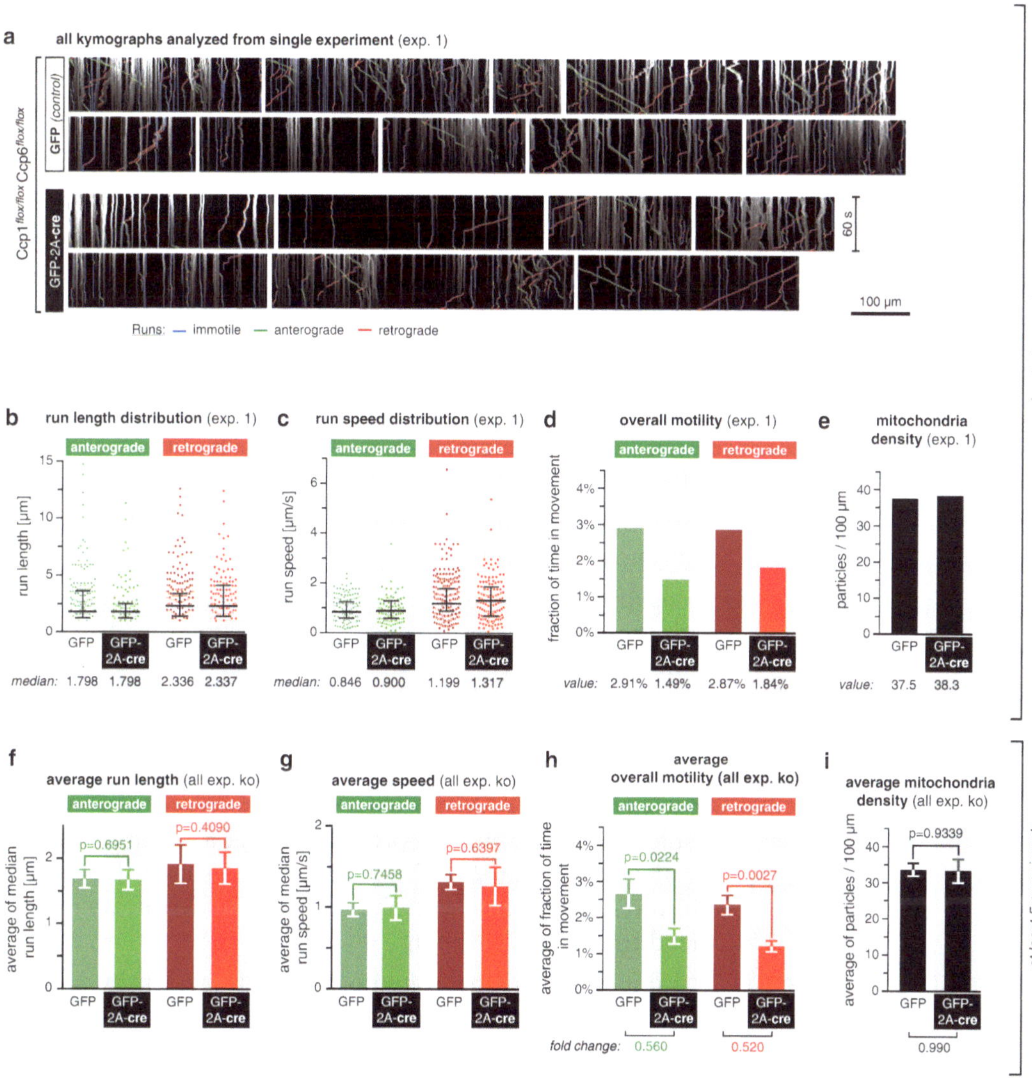

Fig. 4 A representative example of data generated from a mitochondria transport experiments from ref. 14. (**a**) All color-coded kymographs of mitochondria movements in neurons with normal (GFP) and increased polyglutamylation (GFP-2A-Cre) of a single experiment. Runs are defined as immotile (blue), anterograde (green), or retrograde (red) lines. (**b**) A scatter dot plot of run length distributions in different experimental conditions. The median values are shown below the graphs. (**c**) A scatter dot plot of run speed distributions in different experimental conditions. The median values are shown below the graphs. (**d**) Bar graphs of overall motilities of mitochondria in different experimental conditions. It represents the fraction of time the organelles spent in movement throughout the imaging time. (**e**) Bar graphs of mitochondria densities in analyzed neurons. (**f–i**) Bar graphs of mean ± SEM values from seven independent experiments (one experiment: **a–e**), showing run lengths (**f**), run speeds (**g**), overall motility (**h**), and mitochondria densities (**i**). The fold changes between control and hyperglutamylation conditions are shown below the graphs. Statistical significance was tested using unpaired *t*-test. Green: anterograde movement, red: retrograde movement

the respective values (obtained in **step 17**) from all the experiments (Figs. 1 and 4f–i). Calculate the *p*-values using an unpaired *t*-test, or another appropriate statistical method, to determine to which extend the differences between control and knockout conditions are statistically significant.

4 Notes

1. It is crucial to control the atmosphere of the imaging chamber as it might affect the axonal transport. If CO_2-testing equipment is not available, observe the color of the cell culture medium. At sufficient CO_2 levels, it should be between orange and red color, whereas too low CO_2 levels lead to a rather pink color.
2. This step of washing the neurons can be skipped if they are not transduced with lentivirus.
3. Using higher concentrations or incubating for longer times with the dye can be deleterious to mitochondria. Thus, perform test experiments to determine a suitable concentration of dye at which a good signal is observed without affecting the dynamics of mitochondria.
4. We prefer to use the $\times 60$ over the $\times 100$ objective, because it allows to image a larger field without compromising on the resolution. As live imaging produces a huge amount of data, we set the binning at 2×2 to reduce the size of the images. However, if the camera resolution is not sufficient, we recommend to use 1×1 binning.
5. It is important that neurons are seeded at low density to avoid that axons from different neurons grow on each other, which is what they like to do. *See* Chap. 19 for details about neuron culture and seeding for transport experiments. As a general rule, avoid imaging neurons with ambiguous morphology (Fig. 2).
6. The focus planes of bright field and fluorescence are often different; therefore always check that the vesicles/mitochondria are in focus before acquiring.
7. The signal from the vesicles reduces during acquisition, especially when imaging at higher frame rates. Thus, when adjusting the laser settings ensure that it is strong enough to allow detecting the cargoes throughout the imaging time. If needed, use a spare dish of neurons to adjust laser settings before starting the actual measurements.
8. The rate of acquisition depends primarily on the transport cargo. For mitochondria, lysosomes, and endosomes that undergo fast axonal transport, it is suggested to image at least

three frames per second. For such high-speed imaging, we recommend to use the "stream mode" or equivalent settings in which the captured images are temporarily stored in the random access memory (RAM) of the computer. Only after the entire imaging is completed (1 min in our case), the images will be saved on the local hard disk. This avoids any delays in the acquisition, thus allowing to capture each frame with high precision in timing.

9. Alternatively, image stack file can be opened by using "File > Import > Image Sequence" menu in ImageJ.
10. The values are specific for each microscope, dependent on the objective and the camera installed. Ask your microscope facility manager or calculate the pixel value by using the following formula:

$$\text{Image pixel size} = \frac{\text{Camera pixel size} \times \text{Binning}}{\text{Objective maginification} \times \text{Lens magnification} \times \text{C mount}}$$

11. Start tracing the axon ~15 μm distal from the cell body to avoid the axon initial segment, where transport is atypical [31, 32].
12. For our imaging settings, choosing "width" as 15 pixels was sufficient to cover the diameter of the axon. This might need to be adapted based on the resolution of the camera installed.
13. If necessary, saved tracings can be loaded on to the kymograph later. This is very useful to modify the tracings later on.
14. For counting the particles, we drew a horizontal line in about the middle of the kymograph (~30 s) and counted all particles crossing (Fig. 3g).
15. Data can be visualized in different ways. For instance, the distribution of individual events (run length or run duration) in defined ranges can be represented in histogram plots. These plots are very useful, especially when you expect changes to occur in a particular speed category. This information might not be immediately visible in the dot plots with medians: It is possible that run length/time change only in a specific parameter category between control and treated neurons.

Acknowledgments

This work was supported by the ANR-10-IDEX-0001-02, the LabEx CelTisPhyBio ANR-11-LBX-0038. CJ is supported by the Institut Curie, the French National Research Agency (ANR) awards ANR-12-BSV2-0007 and ANR-17-CE13-0021, the

Institut National du Cancer (INCA) grant 2014-PL BIO-11-ICR-1, and the Fondation pour la Recherche Medicale (FRM) grant DEQ20170336756. MMM is supported by the EMBO short-term fellowship ASTF 148-2015 and by the Fondation Vaincre Alzheimer grant FR-16055p, and SB by the FRM grant FDT201805005465. We thank C. Alberti, E. Belloir, F. Bertrand, V. Dangles-Marie, I. Grandjean, C. Caspersen, H. Hermange, A. Thadal, G. Buhagiar, C. Serieyssol, S. Gadadhar, and M. Sittewelle (Institut Curie) for technical assistance. We are grateful to M.-N. Soler, C. Lovo, and L. Besse from the PICT-IBiSA@Orsay Imaging Facility of the Institut Curie supported by the ANR through the "Investment for the future" program (France-BioImaging, ANR-10-INSB-04), and to N. Manel (Institut Curie, Paris) for material and advice for the lentivirus production. We would like to thank F. Cordelières (Bordeaux Imaging Center, France) for the KymoToolBox plug-in, as well as M. Brill (Technical University Munich, Germany), F. Del Bene, V. Marthiens (Institut Curie), and C. González-Billault (University of Chile, Santiago, Chile) for instructive discussions and advice.

References

1. Vale RD (2003) The molecular motor toolbox for intracellular transport. Cell 112 (4):467–480
2. Franker MAM, Hoogenraad CC (2013) Microtubule-based transport—basic mechanisms, traffic rules and role in neurological pathogenesis. J Cell Sci 126(Pt 11):2319–2329. https://doi.org/10.1242/jcs.115030
3. Millecamps S, Julien J-P (2013) Axonal transport deficits and neurodegenerative diseases. Nat Rev Neurosci 14(3):161–176. https://doi.org/10.1038/nrn3380
4. Brady ST, Morfini GA (2017) Regulation of motor proteins, axonal transport deficits and adult-onset neurodegenerative diseases. Neurobiol Dis 105:273–282. https://doi.org/10.1016/j.nbd.2017.04.010
5. Kneynsberg A, Combs B, Christensen K, Morfini G, Kanaan NM (2017) Axonal degeneration in tauopathies: disease relevance and underlying mechanisms. Front Neurosci 11:572. https://doi.org/10.3389/fnins.2017.00572
6. Janke C (2014) The tubulin code: Molecular components, readout mechanisms, and functions. J Cell Biol 206(4):461–472. https://doi.org/10.1083/jcb.201406055
7. Magiera MM, Singh P, Janke C (2018) SnapShot: functions of tubulin posttranslational modifications. Cell 173(6):1552–1552.e1. https://doi.org/10.1016/j.cell.2018.05.032
8. Janke C, Rogowski K, Wloga D, Regnard C, Kajava AV, Strub J-M, Temurak N, van Dijk J, Boucher D, van Dorsselaer A, Suryavanshi S, Gaertig J, Eddé B (2005) Tubulin polyglutamylase enzymes are members of the TTL domain protein family. Science 308 (5729):1758–1762. https://doi.org/10.1126/science.1113010
9. van Dijk J, Rogowski K, Miro J, Lacroix B, Eddé B, Janke C (2007) A targeted multienzyme mechanism for selective microtubule polyglutamylation. Mol Cell 26(3):437–448. https://doi.org/10.1016/j.molcel.2007.04.012
10. Rogowski K, van Dijk J, Magiera MM, Bosc C, Deloulme J-C, Bosson A, Peris L, Gold ND, Lacroix B, Bosch Grau M, Bec N, Larroque C, Desagher S, Holzer M, Andrieux A, Moutin M-J, Janke C (2010) A family of protein-deglutamylating enzymes associated with neurodegeneration. Cell 143(4):564–578. https://doi.org/10.1016/j.cell.2010.10.014
11. Tort O, Tanco S, Rocha C, Bieche I, Seixas C, Bosc C, Andrieux A, Moutin M-J, Xavier Aviles F, Lorenzo J, Janke C (2014) The cytosolic carboxypeptidases CCP2 and CCP3 catalyze posttranslational removal of acidic amino

acids. Mol Biol Cell 25(19):3017–3027. https://doi.org/10.1091/mbc.E14-06-1072
12. Kimura Y, Tsutsumi K, Konno A, Ikegami K, Hameed S, Kaneko T, Kaplan OI, Teramoto T, Fujiwara M, Ishihara T, Blacque OE, Setou M (2018) Environmental responsiveness of tubulin glutamylation in sensory cilia is regulated by the p38 MAPK pathway. Sci Rep 8(1):8392. https://doi.org/10.1038/s41598-018-26694-w
13. Ikegami K, Mukai M, Tsuchida J-I, Heier RL, Macgregor GR, Setou M (2006) TTLL7 is a mammalian beta-tubulin polyglutamylase required for growth of MAP2-positive neurites. J Biol Chem 281(41):30707–30716
14. Magiera MM, Bodakuntla S, Ziak J, Lacomme S, Marques Sousa P, Leboucher S, Hausrat TJ, Bosc C, Andrieux A, Kneussel M, Landry M, Calas A, Balastik M, Janke C (2018) Excessive tubulin polyglutamylation causes neurodegeneration and perturbs neuronal transport. EMBO J 37(23):e100440. https://doi.org/10.15252/embj.2018100440
15. Magiera MM, Singh P, Gadadhar S, Janke C (2018) Tubulin posttranslational modifications and emerging links to human disease. Cell 173 (6):1323–1327. https://doi.org/10.1016/j.cell.2018.05.018
16. Ikegami K, Sato S, Nakamura K, Ostrowski LE, Setou M (2010) Tubulin polyglutamylation is essential for airway ciliary function through the regulation of beating asymmetry. Proc Natl Acad Sci U S A 107(23):10490–10495. https://doi.org/10.1073/pnas.1002128107
17. Wu H-Y, Wei P, Morgan JI (2017) Role of cytosolic carboxypeptidase 5 in neuronal survival and spermatogenesis. Sci Rep 7:41428. https://doi.org/10.1038/srep41428
18. Mullen RJ, Eicher EM, Sidman RL (1976) Purkinje cell degeneration, a new neurological mutation in the mouse. Proc Natl Acad Sci U S A 73(1):208–212
19. Giordano T, Gadadhar S, Bodakuntla S, Straub J, Leboucher S, Martinez G, Chemlali W, Bosc C, Andrieux A, Bieche I, Arnoult C, Geimer S, Janke C (2019) Loss of the deglutamylase CCP5 perturbs multiple steps of spermatogenesis and leads to male infertility. J Cell Sci 132(3). https://doi.org/10.1242/jcs.226951
20. Konno A, Ikegami K, Konishi Y, Yang H-J, Abe M, Yamazaki M, Sakimura K, Yao I, Shiba K, Inaba K, Setou M (2016) Ttll9−/− mice sperm flagella show shortening of doublet 7, reduction of doublet 5 polyglutamylation and a stall in beating. J Cell Sci 129 (14):2757–2766. https://doi.org/10.1242/jcs.185983
21. Bosch Grau M, Masson C, Gadadhar S, Rocha C, Tort O, Marques Sousa P, Vacher S, Bieche I, Janke C (2017) Alterations in the balance of tubulin glycylation and glutamylation in photoreceptors leads to retinal degeneration. J Cell Sci 130:938–949. https://doi.org/10.1242/jcs.199091
22. Gilmore-Hall S, Kuo J, Ward JM, Zahra R, Morrison RS, Perkins G, La Spada AR (2019) CCP1 promotes mitochondrial fusion and motility to prevent Purkinje cell neuron loss in pcd mice. J Cell Biol 218(1):206–219. https://doi.org/10.1083/jcb.201709028
23. Shashi V, Magiera MM, Klein D, Zaki M, Schoch K, Rudnik-Schoneborn S, Norman A, Lopes Abath Neto O, Dusl M, Yuan X, Bartesaghi L, De Marco P, Alfares AA, Marom R, Arold ST, Guzman-Vega FJ, Pena LD, Smith EC, Steinlin M, Babiker MO, Mohassel P, Foley AR, Donkervoort S, Kaur R, Ghosh PS, Stanley V, Musaev D, Nava C, Mignot C, Keren B, Scala M, Tassano E, Picco P, Doneda P, Fiorillo C, Issa MY, Alassiri A, Alahmad A, Gerard A, Liu P, Yang Y, Ertl-Wagner B, Kranz PG, Wentzensen IM, Stucka R, Stong N, Allen AS, Goldstein DB, Schoser B, Rosler KM, Alfadhel M, Capra V, Chrast R, Strom TM, Kamsteeg E-J, Bonnemann CG, Gleeson JG, Martini R, Janke C, Senderek J (2018) Loss of tubulin deglutamylase CCP1 causes infantile-onset neurodegeneration. EMBO J 37(23):e100540. https://doi.org/10.15252/embj.2018100540
24. Sirajuddin M, Rice LM, Vale RD (2014) Regulation of microtubule motors by tubulin isotypes and post-translational modifications. Nat Cell Biol 16(4):335–344. https://doi.org/10.1038/ncb2920
25. Ikegami K, Heier RL, Taruishi M, Takagi H, Mukai M, Shimma S, Taira S, Hatanaka K, Morone N, Yao I, Campbell PK, Yuasa S, Janke C, Macgregor GR, Setou M (2007) Loss of alpha-tubulin polyglutamylation in ROSA22 mice is associated with abnormal targeting of KIF1A and modulated synaptic function. Proc Natl Acad Sci U S A 104 (9):3213–3218. https://doi.org/10.1073/Pnas.0611547104
26. Maas C, Belgardt D, Lee HK, Heisler FF, Lappe-Siefke C, Magiera MM, van Dijk J, Hausrat TJ, Janke C, Kneussel M (2009) Synaptic activation modifies microtubules underlying transport of postsynaptic cargo. Proc Natl Acad Sci U S A 106(21):8731–8736. https://doi.org/10.1073/pnas.0812391106
27. Ghiretti AE, Thies E, Tokito MK, Lin T, Ostap EM, Kneussel M, Holzbaur ELF (2016)

Activity-dependent regulation of distinct transport and cytoskeletal remodeling functions of the dendritic kinesin KIF21B. Neuron 92 (4):857–872. https://doi.org/10.1016/j.neuron.2016.10.003

28. Olenick MA, Dominguez R, Holzbaur ELF (2019) Dynein activator Hook1 is required for trafficking of BDNF-signaling endosomes in neurons. J Cell Biol 218(1):220–233. https://doi.org/10.1083/jcb.201805016

29. Zala D, Hinckelmann M-V, Yu H, Lyra da Cunha MM, Liot G, Cordelieres FP, Marco S, Saudou F (2013) Vesicular glycolysis provides on-board energy for fast axonal transport. Cell 152(3):479–491. https://doi.org/10.1016/j.cell.2012.12.029

30. Schneider CA, Rasband WS, Eliceiri KW (2012) NIH image to ImageJ: 25 years of image analysis. Nat Methods 9(7):671–675

31. Albrecht D, Winterflood CM, Sadeghi M, Tschager T, Noe F, Ewers H (2016) Nanoscopic compartmentalization of membrane protein motion at the axon initial segment. J Cell Biol 215(1):37–46. https://doi.org/10.1083/jcb.201603108

32. Kuijpers M, van de Willige D, Freal A, Chazeau A, Franker MA, Hofenk J, Rodrigues RJC, Kapitein LC, Akhmanova A, Jaarsma D, Hoogenraad CC (2016) Dynein regulator NDEL1 controls polarized cargo transport at the axon initial segment. Neuron 89 (3):461–471. https://doi.org/10.1016/j.neuron.2016.01.022

Chapter 21

Methods to Measure Perinuclear Actin Dynamics During Nuclear Movement in Migrating Cells

Cátia S. Janota, Francisco J. Calero-Cuenca, and Edgar R. Gomes

Abstract

The nucleus is specifically positioned within a cell in diverse biological contexts. There are multiple connections between the nuclear envelope and the cytoskeleton and these connections are involved in nuclear positioning. During cell polarization prior to cell migration, nuclear envelope proteins bind to the actin cytoskeleton and get organized into linear arrays, known as transmembrane actin-associated nuclear (TAN) lines to move the nucleus away from the leading edge. Here we describe methods to study perinuclear actin dynamics, including measurement of the thickness of actin cables coupled to TAN lines, measurement of the number of perinuclear actin cables, and ablation of perinuclear actin cables. These methods are used to identify mechanisms of nuclear positioning.

Key words Perinuclear actin, Stress fibers, Laser ablation, Nuclear positioning, Nuclear movement, Cell migration, Polarization

1 Introduction

The nucleus is the biggest cellular organelle and it is limited by an impressively dynamic and regulated barrier called nuclear envelope (NE), which separates the cytoplasm from nucleoplasm. The position of the nucleus is important in different biological contexts, such as fertilization, differentiation, cell division, or migration, and relies mostly on nucleo-cytoskeletal interactions [1, 2]. The NE is comprised of an outer nuclear membrane (ONM) with KASH proteins that bind to the inner nuclear membrane (INM) SUN proteins, forming the linker of nucleoskeleton and cytoskeleton (LINC) complex [3, 4]. The LINC complex mediates the interaction between the cytoskeleton (actin, microtubules, and intermediate filaments) and the nuclear envelope for nuclear movement and positioning, but also to control nuclear shape and stiffness, cytoskeleton organization, mechanotransduction, gene expression, chromosome positioning, DNA repair, and cell migration [5]. Furthermore, disruptions of nucleo-cytoskeletal interactions could

Helder Maiato (ed.), *Cytoskeleton Dynamics: Methods and Protocols*, Methods in Molecular Biology, vol. 2101, https://doi.org/10.1007/978-1-0716-0219-5_21,

result in developmental and physiological defects, such as brain development diseases, muscle disorders, cancer, and deafness [4, 6].

Multiple models have been used to study nuclear movement and positioning [7], such as *A. thaliana* [8], *D. melanogaster* egg chamber [9], *M. musculus* [10], *C. elegans* [11], and, among others, wounded monolayers of fibroblasts [12]. Different laboratories have been using a modified version of the wound assay to study nuclear positioning in NIH3T3 fibroblasts since this model is simple to manipulate in order to identify new mechanisms of nuclear positioning in mammals, decouples cell migration from nuclear positioning, and is compatible with high-resolution live imaging of multiple cell components involved in nuclear movement [12–18]. The wounding assay is based on removing serum from culture for a long period to eliminate any polarization cues prior to inducing a wound in the monolayer. After wounding, cells are stimulated with lysophosphatidic acid (LPA) to trigger cell polarization without cell migration. During cell polarization, the nucleus moves away from the leading edge. This movement is driven by the movement of perinuclear dorsal actin cables, i.e., actin cables on the top of the nucleus, and it is powered by actin retrograde flow [14]. Perinuclear actin cables that move away from the leading edge reach the top of the nucleus and interact with KASH proteins (Nesprins), part of LINC complexes (with SUN proteins) forming linear arrays of LINC complex proteins known as transmembrane actin-associated (TAN) lines [13–15]. The moving dorsal actin cables are probably similar to the actin cap observed on the dorsal side of immobile nuclei [19]. Here we describe methods to study perinuclear actin cables involved in nuclear movement.

2 Materials

2.1 Cell Culture

1. NIH3T3 fibroblasts (*see* **Note 1**).
2. Growth medium: DMEM, 10 mM HEPES, penicillin/streptomycin at 500 units/mL, and 10% bovine calf serum.
3. Serum-free medium for starvation: DMEM, 10 mM HEPES and penicillin/streptomycin at 500 units/mL.
4. Phosphate-buffered saline (PBS).
5. Tryple Express (Life Technologies).
6. LPA 1 mM (*see* **Note 2**).
7. Lentivirus expressing LifeAct-mCherry [20].
8. Polybrene.
9. Puromycin 2.5 μg/mL.
10. Acid-washed square 22 × 22 mm coverslips #1.5 (*see* **Note 3**).

11. Acid-washed round 25 mm coverslips #1.5.
12. 6-Well culture plate, 35, 60, and 100 mm culture dishes.
13. Fluorodishes 25 mm (World Precision Instruments).

2.2 Microinjection

1. Plasmid encoding GFP-mini-Nesprin-2G [14].
2. P-97 Micropipette Puller (Sutter Instrument) equipped with a 2.5 × 2.5 mm box filament (Sutter Instrument).
3. Borosilicate capillary glass OD 1 mm × ID 0.5 mm (Sutter Instrument).
4. Microinjection buffer: 10 mM HEPES and 140 mM NaCl.
5. Microinjector (Xenoworks, Sutter Instrument) coupled to an inverted microscope.

2.3 Immunofluorescence

1. PBS.
2. Fixation buffer: 4% Paraformaldehyde (PFA) diluted from stock solution of 32% PFA in PBS.
3. Permeabilization buffer: 0.5% Triton-100, diluted in PBS.
4. Blocking buffer: 10% Goat serum diluted in PBS.
5. Anti-GFP antibody.
6. Phalloidin 555 (Thermo Fisher).
7. DAPI 1 μg/mL.
8. Microscope Slides Menzel-Gläser Superfrost Plus 1 mm.
9. Menzel-Gläser coverslips #1.5.
10. Fluoromount-G (Southern Biotech).
11. Parafilm.
12. Tissue paper.

2.4 Imaging of Actin Cables Coupled to TAN Lines

1. Zeiss Cell Observer wide-field inverted microscope controlled by ZEN Blue Edition.
2. sCMOS camera (e.g., ORCA-flash4.0 V2, Hamamatsu).
3. 63× Plan-Apochromat Oil Objective (NA = 1.4).
4. LED light source Colibri2 (Zeiss).
5. Filter cube FS38HE excitation 450–490 nm and emission 500–550 nm.
6. Filter cube FS43HE excitation 538–562 nm and emission 570–640 nm.

2.5 Imaging of Perinuclear Actin Cables

1. Zeiss Cell Observer spinning disk confocal inverted microscope equipped with 37 °C chamber, 5% CO_2, for live-cell microscopy. Controlled by Zen Blue Edition.
2. EM-CCD camera (e.g., Evolve 512, Photometrics).

3. Spinning disk confocal scanner (CSU-x1, Yokogawa).
4. 63x Plan-Apochromat Oil Objective (NA = 1.4).
5. LED light source Colibri2 (Zeiss).
6. Solid-state laser 405 nm, maximum power 50 mW.
7. Solid-state laser 561 nm, maximum power 75 mW.
8. BP 450/50425-475 nm emission filter.
9. BP 600 50575-625 nm emission filter.

2.6 Live Imaging for Actin Laser Ablation

Zeiss LSM 880 point-scanning confocal inverted microscope equipped with 37 °C chamber, 5% CO_2, and Definite Focus for live-cell microscopy.

1. Detector Airyscan 32 channel area (resolution of 140 nm laterally and 400 nm axially, at 488 nm).
2. GaAsP detector (photomultiplier 45% QE).
1. Chamber/holder for coverslips for live imaging (Live Cell Instrument Co.) (*see* **Note 4**).
2. 63× Plan-Apochromat Oil NA, working distance 0.19 mm (NA = 1.4).
3. Argon laser unit 488 nm with a maximum power of 25 mW.

2.7 Data Analysis

ImageJ FIJI software (available at https://fiji.sc/) [21].

3 Methods

There are several aspects of perinuclear actin dynamics that are involved in nuclear movement during cell polarization in a wounding assay. Perinuclear actin is composed of ventral and dorsal actin cables, and the movement of dorsal actin cables is driven by actin retrograde flow. Three methods to study perinuclear actin dynamics are described here: (1) quantification of the thickness of actin cables coupled to TAN lines; (2) quantification of perinuclear actin cable density; and (3) measurement of perinuclear actin cable dynamics by laser ablation. The three methods are based on the wounding assay, briefly described in this chapter. Readers should refer to previously published methods manuscripts for a detailed description of the wounding assay to study nuclear movement [22, 23].

3.1 Wound Assay

1. Grow a 10 cm dish of WT NIH3T3 fibroblasts to 80% confluency.
2. Place one acid-washed coverslip inside each well of a 6-well plate.

3. Wash the 10 cm dish with sterile PBS, add 1 mL of Triple Express solution, and then transfer to the 37 °C incubator until cells detach (approximately 5 min).
4. Add 2.350 mL of growth medium to each 6-well plate (*see* **Note 5**).
5. Resuspend cells in 9 mL of medium and add 150 μL of cells to each well, so the final volume per well is 2500 μL (*see* **Note 6**).
6. Place 6-well plate in the incubator and wait until cells grow until ~75% confluent (usually 48 h) (*see* **Note 7**).
7. To serum-starve the cells prepare a new 6-well plate (if cells are going to be microinjected, each coverslip should be placed in one 3.5 cm dish) with 2 mL of serum-free medium and three 6 cm dishes with 5 mL of serum-free medium. Use sterile forceps to pick up one coverslip and sequentially dip it in each of the 6 cm dishes to wash the serum from the coverslips. After the third 6 cm dish wash, transfer the coverslip to the new 6-well plate containing serum-free medium (*see* **Note 8**).
8. Wait for 48 h until cells are starved (*see* **Note 9**).
9. Wound the cell monolayer with a 20 μL pipette tip and put the cells back in the incubator. Cells are ready to be stimulated with LPA, usually 15–20 min after wounding. Microinjection can be performed before LPA stimulation as described in Subheading 3.2.

3.2 Quantification of the Thickness of Actin Cables Coupled to TAN Lines

To quantify the thickness of actin cables coupled to TAN lines, TAN lines must be visualized simultaneously with actin. The most efficient method to visualize TAN lines is to express GFP-mini-Nesprin-2G in cells on the wound edge by microinjection prior to LPA stimulation. Actin cables are visualized post-fixation using fluorescently labeled phalloidin. A comprehensive method for microinjection is described elsewhere [24]. This method proceeds after **step 9** of Subheading 3.1.

3.2.1 Microinjection

1. Prepare the plasmid encoding GFP-mini-Nesprin-2G: Dilute 1 μL of plasmid (stock should be at 500–3000 ng/μL) in 100 μL of microinjection buffer, in a 500 μL tube.
2. Centrifuge at 4 °C at 13,000*g* for 15 min, and collect the supernatant (about 80 μL) into a new centrifuge tube. Repeat centrifugation and collect the supernatant.
3. Wound the starved cell monolayer in one 35 mm dish and put the cells back in the incubator for 20 min.
4. Back-load the plasmid mix into a glass capillary previously prepared (*see* **Note 10**) and insert it in the microinjection arm.

5. Take cells from the incubator and place them on the microscope. Inject the nuclei of cells at the wound edge (*see* **Note 11**).
6. Wait for 2 h for protein expression (standard time). Depending on the construct and the desired level of expression the time of expression might need to be adjusted.
7. After expression period, stimulate the cells with 10 μM LPA.
8. After 50 min of LPA stimulation, fix cells in 4% PFA for 10 min (*see* **Note 12**).

3.2.2 Immunofluorescence

1. Transfer the coverslips and medium that are inside 3.5 cm dishes into 6-well plates to simplify the next steps.
2. Remove medium, add 1 mL of permeabilization buffer per well, and place the dish in an orbital shaker for 8 min.
3. Add 1 mL of blocking buffer per well for 30 min.
4. Prepare a humidity chamber for immunostaining incubations. Use a 15 cm cell culture dish and place it upside down. Cover the bottom with water-soaked tissue paper. Place parafilm on top of it.
5. Dilute the anti-GFP antibody 1:1000 in 10% goat serum.
6. Place 50 μL of the antibody dilution on parafilm and place gently the coverslip with the cells facing the parafilm, avoiding air bubbles trapped between the coverslip and the parafilm.
7. Incubate for 1 h at RT or overnight at 4 °C.
8. Add PBS to the side of the coverslip to allow the detachment of the coverslip from the parafilm. Transfer the coverslips into a 6-well plate filled with PBS, with the cells facing up.
9. Wash 3 × 10 min with PBS.
10. Prepare the secondary antibody mix in 10% goat serum: phalloidin 555 (1:200), Alexa 488 anti-chicken (1:800), and DAPI (1:100.000 from the stock; a working aliquot of 1:100 from the stock could be useful).
11. Incubate with the secondary antibody mix for 1 h at RT.
12. Wash 3 × 10 min with PBS (*see* **Note 13**).
13. Pick up the coverslip with the forceps and remove the excess of PBS by touching the side of a coverslip to a tissue paper.
14. Put a 25 μL of Fluoromount-G drop on a glass slide.
15. Place the coverslip on top of the Fluoromount-G, with cells facing down, avoiding air bubbles.
16. Let the sample dry overnight by storing it at 4 °C or seal it with nail polish to image on the same day.

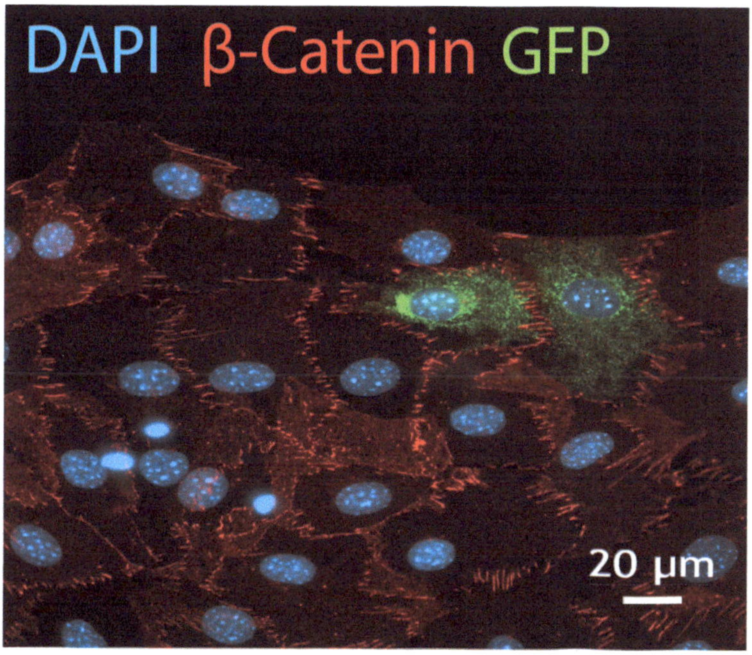

Fig. 1 Representative image of a wound-edge monolayer of fibroblasts immunostained for β-catenin to visualize cell boundaries, GFP to visualize transfected cells, and DAPI to visualize the nucleus. The β-catenin antibody should be included in the primary antibody mix and used at 1:400. Note that the GFP-positive cell on the right is at the wound edge, whereas the GFP-positive cell on the left it is not at the wound edge

3.2.3 Imaging

1. Add a drop of oil to the 63× objective.
2. Place the slide with the coverslip facing the objective (in an inverted microscope the coverslip is facing down).
3. Use the binoculars to identify GFP-positive cells and orientation of the wound edge. Add its XYZ coordinates to the list of positions to acquire. Cell selection should fulfill specific requirements (Fig. 1) (*see* **Note 14**).
4. Select the 555, 488, and 405 nm filters.
5. Select a Z stack of 2.5 μm with a z-step of 0.27 μm, to include all information from the center to the highest Y point of the cell. This is the best way to include an ideal plane of co-localization.
6. Acquire all positions in a single file.

3.2.4 Analysis

1. Open the image in ImageJ FIJI.
2. Find a cell that is positive for GFP-mini-Nesprin-2G. The co-localization of the red and the green channels must be clear (ignore cells that express too high levels of GFP-mini-Nesprin-2G where individual TAN lines are not visible).

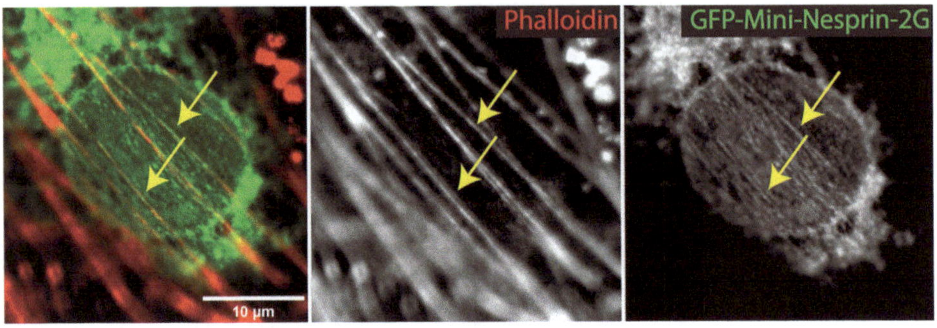

Fig. 2 Identification of TAN lines. Image of a leading-edge fibroblast expressing GFP-mini-Nesprin-2G in green and filamentous actin labeled with phalloidin. The scale bar is 10 µm

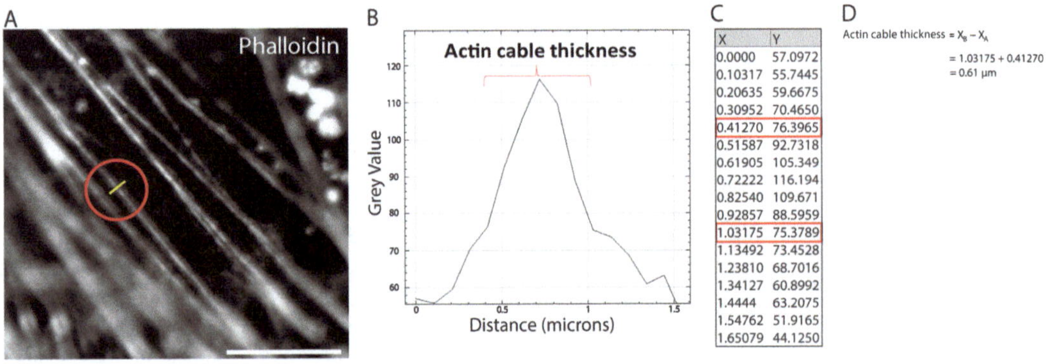

Fig. 3 Quantification of actin cables coupled to TAN line thickness using ImageJ FIJI. (**a**) Representation of a perpendicular line to an actin cable labeled with phalloidin in ImageJ FIJI. (**b**) Plot representing the fluorescence intensity along that drawn straight line. (**c**) The table generated from plot in (**b**), where it is possible to identify the beginning and the end of the actin cable. (**d**) How to measure the thickness of an actin cable using FIJI. The scale bar is 10 µm

3. Open the channels tool (Image > Color > Channels Tool) and select the green and the red channels.
4. Look for TAN lines (green channel) that co-localize with dorsal actin cables (red channel). In Fig. 2 there is an example of the co-localization of an actin cable and a linear array of Nesprin-2G.
5. Select only the red channel.
6. Draw a straight line perpendicular to the actin cable (as depicted in Fig. 3a).
7. Press Control + K to open the plot reflecting the fluorescence intensity along that line (as the one depicted in Fig. 3b). The two points that are at the bottom of the curve, when the slope of the curve starts to increase exponentially, represent the actin cable beginning and end, respectively. The length between the two X coordinates corresponds to the actin filament thickness.

8. Click on “List” to open a table with the fluorescence as a function of the X coordinates (as depicted in Fig. 3c) and copy the X coordinates that limit the beginning and the end of the curve (as in the example in Fig. 3b).
9. Subtract to the bigger coordinate the smaller one and the result will be the actin cable thickness (as shown in Fig. 3d).

3.3 Quantification of Perinuclear Actin Cable Density

The quantification of perinuclear actin cables is similar to the protocol for quantification of thickness of actin cables coupled with TAN lines. For this, microinjection is not required since visualization of TAN lines is not required.

3.3.1 LPA Stimulation

1. Use a 20 μL pipette tip to wound the cell monolayer and wait for 15 min to add 10 μM of LPA.
2. After 50 min of stimulation, fix cells in 4% PFA for 10 min.

3.3.2 Immunofluorescence

Repeat what is described in Subheading 3.2.2 using phalloidin 555 and DAPI, thus skipping **steps** from **5** to **8**. Also do not use Alexa 488 anti-chicken in **step 10**.

3.3.3 Imaging

1. Add a drop of oil to the ×63 objective.
2. Place the slide with the coverslip facing the objective (in an inverted microscope the coverslip is facing down).
3. Use binoculars to identify the wound edge.
4. Select a Z stack that covers totally the height of the cell at the wound edge (around 4 μm).
5. Select the 555 nm and the 405 nm filters.
6. Select all the XYZ coordinates with positive cells, so all the images are saved in a single file. Select at least 30 fields of view with at least one cell.
7. Acquire a Z stack with a z-step of 0.42 μm for each selected field of view.

3.3.4 Analysis

1. Open the image in ImageJ FIJI.
2. Select the lowest Z plane where the ventral actin cables are in focus and count the number of cables under the nucleus region (Fig. 4, left).
3. Select the highest Z plane where the dorsal actin cables are in focus and count the number of cables above the nucleus region (Fig. 4, right).

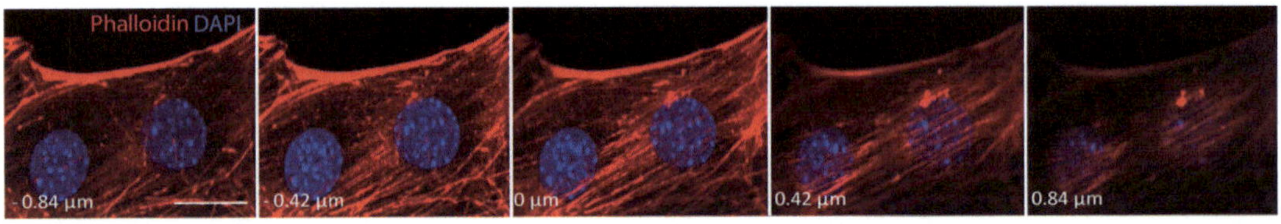

Fig. 4 Spinning disk confocal Z stack with a z-step size of 0.42 µm, from the ventral side (left) toward the dorsal side (right). Phalloidin in red is labeling filamentous actin and DAPI in blue is labeling the nucleus. The scale bar is 20 µm

3.4 Measurement of Perinuclear Actin Cable Dynamics by Laser Ablation

A cell line stably expressing LifeAct-mCherry is very useful to perform perinuclear actin laser ablation, since most of the cells are labeled. NIH3T3 fibroblasts were infected with lentivirus carrying LifeAct-mCherry produced in HEK 293T (pLALI backbone).

3.4.1 Generation of a Stable Cell Life Expressing LifeAct-mCherry

Plate WT NIH3T3 cells at a confluence of 40% on two wells of a 6-well plate (*see* **Note 1**).

1. After 24 h add 10 µL of lentiviral suspension containing LifeAct-mCherry to one well and 10 µL of lentiviral suspension containing a mock DNA sequence to a second well as a control that does not affect protein expression (*see* **Note 15**).
2. Add 1 µL of polybrene 8 µg/mL to each well.
3. Wait for 24 h and then change medium to growth medium (*see* **Note 16**).
4. Wait for 24 h and transfer cells to a 10 cm dish with growth medium complemented with 2.5 µg/mL puromycin to select transfected cells (*see* **Note 17**).
5. Wait until all cells infected with lentivirus containing the mock sequence die.
6. Take LifeAct-expressing fibroblasts to the microscope to check if they are correctly expressing.
7. Start a wound assay with the new stable cell line to verify if these cells orient the centrosome and move their nucleus backwards.

3.4.2 LPA Stimulation and Laser Ablation

1. Begin a wound assay with the stable cell line generated above as described in Subheading 3.1 up to **step 6**, included.
2. After 48 h of starvation, place the coverslip in the chamber/holder, and transfer the same medium where cells were starved into the holder (*see* **Note 18**). As an alternative, the plating in **step 2** of Subheading 3.1 could be done directly in a fluorodish.
3. Turn on the microscope and set up the incubator to 37 °C and the CO_2 to 5% (as an alternative to the CO_2, use 20 mM Hepes in a sealed holder). Wait for at least 30 min for the stabilization of the temperature of the whole system.

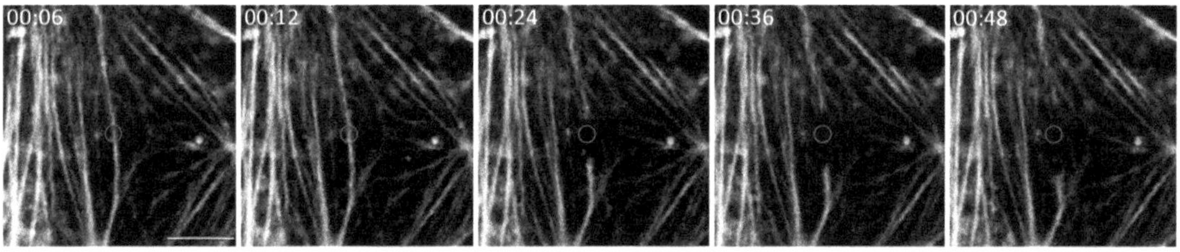

Fig. 5 NIH3T3 fibroblasts expressing LifeAct-mCherry at the leading edge of a wound assay. Representative images of a time-lapse video of actin laser ablation. The ablation was performed on the indicated circle at the third time point (00:24), so only after that it is possible to observe the actin cable being severed. The scale bar is 5 μm. Time is mm:ss

4. Transfer cells to the microscope and do the wounding with a 20 μL pipette tip.
5. Through the binoculars select cells with high levels of expression of LifeAct-mCherry and add these positions to the list of XYZ positions to acquire (select up to six positions so it is possible to acquire all the positions within the first 30 min of LPA).
6. Add 10 μM of LPA.
7. Acquire all positions.
8. Set up the ablation properties to three interactions at bleaching speed 1 with the laser 405 at 100%. Acquire two images before ablation and three images after ablation every 6 s (this could be extended if the cells are not damaged due to phototoxicity, in order to follow the ablation for a longer period). The scanning acquisition must be 6 to protect the cells from phototoxicity. Figure 5 shows an example of the ablation of an actin cable.
9. Perform the ablation in all the selected positions.
10. Ten minutes after the ablation acquire a Z stack of the 6 positions to make sure that they still do not have newly formed actin cables under the nucleus (*see* **Note 19**).
11. Start a time-lapse movie 45 min after LPA addition for each selected position. Acquire a Z stack every 15 min for 1 h, 15 min.

3.4.3 Analysis

1. Open images on ImageJ FIJI.
2. Search for cells that did not recover the actin cables (*see* **Note 20**).
3. Track nuclear movement over time (Plugins—Tracking—Manual Tracking).
4. Correlate nuclear movement with absence of actin cables.

4 Notes

1. NIH3T3 fibroblasts should be kept in culture up to 3 weeks after thawing and never above 90% confluency.
2. To prepare 1 mM LPA stock solution first prepare 4 mL of a solution containing 10 mg/mL BSA (40 mg of tissue culture grade), 100 mM NaCl (133 Ml of 3 M stock or 80 μL from 5 M stock), and 10 mM Hepes (40 μL of 1 M stock) [22]. Filter this solution with a syringe attached to a 0.22 μm filter. Pipette 2.3 mL to the 1 mg LPA flask. Mix very, very well (5 min) pipetting up and down until all the small particles of LPA are not visible. Prepare 50 μL aliquots and snap freeze in liquid nitrogen.
3. To prepare acid-washed coverslips: place the coverslips in porcelain coverslip racks (Thomas Scientific), then place them in a beaker filled up with 1 N HCl for 10 min, and wash three times during 10 min with running water, followed by a rinse in ddH_2O and a final wash with 95% ethanol. Beaker with coverslips is covered with aluminum foil and sterilized in an oven at 100 °C during 60 min and stored sterile.
4. The imaging of live cells can also be performed using glass-bottom dishes, such as fluorodishes (World Precision Instruments).
5. Knockdown of proteins by siRNA is performed at this step. Add 5 μL of 20 μM of siRNA to 250 mL of Opti-MEM and 5 μL of Lipofectamine RNAiMAX Transfection reagent to 250 μL of Opti-MEM Serum-Reduced Medium (31985-047, Invitrogen). Add the second mix to the first to one, wait for 20 min, and add it to the well (on top of 150 μL of cells and 1850 μL of complete medium). The final volume should be 2500 mL. After 24 h substitute this medium with growth medium.
6. If the treatment with siRNA reduces cell proliferation, adjust the density of cells used to obtain the desired confluency upon cell starvation, a key step on this protocol (*see* **Note 8**). Note that this is the correct amount of cells considering the rate of proliferation of WT NIH3T3 fibroblasts; please try a different amount of cells if working with a different cell type.
7. If a protein that affects cell cycle is being depleted, the starvation should be done when the confluence is about 85%.
8. The most critical step in the wound assay is the cell confluence when cells are starved. If the cells are not confluent enough (less than 75%) the monolayer will not be homogenous after 48 h (and wound-edge cells will not move their nucleus as efficiently). Moreover, if cells are above 75% confluency at the

time that starvation begins or if starvation is unsuccessful (for example because the washes were not properly done or it is needed to increase the starving period), it is more likely that cells polarize, even without LPA stimulation or when Nesprin-2G is depleted from the system. To confirm that the starvation was successful, cells should have a minimal number of actin cables 48 h after starvation and in the absence of LPA treatment. If cells have actin cables after starvation or are still positioning their nucleus in the absence of the protein that serves as a positive control (Nesprin-2G), increase starving time.

9. At this point, the confluence of cells is of extreme importance. Overconfluence or underconfluence will impact negatively the nuclear positioning.
10. The length of the glass capillary tip is a critical step to microinject successfully. The program of microinjection is based on five parameters: heat, pressure, pull, time, and velocity. The program we have best results with is as follows: heat is dependent on the Ramp test of each filament box, pressure 470, pull 100 (3 lines), velocity 100 (3 lines), delay 10 (2 first lines) and 25 (last line), and pressure 250 (3 lines). The program should do the 3 lines only once. Since the diameter of the pipette tip is very small and fragile, store them inside a container that can be closed and the tip should not touch anything.
11. Increase the time of expression or increase the amount of DNA if cells are not expressing as much as desired.
12. If one wants to visualize microtubules, please note that fixation should be done with methanol at 20 °C for 10 min.
13. Be careful to avoid the coverslip from drying up during all this process.
14. Choosing cells for quantification is a very important step. Given that the wound is done with a tip, it might not be uniform throughout the entire coverslip. Please look for regions where the wound edge is a straight line, where there are no prominent cells and all cells have a square-like morphology, having a single leading edge, as in Fig. 1.
15. Do a viral titration within a range of 2–15 μL of viral suspension to find the best infection concentration. In alternative to lentiviral infection of adherent cells, it is possible to transfect transiently with Lipofectamine LTX Reagent with Plus Reagent or microinject (Invitrogen).
16. To optimize the amount of plated cells, test different number of cells to prevent an overconfluent monolayer after the infection period (48 h). When using an 80% confluent dish, one could test within a range from 125 to 250 μL of cell

suspension. If cells are close to 80% of confluence, one can pass the cells to a 10 cm dish already with medium supplemented with puromycin, instead of just changing the medium.

17. Some cell types require a different concentration of puromycin. If the survival rate after 48 h of puromycin is low, expand cells an extra 1–2 days before adding puromycin to the medium. Alternatively, test different concentrations of puromycin.
18. Starved cells are very sensitive and they will die if fresh starving medium is used. Therefore, it is important to use the same starving medium where cells were in the previous 48 h.
19. It could happen that the actin cable is not severed with a single ablation step; in this case repeat this step.
20. Depending on the efficiency of the laser ablation, sometimes actin cables are reformed. Therefore, it is not possible to study the role of actin cables on nuclear movement. These cells should not be analyzed.

Acknowledgments

This work was supported by the European Research Council, EMBO installation, AIM France, and LISBOA-01-0145-FEDER-031075, project co-funded by FEDER through POR Lisboa 2020—Programa Operacional Regional de Lisboa, PORTUGAL 2020, and Fundação para a Ciência e a Tecnologia.

References

1. Gundersen GG, Worman HJ (2013) Nuclear positioning. Cell 152:1376–1389
2. Bone CR, Starr DA (2016) Nuclear migration events throughout development. J Cell Sci 129:1951–1961
3. Crisp M, Liu Q, Roux K et al (2006) Coupling of the nucleus and cytoplasm: role of the LINC complex. J Cell Biol 172:41–53
4. Gant Luxton GW, Starr DA (2014) KASHing up with the nucleus: novel functional roles of KASH proteins at the cytoplasmic surface of the nucleus. Curr Opin Cell Biol 0:69–75
5. Thorpe SD, Lee DA (2017) Dynamic regulation of nuclear architecture and mechanics—a rheostatic role for the nucleus in tailoring cellular mechanosensitivity. Nucleus 8:287–300
6. Starr DA, Fridolfsson HN (2010) Interactions between nuclei and the cytoskeleton are mediated by SUN-KASH nuclear-envelope bridges. Annu Rev Cell Dev Biol 26:421–444
7. Meier I (2016) LINCing the eukaryotic tree of life—towards a broad evolutionary comparison of nucleocytoplasmic bridging complexes. J Cell Sci 129:3523–3531
8. Wada M (2018) Nuclear movement and positioning in plant cells. Semin Cell Dev Biol 82:17–24
9. Bernard F, Lepesant J-A, Guichet A (2018) Nucleus positioning within Drosophila egg chamber. Semin Cell Dev Biol 82:25–33
10. Roman W, Gomes ER (2018) Nuclear positioning in skeletal muscle. Semin Cell Dev Biol 82:51–56
11. Bone CR, Tapley EC, Gorjánácz M et al (2014) The Caenorhabditis elegans SUN protein UNC-84 interacts with Lamin to transfer forces from the cytoplasm to the nucleoskeleton during nuclear migration. Mol Biol Cell 25:2853–2865
12. Borrego-Pinto J, Jegou T, Osorio DS et al (2012) Sampl is a component of TAN lines and is required for nuclear movement. J Cell Sci 125:1099–1105

13. Gomes ER, Jani S, Gundersen GG (2005) Nuclear movement regulated by Cdc42, MRCK, myosin, and actin flow establishes MTOC polarization in migrating cells. Cell 121:451–463
14. Luxton GWG, Gomes ER, Folker ES et al (2010) Linear arrays of nuclear envelope proteins harness retrograde actin flow for nuclear movement. Science 329:956–959
15. Chang W, Folker ES, Worman HJ et al (2013) Emerin organizes actin flow for nuclear movement and centrosome orientation in migrating fibroblasts. Mol Biol Cell 24:3869–3880
16. Kutscheidt S, Zhu R, Antoku S et al (2014) FHOD1 interaction with nesprin-2G mediates TAN line formation and nuclear movement. Nat Cell Biol 16:708–715
17. Zhu R, Antoku S, Gundersen GG (2017) Centrifugal displacement of nuclei reveals multiple LINC complex mechanisms for homeostatic nuclear positioning. Curr Biol 27:3097–3110. e5
18. Palazzo AF, Joseph HL, Chen Y-J et al (2001) Cdc42, dynein, and dynactin regulate MTOC reorientation independent of Rho-regulated microtubule stabilization. Curr Biol 11:1536–1541
19. Khatau SB, Hale CM, Stewart-Hutchinson PJ et al (2009) A perinuclear actin cap regulates nuclear shape. Proc Natl Acad Sci 106:19017–19022
20. Baumlin-Schmid N, Salathe M, Fregien NL (2016) Optimal lentivirus production and cell culture conditions necessary to successfully transduce primary human bronchial epithelial cells. J Vis Exp 113:e54176
21. Schindelin J, Arganda-Carreras I, Frise E et al (2012) Fiji: an open-source platform for biological-image analysis. Nat Methods 9:676–682
22. Chang W, Antoku S, Gundersen GG (2016) Wound-healing assays to study mechanisms of nuclear movement in fibroblasts and myoblasts. In: Shackleton S, Collas P, Schirmer EC (eds) The nuclear envelope: methods and protocols. Springer, New York, pp 255–267
23. Gomes ER and Gundersen GG (2006) Real-time centrosome reorientation during fibroblast migration. In: Methods in enzymology. Academic Press, New York, pp 579–592
24. Gueroussov S, Tarnawsky SP, Cui XA et al (2010) Analysis of mRNA nuclear export kinetics in mammalian cells by microinjection. J Vis Exp 46:e2387

Correction to: Visualization and Functional Analysis of Spindle Actin and Chromosome Segregation in Mammalian Oocytes

Binyam Mogessie

Correction to:
Chapter 17 in: Maiato. H. (ed.), *Cytoskeleton Dynamics: Methods and Protocols*, Methods in Molecular Biology, vol. 2101, https://doi.org/10.1007/978-1-0716-0219-5_17

Chapter 17 "Visualization and Functional Analysis of Spindle Actin and Chromosome Segregation in Mammalian Oocytes" was previously published non-open access. It has now been changed to open access under a CC BY 4.0 license and the copyright holder has been updated to "The Author(s)." This book has also been updated with this change.

The updated online version of this chapter can be found at
https://doi.org/10.1007/978-1-0716-0219-5_17

Helder Maiato (ed.), *Cytoskeleton Dynamics: Methods and Protocols*, Methods in Molecular Biology, vol. 2101,
https://doi.org/10.1007/978-1-0716-0219-5_22, © Springer Science+Business Media, LLC, part of Springer Nature 2020

INDEX

A

AAA ATPases 27
Actin, v 53–73, 123, 135–144, 267–294, 298, 301, 302, 306, 310, 311, 327, 331, 371–384
Actomyosin 303, 317
Acute drug treatments 299, 301, 303, 305, 310, 313
Affinity-based purification 23
α-tubulin 1, 2, 6, 150, 157, 158, 182, 237, 238, 241, 243, 249, 255, 256, 259, 334, 339, 345
Alzheimer's disease 78
Anaphase
  A 192
  B 192
Aneuploidy 149, 236, 267, 268, 286, 297
Arabidopsis thaliana
Asters 55, 57–59, 64, 65, 67, 68, 72, 73, 167–169, 175, 303
Augmin 163, 164
Automated cell segmentation 138, 140
Axonal transport, v 335, 353–367
Axonemes 30, 33, 193, 195, 197–200, 202, 206

B

Bacillus subtilis
Bacteria, v 4, 11, 20, 57, 80, 102, 118, 123, 125, 126, 129, 130, 135, 143, 315, 316, 319, 331, 336, 337
Bacterial cytoskeleton 123–132
Baculovirus 2, 97, 98
β-tubulin 1, 2, 6, 14, 235, 327, 329, 354
Biosensors 98–104, 109, 110, 113, 118
Biotin 2, 3, 30, 58, 80, 252, 261
BirA 3, 5, 11, 12
Bradford assay 30, 45, 102

C

Caenorhabditis elegans
  4-cell embryo 298, 300, 303, 304, 306, 307, 310–312, 318
Cancer 57, 149, 219, 221, 236, 372
Cell
  confinement 55, 252, 260, 262
  culture 4, 11, 129, 149, 152, 157, 165, 170, 181, 182, 218, 219, 221, 237, 244, 249, 253, 261, 328–331, 338, 340–342, 348, 357, 358, 364, 366, 372, 376
  cycle 148, 149, 152, 154, 157, 159, 183, 382
  immortalization 249, 253
  migration 53, 147, 179, 226, 327, 371, 372
  shape 1, 147, 212, 327
  transformation 126, 152
Cell-free assays 77
Centrosome
  centriole
    distal appendages 147, 148
    subdistal appendages 147, 148
  pericentriolar material (PCM) 147, 148, 168
CFP/YFP 99, 101, 110
Channelrhodopsins 212
Chlamydomonas
Chromatin 163, 164, 169, 170, 175
Chromosomes 212, 235, 236, 239, 244, 245, 247–265, 267–294, 303, 371
Chromosome spreads 262
Cilia 1, 28, 147, 354
CLASP 35
CLIP170 180
Contractile rings (CR), v 297–300, 304, 306, 308–310, 312–315, 319–321
CPAP 154, 157, 159
Cre-recombinase 330, 335, 336, 339, 340, 345, 358
CRISPR-Cas9 .... 213, 299
Cytokinesis, v 297–321
Cytolinkers 54, 55, 65, 68
Cytoskeleton, v 30, 42, 54, 56, 73, 123–132, 192, 194, 195, 197, 211, 235, 251, 268, 270, 284, 327, 371
Cytosolic carboxy peptidase (CCP) 328

D

DNA replication 148
Down's syndrome 267
Drosophila melanogaster

Helder Maiato (ed.), *Cytoskeleton Dynamics: Methods and Protocols*, Methods in Molecular Biology, vol. 2101, https://doi.org/10.1007/978-1-0716-0219-5, © Springer Science+Business Media, LLC, part of Springer Nature 2020

E

EB1 ......180, 213, 221, 229
EB3 ......54–57, 64, 65, 68, 73, 171, 173, 176, 213, 221, 229, 336, 341
Escherichia coli
Electron tomography (ET) 192, 194–196, 201, 206, 208
Electrophoresis 3, 6, 112, 257, 333, 334, 346
Emulsion droplets 53–73

F

Fiji 34, 46, 138, 139, 143, 159, 166, 174, 182, 187, 240, 287, 308, 309, 374, 377–379, 381
Filaments 53–73, 135, 141, 142, 192, 212, 268, 270, 274, 278, 283, 293, 298, 302, 310, 311, 327, 371, 373, 378, 383
Fixation 157, 164, 168, 169, 174, 183, 185, 192, 194, 196, 200, 257, 262, 264, 289, 294, 373, 383
Flagella 198
FLIM-FRET 95, 99
Fluorescence activated cell sorting (FACS) 229, 250, 255, 264
Fluorescence dissipation after photoactivation (FDAPA) 236, 237, 241
Fluorescence in situ hybridization (FISH) 261, 263
Fluorochromes 94, 253, 264
Förster resonance energy transfer (FRET) 93–120
Freeze-substitution 194, 196, 200

G

γ-tubulin 159, 163
γ-tubulin ring complex (γTuRC) 163
GDP 27, 30, 32
Genotyping 328, 333, 341, 344
GFP 7, 8, 21, 22, 68, 95, 96, 99, 138, 214, 220, 222, 224, 229, 298, 300, 306, 313, 317, 319, 335, 336, 338–341, 345, 348, 356, 357, 360, 365, 377
Gibson assembly 4, 6, 10, 22
Glutamylases 328, 329
GMPCPP 30, 31, 36, 42–44, 47, 79, 80, 97, 114, 117, 118, 193, 195
Golgi 163, 175, 179–188
Golgi-derived microtubules (GDMTs) 179–181, 183, 184, 186, 187
GTP-islands 28, 35
Guanosine 5-triphosphate (GTP) 1, 27–30, 33, 42, 56, 59, 67, 81, 85, 195, 197

H

Halo tag 138
HEK293T cells 4, 5, 11, 15, 16, 19–25
hTERT 149, 152, 249, 253

I

ImageJ 46, 82, 83, 87, 143, 159, 166, 168, 172, 175, 176, 186, 187, 240, 287, 358, 361, 367, 374, 377–379, 381
Immunofluorescence 149, 150, 157, 168, 181, 185, 256, 257, 270, 289, 373, 376, 379
Immunoprecipitation 5, 13
Indian muntjac 247–265
Inner nuclear membrane (INM) 371
Interphase 1, 23, 54, 159, 164, 167–169, 174, 175, 180, 201, 211, 212
Interpolar microtubules 236
Intracellular transport 1, 27, 39, 327, 335, 353

K

Katanin 24, 27–36
Kinesin
-1 94
-5 94
-7 94
-13 96
-14 ......96, 104
Kinetochore microtubules 192, 236, 237, 241, 242
Knockouts 211, 221, 286, 299, 328–330, 335, 345, 354, 358, 364, 366
Köhler illumination 43
Kymographs 46, 47, 64, 65, 82, 87, 88, 136–139, 142, 172–174, 176, 177, 237, 240–243, 245, 309, 314, 358, 361–365, 367

L

Laser microsurgery 252, 259, 260, 299, 303, 306, 314, 320
Lentivirus 254, 255, 330, 331, 338–341, 344, 348, 350, 356, 358, 359, 366, 368, 372, 380
Linker of nucleoskeleton and cytoskeleton (LINC) complexes 371, 372
LOVTRAP 213, 216, 219
LOV2/Zdk1 216
LPA stimulation 375, 376, 379, 380, 383

M

Mass spectrometry 6, 15, 24
MCAK 35, 96, 98, 110, 112, 113, 120

Mean-squared displacement (MSD) 136, 137, 139–141
Mechanotransduction 371
Meiosis, v 268, 271, 274, 282–286, 288, 293, 305, 312, 318
Microfluidics 57, 59–64, 67, 70, 125–127, 129, 130
Microinjection 269, 270, 272–282, 290–293, 315, 316, 320, 321, 373, 375, 379, 383
Microscopy
darkfield 39–49
differential interference contrast (DIC) 39, 40, 130, 182, 320
electron 28, 158, 191, 195–197, 200, 201, 251, 268, 297
fluorescence 28, 40, 45, 49, 60, 62, 63, 135, 157, 164, 165, 215, 221, 236
phase contrast 130, 138, 195
spinning disc confocal 63
STED 251, 258, 259
super-resolution 158, 160, 182, 270, 274, 282
time-lapse 170, 171
total internal reflection fluorescence (TIRF) 20, 28, 40, 49, 67, 77–89, 123–132, 135–144, 164
Microtubule
depolymerization 1, 2, 35, 120, 164, 166, 174
dynamic instability 2, 193
half-life 235–245
lattice 1, 27–36, 40, 68, 73, 163
minus ends 2, 183, 187
nucleation, v 19, 36, 147, 163–177, 179–188, 193
plus ends 2, 35, 53, 54, 164, 177, 186, 193, 213, 218
poleward flux 235–245
polymerization 193, 194, 196
regrowth assay 164, 166
severing 19, 24, 27–36
Microtubule associated proteins (MAPs) 19–25, 40, 77, 78, 80
Microtubule-Organizing Center (MTOC) 147, 163, 164, 180
Microtubule plus-end-tracking proteins (+TIPs) 180, 188, 213, 229
Mitochondria 355, 357, 358, 362–366
Mitosis, v 1, 23, 93–120, 148, 154, 163, 164, 168, 169, 236, 247–265
Mitotic spindles, v 147, 148, 201, 211, 235–245, 248, 260, 303, 327
Molecular motors 78
Mouse 150, 164, 165, 168, 170, 182, 197, 251, 268, 271–274, 281–286, 289–291, 328, 329, 332, 335, 341, 343, 354, 355, 357, 364
Mouse primary hippocampal neuron cultures 332, 341
Mouth pipetting 273, 274, 277, 279, 282, 289, 290, 292, 294, 304, 307, 310, 317, 319
MreB 123–125, 130, 135, 137, 138, 140–142, 144
mRNA
synthesis 269
Muscle disorders 372

N

Negative stain 28, 106
Neuron 2, 28, 164–166, 170, 171, 173, 175, 176, 179, 212, 213, 327–349, 353–361, 364–367
Neuronal cell culture 342
Nuclear movement 371–384
Nucleo-cytoskeletal interactions 371
Nucleoskeleton 371

O

Oligomer 78, 79, 83, 86, 89, 202, 204
Oocyte
culturing 269, 272
isolation 268, 271, 272
maturation 269, 271, 272, 278
transfer 271, 273, 282, 286, 290, 294
Optogenetics, v 211–232
Overexpression 2, 20, 150, 152, 154, 157, 159, 176, 293, 328, 329

P

PA-GFP-α-tubulin 237, 238, 241, 243
Parthenogenetic activation of mouse eggs 286
P2A self-cleavable peptide 2
Perinuclear actin 371–384
Pharmacological drug addition 284
Photoactivation 214, 216, 217, 220, 222, 225, 226, 229, 235–245
Photobleaching 40, 46, 81, 83–85, 87, 89, 95, 99, 138, 143, 144, 236, 240, 241
Photodissociation 213–218, 221, 224, 226–229
Photoinactivation 221, 222
Photomask 61, 69
Phototoxicity 216, 218, 224, 293, 381
Plasma Cleaner 35, 82, 89

Plk4 .... 148–150, 152, 153, 159
Polycistronic 2
Polydimethylsiloxane (PDMS) 57, 59–63, 69, 70, 72, 252, 260, 261
Polymerase chain reaction (PCR) 3, 6–10, 253, 261, 265, 315, 320, 330, 336, 337, 344, 345, 349
Protein-protein interactions 93–120
Protofilaments 1, 202, 205, 235

R

Rhodamin tubulin 59, 63, 64, 67, 68, 73
Rho-GTPase 54
RNA interference (RNAi) 166, 211, 213, 221, 250, 255, 256, 299, 303, 304, 306, 311, 312, 315, 316, 318, 319, 321, 329

S

Sea urchin 30, 195
Self-organization 55
shRNA 176, 219
Single molecule 77–89, 144, 216
Single-particle tracking (SPT) 136, 139, 140, 142–144
SiR-tubulin 57, 63, 71, 251, 252, 259–262, 265
SNAP tag 138, 292
Spastin 27–36
Spectral response 214, 215
Spindle actin 267–294
Spindle assembly checkpoint (SAC) 236
Streptavidin 5, 13
Sumo 3
SUN proteins 372

T

Tau 77–89
Taxol 28, 36, 97, 114, 115, 245, 302
Temperature-sensitive mutations 299–301, 305, 309
Tetrahymena
Transduction of neurons 335, 341, 344, 345
Transfection 221, 229, 249, 253–256, 331, 338–340, 347, 382
Transmembrane actin-associated nuclear (TAN) lines 372–375, 377–379
Tubulin
  acetylation 328, 354
  C-terminal tail 2
  detyrosination 328
  heterodimer 1, 39
  isotypes 2
  polyglutamylation 328, 335, 345, 346, 354
  post-translational modifications (PTMs) 328, 335, 354, 355, 357, 358
Tubulin associated proteins (TAPs) 3, 6, 14, 16, 17
Tubulin-specific folding cofactors (TBCs) 2
Tubulin tyrosine ligase (TTL) 220, 226, 228, 230, 328
Tubulin tyrosine ligase like (TTLL) proteins 328

U

Ultrastructural analysis 191–208

V

Vectors 4, 10, 21, 22, 24, 249, 311, 315, 316, 330, 331, 335–339, 347, 348
Viral transduction 249, 254

W

Western blotting 112, 152, 157, 250, 256
Wound assay 372, 374, 380–382

X

Xenopus egg extracts 55

MIX
Papier aus verantwortungsvollen Quellen
Paper from responsible sources
FSC® C105338

If you have any concerns about our products,
you can contact us on
ProductSafety@springernature.com

In case Publisher is established outside the EU,
the EU authorized representative is:
Springer Nature Customer Service Center GmbH
Europaplatz 3, 69115 Heidelberg, Germany

Printed by Libri Plureos GmbH
in Hamburg, Germany